Devon Bird R

Number 74: 2002

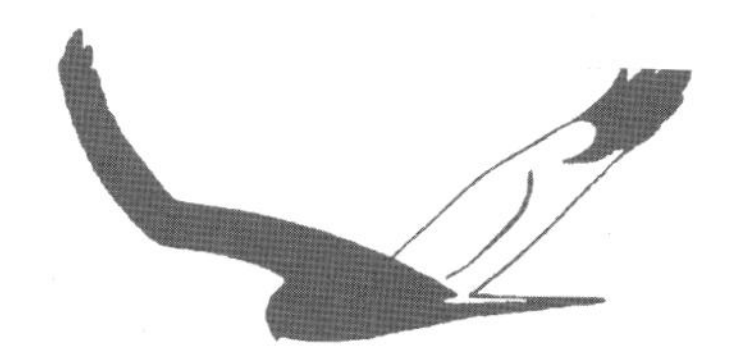

Editors

Ian Farrell	Richard Hibbert	Peter Reay

Section Writers

Roy Adams	Lee Collins	Mark Darlaston
Martin Elcoate	John Fortey	Simon Geary
Richard Hibbert	Cath Jeffs	Matthew Knott
Ivan Lakin	Mike Langman	Bob Normand
Peter Reay	Roger Thornett	Barrie Whitehall

Article & Report Writers

Steve Cooper	Ed Drewitt	Nick Dixon
Simon Geary	Peter Goodfellow	Richard Hibbert
Leonard Hurrell	Ray Jones	Matthew Knott
Mike Langman	Richard Patient	Peter Reay
Phil Stidwill	Roger Swinfen	

Artists & Photographers

Ron Champion	Ade Cooper	Steve Cooper
John Fortey	John Gale	Ren Hathway
Leonard Hurrell	Matthew Knott	Mike Langman
Robert Marshall	Bob Montgomery	Bob Normand
Paul Nunn	Richard Patient	Chris Proctor
Chris Robbins	Tony Taylor	John Walters
Tom Whiley	Steve Young	John Woodland

Society Patrons

Lord Clinton, Clinton Devon Estates; Devon Ashphalte Company;
E & J W Glendinning Ltd; In Focus; Michelmore Hughes;
Natural History Book Service Ltd; Watts Blake Bearne & Co plc

Pied Flycatcher (male) - *Ren Hathway*

ISBN: 0-9540088-2-0

Published by the Devon Bird Watching and Preservation Society
POB 71 Okehampton, Devon EX20 1WF
Email: info@devonbirds.org.uk
www.devonbirds.org.uk
Registered Charity 228966

This publication was produced and distributed by the Devon Bird Watching and Preservation Society with kind and generous support from the Claude and Margaret Pike Woodlands Trust, the Environment Agency, the John Spedan Lewis Foundation, Lord Clinton's Charitable Trust and the Norman Family Charitable Trust.

Designed and typeset by Ian Farrell in InDesign ® in Lino Type Times New Roman
Maps designed and produced by Simon Geary with Philippa Burrell of the Devon Biodiversity Records Centre using MapInfo ®

Printing: Kingfisher Print & Design Ltd. Wills Road, Totnes Industrial Estate, Totnes, Devon TQ9 5XN Tel: 01803 867087

Goshawk - *Steve Young*

Table of Contents

EDITORIAL

Back in 2001, as the **Foot & Mouth Disease (FMD) access restrictions** took a grip and affected all our lives, we wondered if there would be enough records to make a *DBR* possible. At the very least we assumed it would be a thin one. But of course it isn't, and for reasons still not entirely clear, members managed to increase the number of records to an extraordinary record-breaking 48,000! This is not to say, of course, that the FMD restrictions had no impact, and there is much reference to their presumed effects throughout the Report. **When comparing 2001 records with those of other years, it is important that both current and future readers remember that much of the Devon countryside was closed from March to August, and that all national surveys were cancelled during this period. Thus, apart from affecting the distribution, if not the quantity of casual records over this period, there were effectively no BBS or WBBS surveys in 2001 and WeBS, which usually contributes so much to the tables in waterbird species accounts, only provided count data for January, February and September-December.**

For the second year running we would like to take this opportunity to thank the **Claude and Margaret Pike Woodlands Trust**, the **Environment Agency**, the **John Spedan Lewis Foundation**, **Lord Clinton's Charitable Trust** and the **Norman Family Charitable Trust** for their generosity in enabling us to maintain and extend improvements, such as:

- A continued increase in the number of pages within the report; *DBR* 2001 runs to an unprecedented 256 pages in black and white, and eight pages of colour photographs.
- More short reports, information boxes, figures and maps.
- Improvements in the size and quality of charts and maps through the use of more powerful software and analysis tools.
- Improvements in the standardisation and accessibility of information in the species accounts and a fuller treatment of the data and improvements in content.
- The inclusion of a weather report to help interpret movements, breeding success and other aspects of bird behaviour recorded during the year.
- The return of the Site Gazetteer, revised and updated.
- And maintaining last year's increase in the number of copies printed to allow for the free distribution of the Report as an educational and reference tool to selected secondary schools, libraries, universities, research establishments and other organisations (with the kind assistance of the Devon County Council Education Department).

Thanks, as usual, are due to all observers for submitting records, to the data capture team for building the database, and to all section writers, report writers, artists and photographers for their contributions. We are also grateful to the contributions of collated data and records from A J Bellamy (RSPB), Ivan Lakin, David Rogers, Roger Smaldon, Phil Stidwill and Tony Taylor, but are particularly indebted, once again, to Simon Geary and Mike Langman for all their help. Nominally Data Manager and County Recorder respectively, their invaluable assistance and contributions extend well beyond these roles. And, at the end of 2002, it is time to announce, with considerable regret, that Mike Langman has finally had enough of us and will be hanging up his quill. We can only say that Mike has been a veritable fountain of knowledge, enthusiasm and patient criticism, without which we probably wouldn't have survived our first year as Editors. He has consistently delivered the goods with panache, skill and accuracy, and his input, calm good sense and enormous wealth of knowledge and experience will be greatly missed. He does however; assure us that we shall continue to receive his annual flock of pictures to grace these pages, and that he will continue to help collect and collate work from other artists and photographers for the DBR. Mike's various responsibilities, including membership of the Editorial Committee, have been handed over to the new County Recorder, Mike Tyler, whom we warmly welcome on board for 2003.

Ian Farrell, Richard Hibbert, Peter Reay
December 2002

REVIEW OF THE YEAR 2001

Mike Langman

In total, **271 species** were recorded in 2001 (*excluding* Snow Goose and 21 Category E species), compared to 268 in 2000 and 275 in 1999. Two species were added to the Devon list during the year, **Pied-billed Grebe** and **Black-faced Bunting**, as well as a subspecies, **'Siberian' Stonechat**. Another subspecies, **'American' Herring Gull**, though recorded in 1998, has eventually been accepted by BBRC and represents another first for Devon. Adding the two new species, and the newly split Hooded Crow, brings the overall county total to **415 species** (*cf.* slightly underestimated totals in previous *DBR*s); the total includes Egyptian Goose, Ruddy Shelduck, Red-crested Pochard and Golden Pheasant on the assumption that records of these species were of birds from wild breeding populations (*cf.* Snow Goose).

Rarity Highlights

The National (BBRC) Rarities were: **Pied-billed Grebe**, **Fea's/Zino's Petrel** (still under consideration by BBRC), **Night Heron**, **Black Stork**, **Black Duck**, **Lesser Yellowlegs**, **Baird's Sandpiper**, **Franklin's Gull**, **Bonaparte's Gull**, **Alpine Swift**, two **'Siberian' Stonechats**, **Subalpine Warbler**, two **Rose-coloured Starlings**, **Rose-breasted Grosbeak**, **Bobolink** and **Black-faced Bunting**.

Other rarities included: a returning male **Ring-necked Duck**, **Green-winged Teal**, **Surf Scoter**, **Spotted Crake**, six **Kentish Plovers**, **Temminck's Stint**, **Pectoral Sandpiper**, several **Ring-billed Gulls**, **Hoopoes**, **Yellow-browed Warbler**, **Barred Warbler**, **Red-breasted Flycatchers**, **Hooded Crow** and **Common Rosefinch**. Notable events included excellent numbers of summering **Roseate Terns** off Dawlish Warren and good numbers of **Grey Phalaropes** and **Short-eared Owls** during the autumn. **Ruddy Duck** bred in Devon for the first time.

Fig. 1: Monthly Distribution of BBRC Rarities in Devon in 2000 and 2001

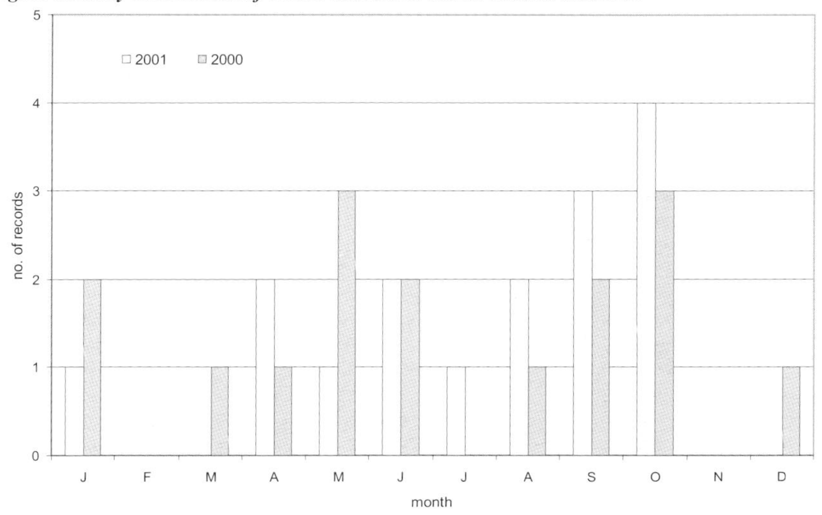

January

Following some gale force SW winds and plenty of rain early in the month, high pressure then dominated, with mainly E winds, which also reached gale force in mid-month. There were more strong SW winds from 22nd as low pressure systems swept in, but the winds soon turned light from the SW towards the end of the month. Birds lingering from 2000 included: the **Black Duck** at Slapton; a drake **Ring-necked Duck** at Burrator; three **Long-tailed Ducks** and a **Velvet Scoter** at Dawlish Warren;

eight **Spoonbills** split equally between Taw/Torridge and Exe Ests; a **Great Grey Shrike** and two **Ring-billed Gulls** in N Devon; and, in S Devon, a **Hooded Crow**, two **Tree Sparrows** and at least one **Whimbrel**. Concerning new arrivals, a good start to the year was an immature **Whooper Swan** on the Axe Estuary, with a **Water Pipit** also there on the marshes. Hope's Nose hosted a **Long-tailed Duck** and 11 **Red-throated** and four **Great Northern Divers**; there was a **Red-necked Grebe** at Plymouth, and at Beesands, away from the normal wintering areas, there was a **Black-necked Grebe.** In N Devon, another **Ring-billed Gull** was discovered. Some poor weather conditions on 2nd produced a **Pomarine Skua** at Berry Head and good numbers of **Great Northern Divers** at several sites. As usual, Hartland watchers came up with a good count of over 100 **Red-throated Divers** on 3rd and a **Great Skua** nearby at Baggy Pt on 4th. Roadford Reservoir did well on 5th with an **Avocet** and a **Scaup** (and another was at Wistlandpound). Elsewhere, Bursdon Moor saw two **Hen Harriers** and a **Merlin** come to roost. Sadly suppressed was the news of a first for the county – a **Pied-billed Grebe** – at the very public-friendly Tamar Lakes from the 6th. A sizeable wintering flock of **Hawfinches** was discovered near Haldon from 7th, while the Kingsbridge Estuary attracted four **Slavonian Grebes** and a **Black-throated Diver**, and three **Water Pipits** were also found wintering on the Otter Estuary. An immature **Whooper Swan** took up residence in N Devon from 8th, when two **Black Redstarts** fed around the cliffs at Ladram Bay. On 9th, a wing-tagged **Red Kite** lingered at North Molton and a **Ring-billed Gull** was found on the Exe Estuary. There was a **Glaucous Gull** at Slapton, and three **Little Stints** at Pottington, on 12th. The weekend of 13-14th brought plenty of good birds, including **Green-winged Teal**, **Smew**, **Bittern**, **Iceland** and **Glaucous Gulls** and a **Black Guillemot**. Two **Jack Snipe** were flushed at Bishop's Tawton, and four **Water Pipits** fed at South Huish on 15th. On 19th, the **Franklin's Gull** was rediscovered roosting in Torbay, and a **Black-throated Diver** was off Abbotsham Cliffs. Two **Red Kites** were seen near Crediton, and two **Firecrests** wintered at Galmpton sewage works from 20th. Dawlish Warren had a **Long-tailed Duck** and five **Mediterranean Gulls** on 22nd, and a **Red-necked Grebe** was in Torbay on 23rd. An **Iceland Gull** was in Torbay on 25th. Elsewhere there were three **Water Pipits** on the River Tavy, a **Lesser Whitethroat** in an Ilfracombe Garden and, from 26th, a much-discussed **Redpoll** at Topsham. A high count of 18 **Slavonian Grebes** was made at Dawlish Warren on 30th, and, to end the month, two **Tree Sparrows** were recorded from an Okehampton Garden.

February

The month started unsettled, but warm, as weather fronts swept in from the Atlantic. Amazingly, the month started with the discovery of a **Robin's** nest with eggs, although the breeding attempt failed. More seasonal was a **Red-necked Grebe** at Roadford. Slapton hosted a **Bittern** and up to three **Firecrests** during the month, and a **Little Gull** was regularly seen at Dawlish Warren. The count of 120 **Great Crested Grebes**, sheltering from gale force SW winds at Babbacombe on 4th, was close to the county record. A **Red Kite** passed over Tiverton on 9th, and in Torbay on 11th there was a count of 35 (rising to 38 on 21st) **Black-necked Grebes** and a **Black-throated Diver**. A **Bittern**, perhaps the Slapton bird, was at Beesands from 12th, and a **Little Gull** and an exceptionally early **Swallow** were at Slapton on 13th. Calm conditions from14th allowed 20 **Red-throated Divers** to be counted at Dawlish Warren on 16th. In Torbay on 17th, there was a **Glaucous Gull** and a count of 16 **Purple Sandpipers**, while **Redpoll** watchers at Topsham saw a **Water Pipit**. On 19th, Slapton held a **Red-necked Grebe** and four **Slavonian Grebes**. A **Ring-billed Gull** was seen roosting in Torbay on 23rd, and a **Serin** was discovered at Prawle Pt from 24th. A **Glaucous Gull** stopped at Thurlestone on 25th, and a **Great Grey Shrike** was noted just in Devon, near Kilkhampton, on 26th. An adult **Bewick's Swan** dropped on to Braunton Marshes on 27th.

March

A series of low pressure systems rolling in from the Atlantic dominated, giving generally mild but very wet weather. A quiet start to the month, with only a **Little Gull** on the Exe Estuary of note, but on 8th the first **Sand Martins** were seen, and a first summer **Ring-billed Gull** was found at Countess Wear; further west an **Iceland Gull** was seen regularly on the Plym Est from 10th. Two **Sandwich Terns** arrived at Dawlish Warren on 12th where up to 17 **Slavonian Grebes** were counted, while up the estuary, four **Spotted Redshanks** roosted at Bowling Green Marsh. **Wheatears** arrived from 15th on light to moderate SE winds, and three **Snow Buntings** scuttled around the coastal path at Ilfracombe. From 17th, good numbers of **Wheatear** and **Chiffchaff** were also noted on the S coast together with several **Firecrests**. The 18th brought flocks of **martins** and the odd **Swallow,** and also a **White Stork** in N Devon, before it moved to Exebridge. A typical **Hoopoe** visit was to a garden in E Devon on 19th. A colour ringed **Kentish Plover** dropped into Dawlish Warren on 22nd, when a **Red-necked Grebe** was in Torbay. A **Red Kite** took up residence near Braunton from 23rd. Two **Glaucous Gulls** on the Plym Estuary, an **Iceland Gull** at Dawlish Warren and a singing **Willow Warbler** were recorded on 24th. As usual, the Teign Estuary produced two adult **Ring-billed Gulls** from 25th, and the next day a pair of **Kentish Plovers** were at nearby Dawlish Warren. Also on 26th, an early **Pied Flycatcher** sheltered in Torbay. Another **Hoopoe** visited a DBWPS member's garden on 27th, and singing **Tree Pipits** were heard at Yarner Wood on 28th. Just when it was thought the **Franklin's Gull** had gone for good, it reappeared at Hope's Nose before regularly commuting between there and the Teign Estuary from 31st.

April

Another unsettled month; generally warm, but with twice the average rainfall as weather fronts moved in and settled over England. The month started very well with a **Chough**, first noted at Kingswear, then seen later in the day at Slapton before becoming a little more reliable at Prawle later in the month. There were still four **Spoonbills** on the Exe Est and two at Penhill, and a **Hoopoe** appeared in Budleigh Salterton on 4th. In N Devon, a **Blue-headed Wagtail** lingered at Velator. The Teign Estuary produced a **Glaucous Gull** from 7th, while on 8th there was a **Marsh Harrier** over Bowling Green Marsh and, even better, a male **Subalpine Warbler** in a garden at Wembury. Hope's Nose was the place to be on 9th, with a **Nightingale** and two **Iceland Gulls**. Light winds turning to the SE produced some good movements of common and scarcer migrants including **Hoopoe**, **Little Ringed Plover** and another **Marsh Harrier** on 10th. Up to three **Garganey** were counted on the Exe Est on 11th, while the 12th brought **Osprey**, **Night Heron** and another **Nightingale** to S Devon, and a **Whooper Swan** to Velator. Good numbers of summer migrants arrived on 13th, including **Grasshopper** and **Sedge Warblers**, **Redstarts**, **Cuckoos** and two **Little Ringed Plovers**. Another **Osprey** was seen on 14th and a **Common Crane** appeared at Slapton. Several more **Iceland Gulls** passed through the county from 15th and the first **Swift** was seen on 17th. **Hobbies** started to return from 20th, and a **Lesser Whitethroat** was noted on 21st. Some strong to gale force S-SW winds brought three **Great Skuas** off Hope's Nose on 22nd, and at Dawlish Warren there were several **Little Terns** and a **Little Gull**. Also there on 24th were three **Great Skuas**, two **Black Terns** and good numbers of **Little Terns**. A **Quail** was a surprise find near Holsworthy on 25th as it hid under a tractor. There was a good count of nearly 300 **Whimbrel** at Bowling Green Marsh on 26th. A **Nightjar** was seen at Stover CP on 27th, on 29th a **Long-tailed Duck** could be seen at Slapton and many sites recorded **Lesser Whitethroat**.

May

A very dry month, which started with settled high pressure and light, mostly N, winds. The first **Roseate Tern**, in what would prove to be an excellent year for the species, turned up at Dawlish Warren on 1st. Birds of prey then became the focus of attention with the first few days of the month recording several **Marsh Harriers**, **Ospreys**, **Red Kites** and the only **Montagu's Harrier** of the year. Other interesting birds included: a **Red-necked Grebe** at Slapton on 4th; a **Hawfinch** on Lundy and a long staying **Blue-headed Wagtail** at Prawle from 5th; a couple of **Curlew Sandpipers** at Dawlish Warren on 8th; and the **Black Guillemot** still lingering in Plymouth. The 10th produced a **Wryneck** at Hatherleigh Moor, and **Black-throated** and two **Great Northern Divers** on the S coast. A **Dotterel** lingered on Lundy on 11th, but warm anticyclonic air on the weekend of 12-13th brought several **Black** and **Roseate Terns**, **Garganeys**, **Wood Sandpipers** and another **Marsh Harrier**. Up to five **Hobbies** were at Exminster Marshes on 15th. The only strong winds of the month were westerlies on 17th. Three more **Roseate Terns** were noted on 18th, while Bowling Green Marsh attracted a **Bonaparte's Gull** on the 19th. One of the most unusual sights of the month must go to two **Razorbills** flying over Marsh Mills roundabout in Plymouth on 20th. The month finished with a **Kentish Plover** and several more **Great Northern Divers** along the S coast, but taking the limelight was a **Temminck's Stint** at Bowling Green Marsh on 31st.

June

Another fairly settled dry month, but with a few thunderstorms. A new **Osprey** was on the Exe Est from 1st, and also there was a **Little Ringed Plover**. On the 2nd, a **Red Kite** flew over Ashclyst Forest and a **Pomarine Skua** passed Prawle Pt. Early morning birding at Bowling Green Marsh on 3rd paid off with a singing **Golden Oriole**. A **Balearic Shearwater** and **Red-throated Diver** were at Dawlish Warren, but the best bird was a probable **Sooty Tern** four miles W of Lundy on 4th. A female **Scaup** at Radford Lake, in Plymouth on 8th was most unexpected, and nearby a **Hoopoe** was found on Plymouth Hoe on 9th. Also on this day were a **Little Stint** at Dawlish Warren and a **Red-backed Shrike** at Grimspound. A **Rose-coloured Starling**, part of a national influx, breakfasted in a Paignton Garden from 11th. Another was briefly at Lee Moor on 14th, when some appalling weather, with strong to gale force S winds, forced a **Black Guillemot** and an adult **Long-tailed Skua** onto the S coast. The annual summer build-up of **Mediterranean Gulls** peaked at Bowling Green Marsh with eight on 16th. Reserve volunteers at Sherpa Marsh were rewarded with a booming **Bittern** from 18th, and five adult **Pomarine Skuas** flew past Teignmouth seafront on 19th. North Devon reserve watchers came up trumps again on 20th; this time it was a **Purple Heron** at Bradiford, shortly followed by a **Red Kite** nearby on 22nd. One or two **Quail** were noted in the South Hams from 24th. Some fresh to strong S winds at the end of the month brought several **skuas**, **Puffins** and **Storm Petrels** close inshore on 29th.

Fea's Petrel - (with Manx Shearwaters) - *Mike Langman*

July

July was generally unsettled and wet, with more than twice the average rainfall. The month started with more **Mediterranean Gulls**, several **Roseate Terns** and a couple of **Little Ringed Plovers**. A **Hawfinch** was an unusual visitor to a bird table in the South Hams from 7th, a day when winds picked up from the SW, and for several days good numbers of **Manx Shearwaters**, **Storm Petrels** and smaller numbers of **Balearic Shearwaters**, **Puffins** and a few **skuas** were brought close to the S coast headlands. The focus of **Mediterranean Gull** distribution moved to the N coast, with perhaps as many as 12 individuals recorded, including several with colour rings. Small numbers of **Storm Petrels**, **Balearic Shearwaters** and more **Roseate Terns** continued to be noted in some unsettled weather conditions. An unseasonal, fast-moving depression headed out of Biscay into the English Channel on the night of 16th, and conditions looked very promising for the following morning. Indeed, watchers braving the stormy SE winds and rain were treated to **Sabine's Gull**, four species of **skua**, **Storm Petrels**, **Puffins**, **Sooty** and

Balearic Shearwaters, **Black** and **Roseate Terns**, with pride of place going to a **Fea's/Zino's Petrel** noted at both Hope's Nose and Berry Head. Although conditions soon calmed down, the weather had unsettled many seabirds, most notably **Balearic Shearwaters**, and regular daily counts from several sites included many birds loafing in groups on the sea. More **skuas**, **Storm Petrels**, **Puffins** and good numbers of **Manx Shearwater** were also seen. Autumn passage was upon us by mid-month as small numbers of **Wood** and **Green Sandpipers**, **Little Stints** and **Curlew Sandpipers** arrived on the marshes and estuaries until the end of the month. A small influx of **Crossbills** was a feature from 23rd. The weather eventually settled down and the month closed with the arrival of two **Ospreys**.

August

The seabird theme continued into early August, when a series of weak depressions moved in from the Atlantic from 2nd, bringing a **Great Shearwater** and good numbers of **Balearic Shearwaters** and a **Black-throated Diver** into Lyme Bay. Similar conditions on 7th produced another two **Great Shearwaters**. Elsewhere, more waders were on the move, with several more **Wood Sandpipers** and **Little Ringed Plover** noted. More Atlantic weather systems from 11th brought **Sooty** and **Cory's Shearwaters** and more **Storm Petrels** along the S coast. Dawlish Warren produced a count of 37 **Curlew Sandpipers** and two **Little Ringed Plovers** on 15th. The N coast saw a movement of 1,400+ **Manx** and two **Sooty Shearwaters** on 16th, and there was an **Osprey** on Roadford Reservoir. Strong S winds on 18th saw some good counts of seabirds, including over 100 **skuas** at Berry Head. In N Devon on the 18th, several **Arctic Terns** were counted, and the returning adult **Ring-billed Gull** was back in Barnstaple. On the S coast, a **Red-backed Shrike** was at Prawle. More **Wood Sandpipers** were seen from 20th, and the weather calmed as high pressure gave light S-SE winds. A **Bittern** skulked at Slapton and an **Osprey** flew over Haldon on 21st. **Crossbills** were still on the move, with 12 at Start Pt on 22nd. A distant **Spotted Crake** was well picked out at Bowling Green Marsh on 23rd, and a **Marsh Harrier** was noted at Braunton Burrows. The 24th saw a **Black-necked Grebe** and an **Osprey** on the Plym Est. A **Wryneck** visited an E Devon garden on 25th, while the 26th brought four **Wood Sandpipers** to one site and a long-staying, but very elusive, **Black Stork** to the edge of Dartmoor. A day visitor to Lundy on 28th was rewarded with an **Isabelline Shrike**. The month finished with more seabirds, this time off the N coast following some strong NW winds, including **Great Shearwater** and **Sabine's Gull** at Westward Ho! and 30 **Balearic Shearwaters**, two **Sooty Shearwaters** and two **Black Terns** off Hartland Pt on 30th.

September

The month was dominated by strong NW winds with very little rain during the first half, and these conditions influenced the rest of the autumn's birds, especially the American vagrants. A **Wryneck** at Aylesbeare Common on the 1st was followed by a **Pectoral Sandpiper** at Tamar Lakes from 3rd and some good counts of **Balearic Shearwaters** off Hartland Pt. A flock of 70 **Ravens** was seen near the Dart Estuary on 6th. A **Grey Phalarope**, the first of a notable influx, rested at Axmouth Marshes from 7th when five **Sooty Shearwaters** passed Hartland Pt. A long-staying **Garganey** could be seen at Slapton, and a strange-looking wader on 8th in Kingsbridge turned out to be a leucistic **Greenshank**. A **Black Guillemot** flew past Hope's Nose on 9th, and Tamar Lakes produced a count of over 20 **Curlew Sandpipers** on 11th. Some strong SW, veering W, winds on 13th brought in some good numbers of seabirds, including three **Pomarine** and 18 **Great Skuas** at Berry Head, and 41 **Balearic Shearwaters** off Hartland Pt; there were also two **Black Terns** at Lower Tamar Lake. Waders were the order of the day on 16th following some persistent fresh NW winds, including 47 **Curlew Sandpipers** at Bowling Green Marsh and another 15 at Tamar Lakes. There were also five **Little Stints** on the Exe Estuary and, best of all, a **Lesser Yellowlegs** at Roadford. Another American wader, this time a long-staying **Baird's Sandpiper**, was at Dawlish Warren from 19th, and with birdwatching efforts being concentrated on the site over the next few days, **Kentish Plover**, 11 **Little Stints**, **Osprey**, **Pale-bellied Brent Goose**, **Black-Necked Grebe** and 11 passing **Coal Tits** were also found. Elsewhere, the **Curlew Sandpiper** influx continued, a **Richard's Pipit** was on Dartmoor on 22nd and several **Merlin** hunted around the S coast. A first-winter **'Siberian' Stonechat** mixed with common **Stonechats** at Start Pt on 26th when also an early **Redwing** was noted at Stover CP. A couple of **Firecrests** were found on 27th, when a **Grey Phalarope** was at Hope's Nose. **Ospreys** and two **Wrynecks** could be seen on 29th. With strong to gale force S-SW winds and rain overnight, the next day brought some excellent seawatching, with the year's best counts of **skuas**, several **Black Terns**, **Grey Phalarope** and a particularly confiding **Sabine's Gull** on the S coast.

October

October was a very warm, but usually wet and windy, month as several Atlantic lows swept in with their weather fronts. Good number of **skuas** could still be seen passing headlands on 1st, together with a few **Little Gulls** and another **Grey Phalarope**, but the bird of the day was an **Alpine Swift** that passed over Berry Head twice during the morning. An **Ortolan Bunting** was present all too briefly at Hope's Nose on 2nd, and another **Grey Phalarope** was there on 4th. Nearby at Berry Head on 5th, among good numbers of **skuas** there were 15 **Pomarine** and a **Long-tailed Skua**, plus six **Sooty Shearwaters**, following another deep Atlantic depression. The next few days produced more **Grey Phalaropes**, but more notably a **Rose-breasted Grosbeak** on Lundy from 6th. At least three **Sabine's Gulls** tried to find shelter in Plymouth on 8th, following one there on 7th. An inland **Grey Phalarope** was noted on a pond at Crediton on the 8th. The wind slackened, and on 9th, a fine day with a light NW wind, birders were out and finding some impressive birds, including: **Bobolink**, **Red-backed Shrike** and no less than 29 **Coal Tits** at Prawle Pt; **Common Rosefinch** on Lundy; **Pomarine Skua** at Hartland Pt; and **Snow Bunting** at Baggy Pt. Prawle also attracted a **Red-breasted Flycatcher** on 10th and there was also a **Red-necked Grebe** in Torbay. Lundy produced another excellent rarity in the form of a **Black-faced Bunting** on 12th. Sites on the S coast had to manage with **Red-backed Shrike**, **Yellow-browed Warbler**, several **Firecrests** and many more **Coal Tits**! The 13th was another great day, with several **Yellow-browed Warblers** (including at least three, and a **Barred Warbler**, on Lundy) two **Wrynecks**, many **Firecrests** and the second **'Siberian' Stonechat** of the year, this time at Lannacombe. Another **Yellow-browed Warbler** was at Start Farm on 14th. The S coast managed a couple of **Sooty Shearwaters** on 15th and another **Grey Phalarope**. The 16th brought an enormous count of 226 **Little Egrets** roosting at Powderham, and a **Red-breasted Flycatcher** at Prawle Pt, where there was another **Red-backed Shrike** the next day. A **Leach's Petrel** flew past Hope's Nose on 17th and there were three **Snow Buntings** at Northam Burrows. A **Richard's Pipit** lingered at Berry Head on 18th, while there was a **Marsh Harrier** at Slapton, a remarkable four **Corn Buntings** at Soar, and a **Long-tailed Skua** at Dawlish Warren. Dawlish Warren completed a fantastic year for **Kentish Plovers** with its final bird from 19th. Calm and bright conditions on 20th helped to produce **Yellow-browed Warbler**, **Black Tern** and many **Firecrests** in the South Hams, a **Red-breasted Flycatcher** on Lundy and two **Short-eared Owls** at Dawlish Warren, followed by several more at other sites over the next few days. Lundy attracted a **Richard's Pipit** from 23rd. A small low-pressure system swept in on 26th to bring some relatively good numbers of seabirds around the S coast, while on Lundy there was a **Common Rosefinch**, a **Barred Warbler** and an **Osprey** on 27th, and there were by now four **Spoonbills** at Penhill Marshes. A **Lapland Bunting** was discovered at Soar on 28th, and five **Bewick's Swans** visited the Axe Estuary. Back on Lundy, there were another two **Yellow-browed Warblers** on 29th and, completing an excellent month, three **Snow Buntings** at Berry Head on 30th.

November

Another unsettled month, but not nearly so wet or windy as October. The focus was still on Lundy with the first few days of the month producing **Little**, **Snow** and **Lapland Buntings**, **Twite**, four **Whooper Swans** and a **Greenland White-fronted Goose**. A **Whooper Swan** popped in at Roadford on 1st. Visible migration by **Woodpigeon** and **Stock Dove** produced some very high counts from S coast sites early in the month, and there were several **Ring Ouzels** at Soar on 3rd. Winter seabirds were on the move on 4th, and flat seas helped birders pick out a group of eight **Velvet Scoters** at Dawlish, and again briefly in Torbay where there were also five **Black-necked Grebes** and a **Long-tailed Duck**. A **Goshawk** flew over Halwell, nr Totnes on 4th, and another was noted on Dartmoor on 5th. Elsewhere, there was a **Red Kite** nr Chudleigh, four **Hen Harriers** roosting at Bursdon Moor and a late **Reed Warbler** at Soar. A skulking **Barred Warbler** was well found on 6th at Dawlish Warren, where there were five **Pale-bellied Brent Geese** on 7th. More **Short-eared Owls** made the most of still, high pressure conditions and were seen at several sites from 10th, when also a couple of late **Swallows** hawked the Ley at Slapton, a **Black-throated Diver** was in Torbay and a late **Redstart** and **Whinchat** could be seen at Berry Head. Over 500 **Stock Doves** were counted at Prawle, together with a late **Red-backed Shrike**, on 11th, and three **Little**

Auks were noted, including one well up the Exe Estuary. Three of each of **Black-throated Diver** and **Red-necked Grebe** could be seen at Dawlish Warren on 12th. News broke of an **Osprey** at Rattery on 13th, but it had probably been present since mid-Oct, and a **Whooper Swan** was noted at Bideford. Back at Dawlish Warren, the next few days brought **Little Auk**, **Long-tailed Duck** and several **Little Gulls** and **Great Northern Divers**. There were by now five **Spoonbill**, 2,000 **Golden Plover** and a couple of **Little Stints** at Penhill Marshes. Two **Scaup** were mixed in with the **Tufted Ducks** at Slapton from 16th, probably brought in on the strengthening E-NE winds. The winds took their toll on 17th as two weak **Little Auks** arrived at Broadsands, one dying on the beach. Slightly calmer weather on 18th produced a **Yellow-browed Warbler** at Slapton, and two **Black-necked Grebes** on the Kingsbridge Estuary. A **Bittern** was spotted at Slapton on 19th when there was a **Ruddy Duck** and a **Scaup** at Roadford, and a **Spoonbill**, two **Red-necked** and a **Black-necked Grebe** at Dawlish Warren. Two **Long-tailed Ducks** could be seen in Torbay on 20th and a **Little Auk** passed nearby Berry Head on 21st, when over 500 **Avocet** were counted on the Exe Estuary. The next few days saw several more wintering **grebes**, **divers** and **ducks** and a **Little Auk** again at Berry Head on 25th. The month closed with a **Spoonbill** at Stoke Gabriel, **Little Auks** at Hope's Nose and Prawle and an **Iceland Gull** at Berry Head.

December

After a depression swept through early in the month, most of December was dominated by high pressure and cool E-NE winds. The 1st produced a couple of flyby **Velvet Scoters** at Prawle, and by now there were two **Long-tailed Ducks** at Dawlish Warren and as many as 16 **Ruddy Ducks** at Slapton. A **Snow Bunting** was at Dawlish Warren on 2nd, when the **Iceland Gull** could be seen at Mansands Beach and two **Velvet Scoters** were in Torbay. Several **Jack Snipe** were found on 8-9th and a very late **Common Tern** was off Dawlish Warren. Prawle produced a **Pomarine Skua** on 9th while in N Devon some 120 **Red-throated Divers** were off Hartland and four **Hen Harrier**, three **Merlin** and two **Short-eared Owls** roosted at Bursdon Moor. Another two **Velvet Scoter** were noted off the N coast on 12th, and two **Ring-billed Gulls** lingered on the Plym Estuary. A very late **Swallow** flew over Dawlish Warren on 13th, and a **Black-throated Diver** was off Jennycliff on 14th. The regularly returning **Ring-necked Duck** was back at Burrator on 15th when **Whooper Swan** and **Iceland Gull** could be seen near Bideford. Two **Water Pipits** were found wintering on the Otter Estuary and several **White-fronted Geese** roosted at Tamar Lakes on 16th. South Huish attracted four **White-fronted Geese** on 17th, and four **Black Redstarts** on 18th. In the Torbay area, several more **Velvet Scoter** could be seen from 19th. A wintering **Whimbrel** was recorded on the Teign Estuary on 21st, and West Charleton Marsh continued to attract **Jack Snipe** and a couple of **Water Pipits**. The year closed with a cold snap and an exciting find, a **Cattle Egret** in E Devon in the snow, but unfortunately this tame, ringed bird was probably an escape. Elsewhere there was a big count of over 1,000 **Great Black-backed Gulls** on the Plym Estuary and a **Water Pipit** on the River Dart.

REVIEW OF THE YEAR – WEATHER

Dr Len Wood, University of Plymouth

The data.

Weather data, obtained from the British Atmospheric Data Centre, are presented for Plymouth in 2001 – a year which was not particularly remarkable in the context of ranking in the last 50 years. Although weather will vary across the county, the Plymouth data should give a reasonable indication of the main features and changes in Devon. Note, however, that rainfall is very altitude dependent, and the annual total increases from the Plymouth value by about 100 mm for each altitude rise of 40 m; thus Princetown, at 414 m above sea level, has an annual total of approximately 1900 mm. The temperature data are broadly representative of coastal Devon, but moorland areas above 400 m may be three or four deg.C colder, especially in winter. Wind data for Plymouth is representative of other locations in Devon, except when there are sea breezes on fine spring and summer days. These onshore winds of approximately 10 knots sometimes penetrate into the county on such days, producing insect-rich updraughts and cumulus clouds, under which birds occasionally feed.

Rainfall.

After three successive wetter than normal years in Plymouth, 2001 was the 11th driest, with a total of 877 mm compared with the climatological 30 year average of 957 mm. Fig. 2 shows that while January, March, April, July and October were wetter than normal, there were some notable dry periods in early (May/June) and late (August/September) summer.

Temperature.

The year ranked as the 16th warmest in Plymouth over the last 50 years, the only remarkable months being October, which was 2.1 deg.C warmer, and December, which was 1.2 deg.C colder, than the average values (Fig. 2). The mean temperatures quoted are an average of the daily maximum and minimum temperatures.

Wind.

The daily wind record is shown in the monthly bar charts (Figs. 3-14). Particularly windy months were January, April, and October. January had strong easterlies in the second week, and westerly gales occurred in the first week of February. There was a notable southerly gale on 30 September followed by southwesterly gales in the first and last weeks of October. Where wind was predominantly in one direction each day, the heading is given at the top of each wind bar, winds of variable direction are not marked and boldly marked directions on the bar charts refer to occasions when the wind gusted to gale force, which is above 34 knots.

Choughs - *Mike Langman*

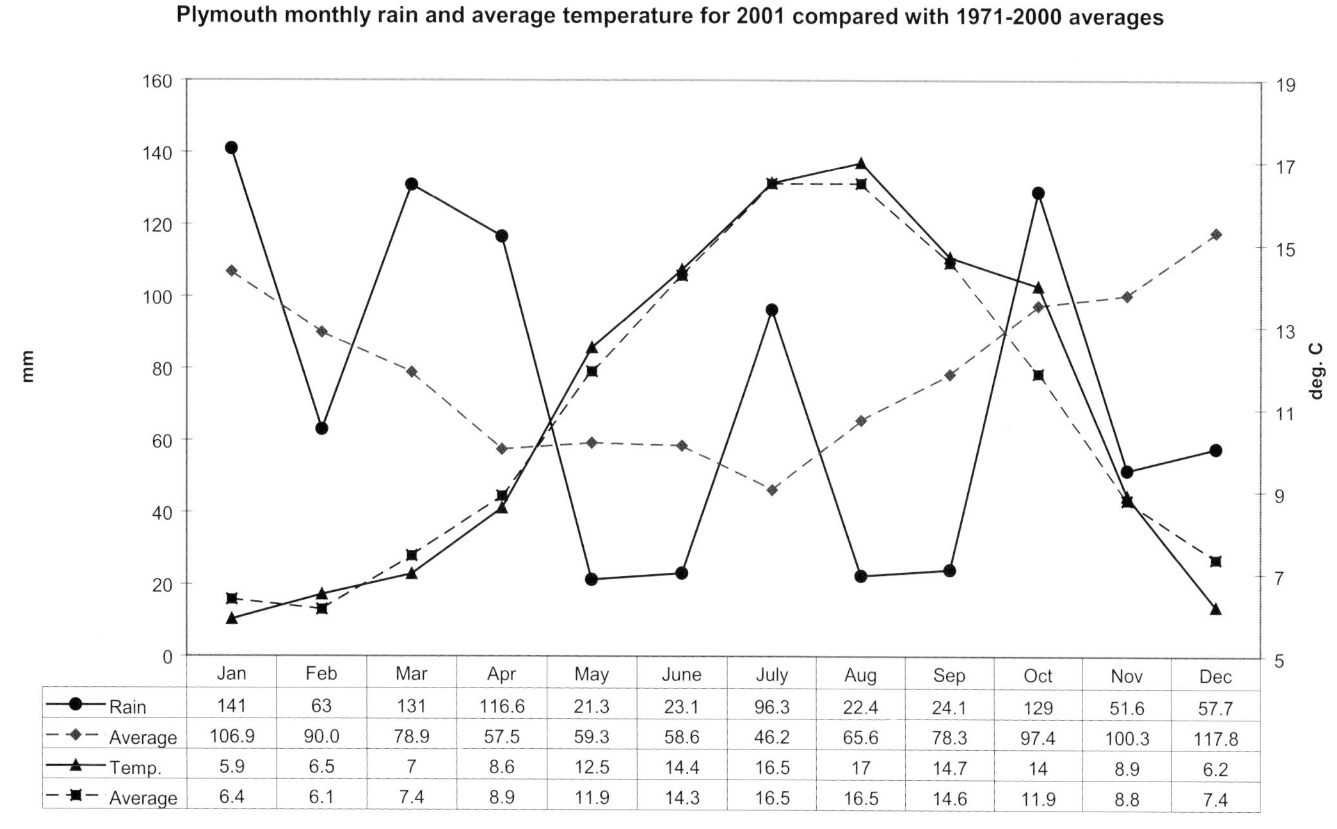

	Jan	Feb	Mar	Apr	May	June	July	Aug	Sep	Oct	Nov	Dec
Rain	141	63	131	116.6	21.3	23.1	96.3	22.4	24.1	129	51.6	57.7
Average	106.9	90.0	78.9	57.5	59.3	58.6	46.2	65.6	78.3	97.4	100.3	117.8
Temp.	5.9	6.5	7	8.6	12.5	14.4	16.5	17	14.7	14	8.9	6.2
Average	6.4	6.1	7.4	8.9	11.9	14.3	16.5	16.5	14.6	11.9	8.8	7.4

Fig. 2 - Rainfall and Temperature

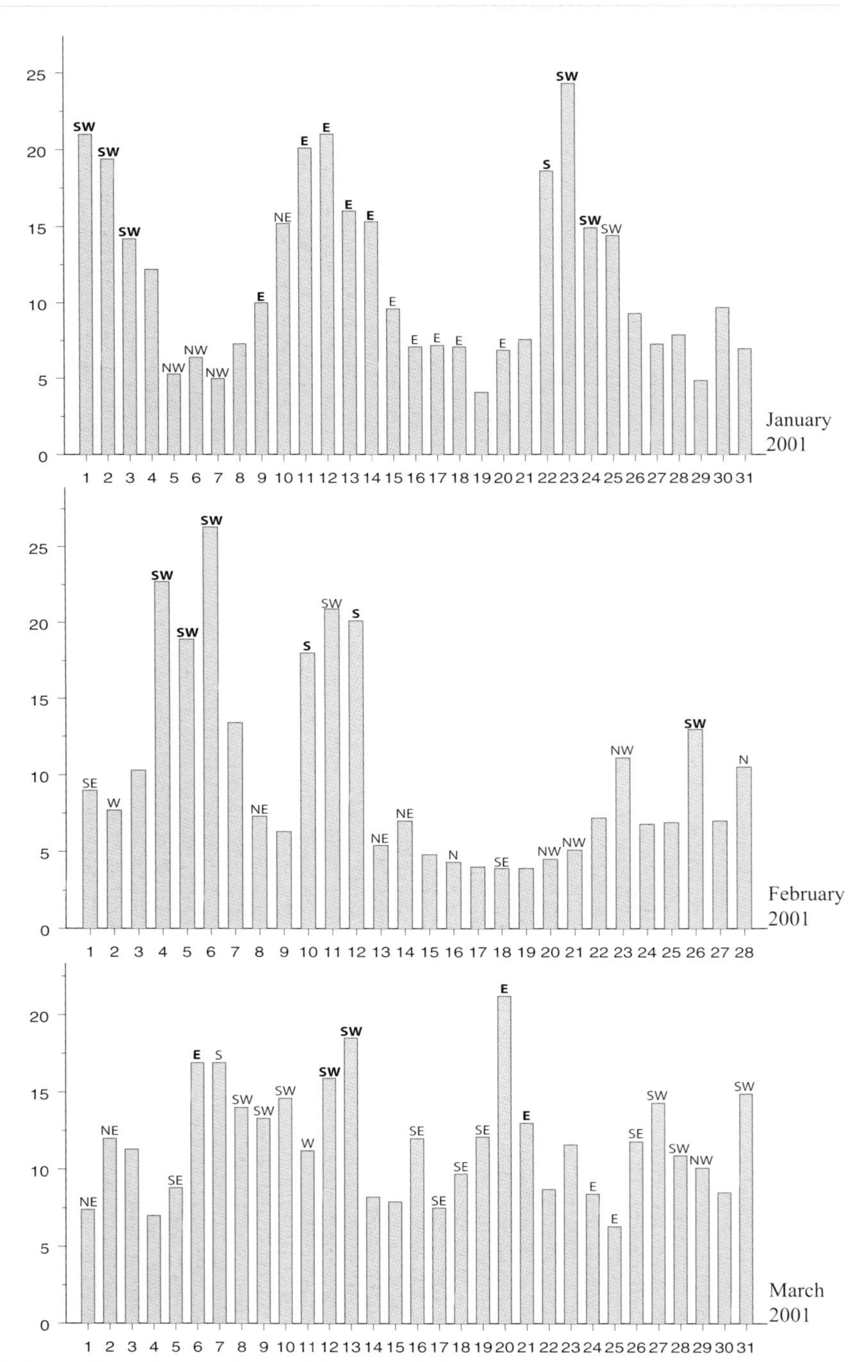

Figs 3, 4 & 5: Mean wind speed - Jan, Feb & Mar 2001

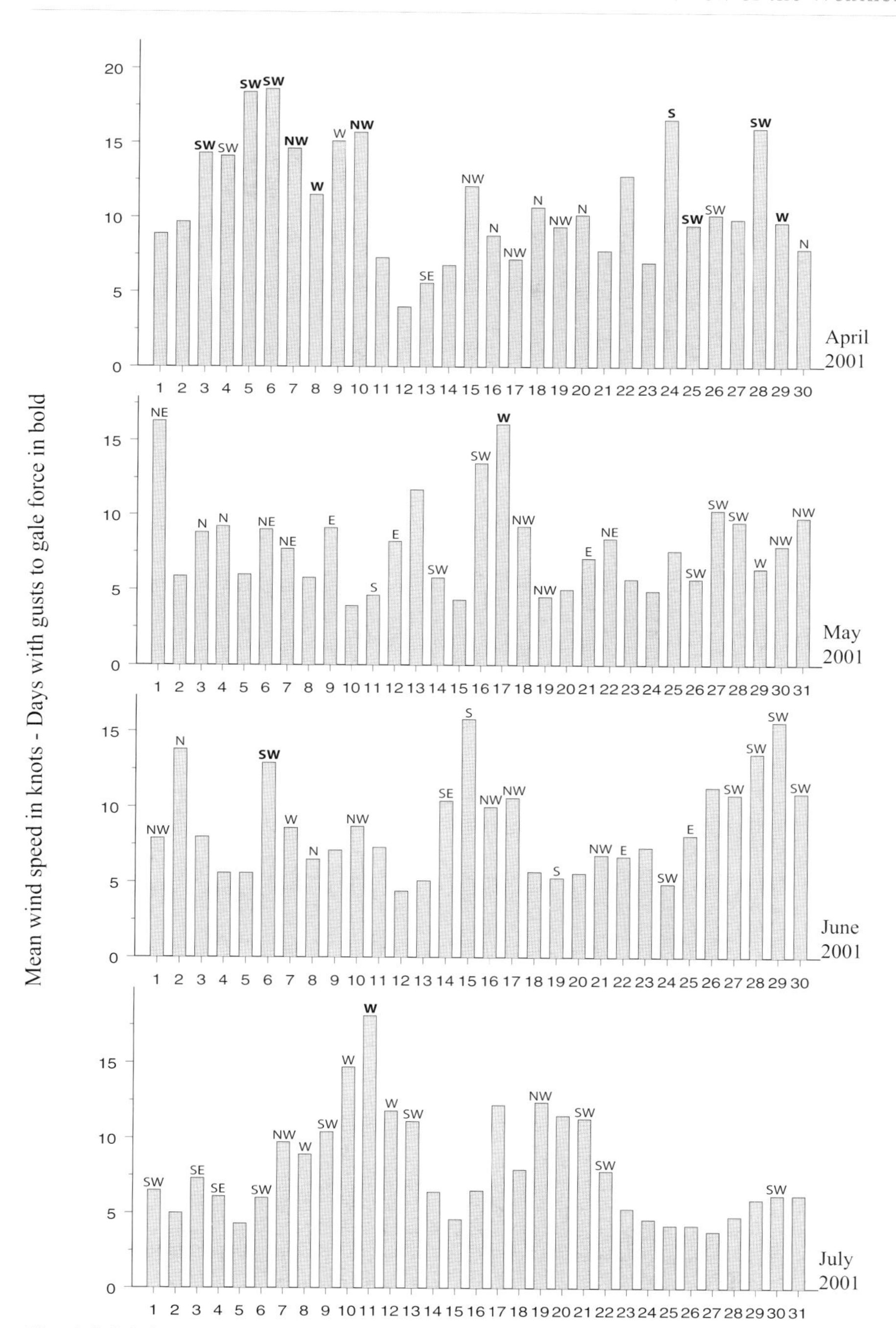

Figs 6, 7, 8 & 9: Mean wind speed - Apr, May, Jun & Jul

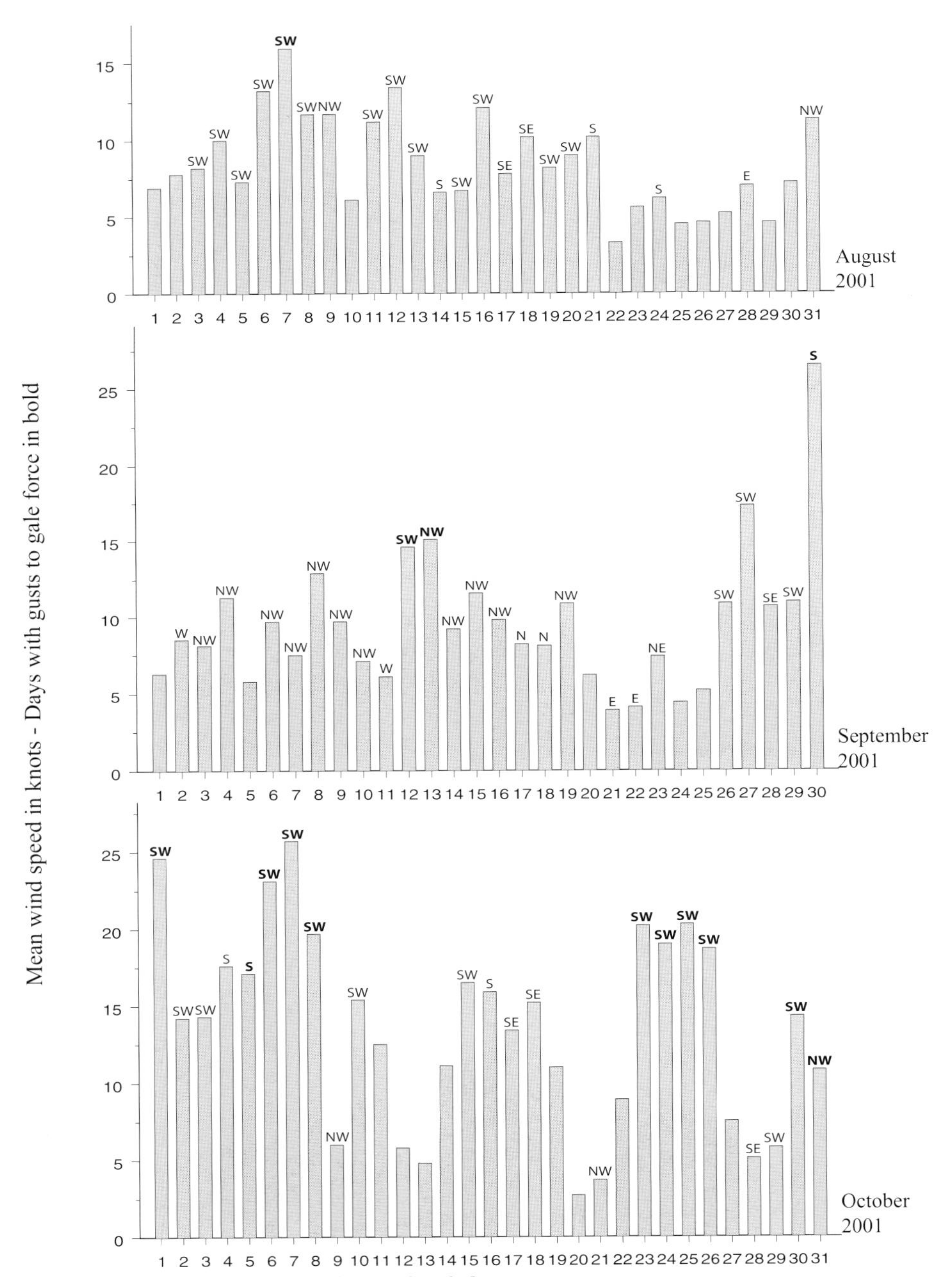

Figs 10, 11 & 12: Mean wind speed - Aug, Sep & Oct

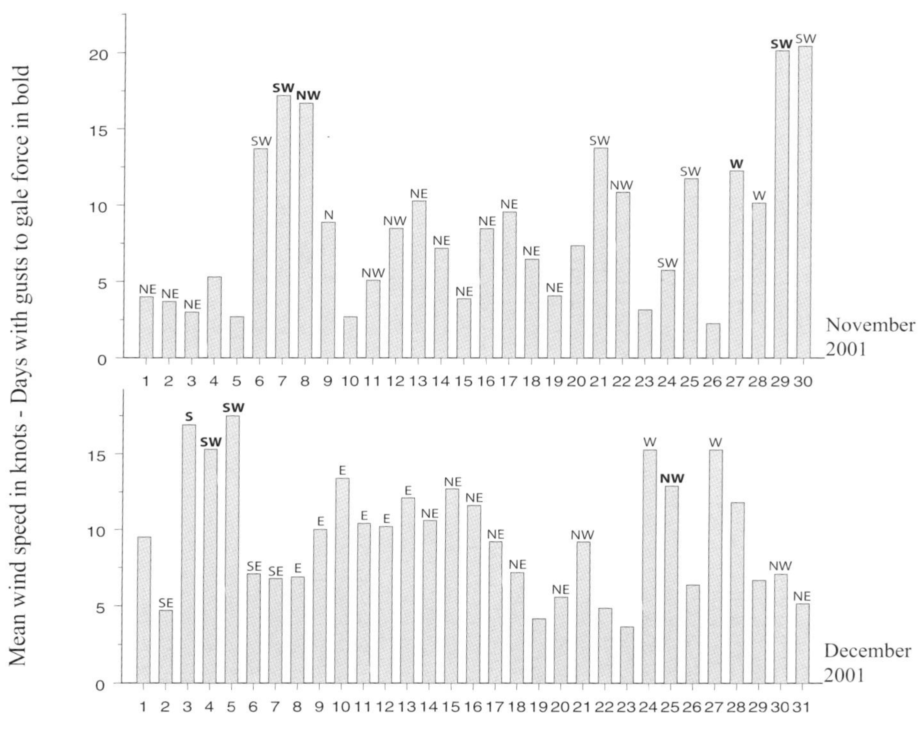

Figs 13 & 14: Mean wind speed - Nov & Dec

Skylark - *Mike Langman*

INTRODUCTION TO THE SYSTEMATIC LIST

The list broadly follows the format of other recent *Devon Bird Reports* (*DBR*s), but the following notes on Sources and Presentation should facilitate the use and interpretation of the individual **species accounts**.

SOURCES

a) **Records** submitted by DBWPS members and others. This is the major source, with *c*.48,000 records held on the database for 2001. This is considerably more than the *c*.32,000 for 2000, and is in fact the highest annual total so far, in spite of the FMD access restrictions. It suggests that the trend of increasing numbers of records submitted (see Fig.1 of *DBR* 1997) has not yet halted.

b) **Bird Guides** and other internet and telephone bird information services.

c) **Local Reports** for Dartmoor (Smaldon, 2002), Exmoor (Butcher, 2002) and Lundy (Taylor, in press).

d) **Data from national surveys**, in particular the Wetland Bird Survey (WeBS) administered by the WWT, and the *Seabird 2000* census administered by JNCC. Data from these are included under individual species accounts where appropriate, and are also summarised, explained and interpreted in separate reports. Note that in 2001 there were no BTO BBS and WBBS data for Devon.

e) **Selected ringing recoveries** relating to 2001 as provided by the Ringing Officer; more detail is given in the Ringing Report (including previously unpublished recoveries for earlier years).

f) **Publications**, including previous *DBR*s, and both local and national reports and publications which have relevance to Devon's birds in 2001. All are listed under References & Sources of Information.

PRESENTATION

a) **Species, sequences and names**. All records of species in Categories A, B and C of the revised British List (BOU, 1998) are included in the main section, even if suspected of being of captive origin. There were no records of species in the revised Category D (where there is doubt that they have occurred naturally in a wild state), but records of Category E species (which have occurred as introductions, transportees or escapes but whose breeding populations (if any) are thought not to be self-supporting) are listed separately under Category E species. For national rarities and DBWPS List A and B species (see Guidance on Record Submission and Devon County List), only records which have been accepted by BBRC or the Devon Birds Records Committee are included. Records which are still under consideration by the relevant records committee are listed in Records Pending and those which have been submitted or published (e.g. on bird 'phone lines), but for which adequate supporting details are still required are given in Records Requiring Further Information.

The sequence and scientific nomenclature follow Voous (1977), as subsequently amended by BOURC reports published in the *Ibis*. The latest decisions have been summarised in *British Birds* (Anon, 2000 & 2002), and are adopted in this report. English names follow *Birds of the Western Palearctic* (*BWP*) (Cramp *et al.*, 1977 *et seq.*) and generally concur with those in common usage (the exceptions being where current BOU taxonomic treatment differs from *BWP*, such as with recent 'splits', in which case the BOU English names are used). Where different, the English names recommended by the BOU (1992) are given on the right hand side of the title line. Records of distinctive subspecies are listed separately at the end of the account of the commonly occurring subspecies.

Within each account, records are generally reviewed either chronologically or alphabetically (by site) in order to aid interpretation. For many species, a summary of the year or review of status is given first.

b) **Abundance and Status**. A brief statement on the abundance and status of each species and subspecies in Devon is given using the definitions listed below, and in general following Rosier (1995).

Abundance	Breeding pairs in Devon	Non breeding pairs in Devon
Rare	Less than annual	Less than three annually
Very Scarce	1–9	3–25
Scarce	10–99	25–249
Not Scarce	100–999	250–2,499
Fairly Numerous	1,000–9,999	2,500–24,999
Numerous	10,000–49,999	25,000–124,999
Abundant	≥50,000	≥125,000

Resident breeder	Breeding species typically making only short-range movements to, for example, feeding or roosting sites (e.g. Nuthatch);
Migrant breeder	Species in which individuals migrate long distances after breeding, typically to Iberia and Africa (e.g. Swallow) i.e. Summer visitors;
Winter visitor	Non-breeding species, spending variable lengths of time in Devon after migration, typically from the N or E (e.g. Brent Goose);
Passage visitor	Species with individuals appearing for relatively short periods during migration from breeding to non-breeding areas and *vice versa* (e.g. Curlew Sandpiper);
Vagrant	Species occurring well outside their normal range, often as a result of exceptional weather conditions (e.g. Red-rumped Swallow), sometimes in irregular irruptions (e.g. Two-barred Crossbill).

For established non-native ('naturalised') species, definitions follow those recommended by Holmes & Stroud (1995), according to the process by which they have become established, i.e.:

Naturalised feral	Domesticated species gone wild (e.g. Feral Pigeon);
Naturalised introduction	Introduced by man (e.g. Red-legged Partridge);
Naturalised re-establishment	Species re-established in area of former occurrence (e.g. Red Kite);
Naturalised establishment	Species which occur in the wild, but breeding established by introduction (e.g. Greylag Goose).
Escapes	All other non-natives, i.e. Category E species.

c) **Conservation Status**. A detailed explanation of Conservation Status and Biodiversity Action Plans is available in *DBR*s for 1997 and 1998, and here only definitions of the terms used are given.

BoCC(Red)	UK Birds of Conservation Concern (Red List Species of *high conservation concern* due to either: being Globally Threatened; a rapid decline in population or range in recent years; or a historical decline and no substantial recent recovery)
BoCC(Amber)	UK Birds of Conservation Concern (Amber List Species of *medium conservation concern* due to either: unfavourable conservation status in Europe; a moderate decline in population or range in recent years; a historical population decline, but substantial recent recovery; being a rare breeder; or having internationally important or localised populations) Anon (2000 and 2002a) and Gregory *et al.* (2002)

SPEC(1)	Species of European Conservation Concern (Category 1: species of global conservation concern because on a world-wide basis classified as Globally Threatened, Conservation Dependant or Data Deficient)
SPEC(2)	Species of European Conservation Concern (Category 2: species whose global populations are concentrated in Europe and which have Unfavourable Conservation Status in Europe);
SPEC(3)	Species of European Conservation Concern (Category 3: species whose global populations are not concentrated in Europe, but which have Unfavourable Conservation Status in Europe);
SPEC(4)	Species of European Conservation Concern (Category 4: species whose global populations are concentrated in Europe, but which have 'Favourable Conservation Status' in Europe);
SPEC(1–4^{w})	Species of European Conservation Concern (categories relate to winter population); Tucker and Heath (1994)
UK BAP	UK Biodiversity Action Plan Priority Species;
SW BAP	SW Biodiversity Action Plan Priority Species;
Devon BAP	Devon Biodiversity Action Plan Priority Species. Anon (1994), Cordrey (1996, 1997) and Devon Biodiversity Partnership (1998)

d) **Tables**. Unless otherwise stated, the figures given in the tables are the highest day total of birds for the period in question (usually a month). Bird-days may be used in situations where individual birds cannot or have not been distinguished, and where numbers are in a state of flux, usually during migration; bird-days are then simply the cumulative daily total for a given period. For some species, it is considered that the number of records or sites gives an appropriate index of abundance. For tables involving waterbird species (i.e. divers, grebes, wildfowl, waders and gulls etc), *WeBS County Totals* are also included, both for 2000 and 2001 in order to facilitate comparisons; sometimes the monthly maxima for individual sites will also be derived from WeBS data, but often these will be from independent counts. Other features of tables are:

Blank (-) = no count/record received or zero count received (usually from WeBS)
Bold figures = nationally important (usually >1% of national population); based on GB thresholds given in Musgrove *et al.* (2001)
Italics = subsites (also indented)
( ) = refers to individual birds known to have also been present the previous month

e) **Observers**. Observers' initials are given for BBRC rarities, DBWPS List A and many B species and other 'rare' and 'very scarce' species. For these, the finder(s), where known, or those reporting the record on the first or last date, or those who provided written descriptions, are acknowledged first; otherwise names are in alphabetic order. Where numerous observers are involved, or the finder(s) are unknown, '*et al.*' (and others) or 'mo' (many observers) are used. Initials are also typically given for unusually early or late dates and other important or notable records such as high counts, or counts of localised colonial or nationally/regionally important breeding species etc. Initials can be deciphered by referring to the List of Observers and Contributors.

f) **Seasons**. Descriptors for the seasons are perhaps not always used consistently, but in general **winter** = Nov-Feb inclusive; **first winter-period** = Jan and Feb; **second winter-period** = Nov and Dec; **spring** = Mar, Apr and May; **summer** = Jun and Jul; and **autumn** = Aug, Sep and Oct. One complication is that the *WeBS year* has no summer, 'winter' is from Nov to Mar inclusive, 'spring' is from Apr to Jun and 'autumn' is from Jul to Oct. While trying to be as consistent as possible, it seems that the seasons may have to be defined slightly differently for different groups of birds.

g) **Sites**. The sites referred to in the text are listed in the Site Gazetteer where further details, including OS Grid References, can be found. In most tables, and in places in the text, some individual locations are grouped together into ecologically meaningful sites (especially where there is known movement between 'sub-sites'). (Although total counts or maxima may be given for sites rather than sub-sites, it is important that observers should continue to be precise in submitting the location of their records and give and Ordnance Survey 6 or 4-figure grid reference wherever possible; this information is retained on the computer database for monitoring and site conservation purposes). For most waterbirds, these sites are based on estuaries, but include counts of birds found on the sea immediately offshore from the estuary, secondary river systems flowing into the main river and bodies of freshwater. Thus counts for the following sub-sites are generally included in their parent site (unless the species is restricted to, or is most abundant at particular sub-sites). In some cases, counts are given for both the parent site and one or more sub-sites in the same table; in these cases the sub-site follows below the parent site and is indicated by appearing indented in italics.

Parent Site	**Sub-site(s) usually included within the Parent Site**
Axe Est	Axmouth/Seaton/Colyford Marshes, and the sea immediately off Seaton.
Avon Est	Meadows around Aveton Gifford, and the sea immediately off Bantham and Burgh Island.
Dart Est	Dartmouth, Kingswear, estuary to tidal limit at Totnes but not inc. town, Stoke Gabriel Mill Pool, and the sea between and including Coombe Pt and Inner Froward Pt
Erme Est	Efford, Flete, Orcheton and Pamflete.
Exe Est	Bowling Green Marsh (Topsham), Clyst Est and valley up to Fishers Mill area, Cockwood, Countess Wear, Dawlish Warren, Exminster Marshes, River Kenn/ Powderham Park, Starcross and the sea between Dawlish and Orcombe Pt.
Kingsbridge Est	Bowcombe Creek, Blanksmill, Collapit Creek, Frogmore Creek, Southpool Creek, Salcombe area and West Charleton Marsh.
Otter Est	Rocks at Otterton Ledge, the sea off Budleigh Salterton and Otter Cliffs.
Plym Est	Chelson Landfill, Saltram Park, Saltram tidal pool (Blaxton Meadow), Hooe Lake and estuary to tidal limit at Coy Pool.
Plymouth Sound	Bovisand, Cattewater, Cawsand Bay (Cornwall), Devils Pt, Hooe Lake, Jennycliff, Mount Batten Bay, Plymouth Hoe and Turnchapel.
Start Bay.	The sea between Strete and Hallsands.
South Huish area	Thurlestone Marsh, South Milton Ley, South Huish Marsh, Thurlestone Sands and immediate sea area.
Tamar Complex	(Devon) Mayflower Marina, Stonehouse Creek, Bull Pt, Ernesettle, Saltash Passage, Upper/Mid-Tamar Est, Tamerton Creek and Tavy Est (Warleigh Pt, Warleigh Marsh, Blaxton Wood/Marsh, Lopwell Dam/Lake and Bere Ferrers)
Taw/Torridge Est	The Taw upstream to Barnstaple including Ashford, Bradiford, Braunton Burrows, Braunton Great Field, Braunton Marsh, River Caen, Chivenor, Crow Neck/Pt, Fremington Pill, Horsey Is, Islay Marsh, Penhill Pt and Marshes, Sherpa Marsh, Velator, Wrafton and Yelland; the Torridge upstream to Landcross including Appledore, East-the-Water, Instow, Northam Burrows, Skern, Westleigh, R Yeo at Landcross; and the sea between Baggy Pt and Westward Ho!
Teign Est	Passage House Inn and Hackney Marshes.

Torbay	The sea between Meadfoot Beach and Brixham Harbour. 'Torbay area' is used to indicate a slightly wider area including the adjacent headlands of Hope's Nose and Berry Hd
Yealm Est	Steer Pt, Cofflete Creek, Puslinch, Newton Ferrers, Noss Mayo, Kitley Pond and the sea area from Mouth Stone to Season P

Treatment of the 'Tamar Estuary Complex' is problematic, since the county boundary with Cornwall divides the estuary system. The complex is treated as a single site for national reporting on WeBS counts (Musgrove *et al.*, 2001) which clearly makes sense in terms of an ecological unit and for the assessment of the importance of populations, and it matches the statutory designation as a candidate SAC ('Plymouth Sound and Estuaries'). The treatment in this report follows that in *DBR*s since 1999 in that, for some of the commoner species, figures are given in the tables for the whole complex (i.e. including Cornish sectors) based on WeBS counts. In addition, counts are generally also given for the two sub-sites which are at least partly in Devon. *Where WeBS County Totals are given, they include Tamar counts only for these two sub-sites, which are*:

- Upper/Mid-Tamar Est (above the Tamar Bridge as far as Halton Quay, but excluding the Tavy) - split between Devon and Cornwall, and including important Devon sites such as Ernesettle and Tamerton Creek.
- Tavy Est (above the Tavy Bridge) - entirely in Devon.

h) Other Conventions and Abbreviations: **Ages/plumages:**

aos	Apparently occupied (nest) site	♂	Male
BAP	Biodiversity Action Plan	♀	Female
BBRC	British Birds Rarities Committee	yg	Young
BBS	Breeding Bird Survey	juv	Juvenile
BOU	British Ornithologists' Union	imm	Immature
BOURC	BOU Records Committee	1s/1w	First summer/winter
BWP	*Birds of the Western Palearctic*	2s/2w	Second summer/winte
BoCC	Birds of Conservation Concern	3s/3w	Third summer/winter
CES	Constant Effort Sites Scheme	1y	First year (juv/1w/1s)
CP	Country Park	2y	Second year (2w/2s)
DBR	Devon Bird Report	3y	Third year (3w/3s)
DNPA	Dartmoor National Park Authority	1cy	First calendar year
DSGp	Dartmoor Study Group	2cy	Second calendar year
Est/Ests	Estuary/Estuaries	3cy	Third calendar year
GBS	Garden Bird Survey	ad	Adult
Hd	Head	bp	Breeding plumage
L	Lake	nbp	Non-breeding plumage
mo	Many observers	pr/prs	pair/pairs
NR/LNR/NNR	Nature Reserve/Local NR/National NR		
Pt	Point		
R	River		

Res	Reservoir(s)		
SAC	Special Area of Conservation		
sp/spp	Species (singular/plural)		
SPEC	Species of European Conservation Concern		
SPA	Special Protection Area		
WBBS	Waterways Breeding Bird Survey		
WeBS	Wetland Bird Survey		
WWT	Wildfowl and Wetlands Trust		

GENERAL OVERVIEW OF DISTRIBUTION MAPS

Twelve species were chosen for mapping in 2001, these being Little Egret, Red-legged Partridge, Grey Partridge, Golden Plover, Lapwing, Jack Snipe, Little Owl, Nightjar, Black Redstart, Cetti's Warbler, Firecrest, and Willow Tit. This suite of species was chosen to represent a combination of breeding, migrant and/or winter distribution. Ten years of data has been used for all species 1992-2001, except Little Egret 1997-2001.

The division of migrant and wintering birds was arbitrary, with winter being Dec-Feb, except where stated. Temporal parameters for breeding birds were based on the breeding cycle described in BWP, i.e. first eggs to latest hatching.

Thematic mapping has been used to indicate relative abundance, whilst allowing the maximum number of records to be shown per map. Class intervals were chosen arbitrarily, so as to emphasise relative abundance between sites.

Grid references were lacking for most records, these have mostly been allocated as per the location name given, based on the Ordnance Survey 1:25000 map series. For supersites and large estuaries the most central 1-km square was used.

Where identified, cold-weather records of Golden Plover, Lapwing and Jack Snipe, together with obvious fly-over records of these and Little Egret have been excluded.

SECTION WRITERS FOR THE SPECIES ACCOUNTS

Divers to Petrels - **Lee Collins**

Gannet to Spoonbill - **Peter Reay**

Wildfowl - **Bob Normand & John Fortey**

Raptors - **Mark Darlaston**

Gamebirds, Rails - **Richard Hibbert**

Crane & Waders - **Ivan Lakin**

Skuas - **Mike Langman**

Gulls - **Simon Geary & Peter Reay**

Terns & Auks - **Martin Elcoate & Peter Reay**

Pigeons to Wagtails - **Richard Hibbert & Peter Reay**

Waxwing to Wheatear - **Simon Geary**

Ring Ouzel to Mistle Thrush - **Richard Hibbert**

Warblers to G Oriole - **Matthew Knott & Peter Reay**

Dartford Warbler - **Roger Thornett**

Shrikes to Buntings - **Barrie Whitehall**

Hawfinch - **Roy Adams**

Cirl Bunting - **Cath Jeffs**

SYSTEMATIC LIST - SPECIES ACCOUNTS

RED-THROATED DIVER *Gavia stellata*

Scarce winter and passage visitor to estuaries and offshore; rare inland. SPEC(3), BoCC(Amber).

The highest counts of the year both emanated from Hartland, with 100 plus on 3 Jan and c.120 on 9 Dec from Shipload Bay, similar to 2000 and well above the qualifying level for national importance (50). Unfortunately Ladram Bay had only two records, with a max count of five, dramatically down on previous years (e.g. 50 in 2000).

Table 1: Monthly maxima at main/well-watched sites:

	J	F	M	A	M	J	J	A	S	O	N	D
Dawlish Warren / Bay	9	26	12	3	1	-	-	-	1	1	6	4
Hartland Point	100+	2	-	-	-	-	-	-	-	1	3*	120
Otter Est / Ladram Bay	5	1	-	-	-	-	-	-	-	-	-	4
Start Bay	4*	1	-	1	1	1	-	-	-	1	2*	2*
Torbay	12(11*)	2	1*	9*	-	-	-	-	1	2	2	15

** refers to passage birds*

First winter-period/spring. With 83 submitted records for Jan & Feb alone, all those quoted are of counts in double figures. As usual, the highest count was from Hartland with 100+ on 3 Jan (ColW), with other large Jan counts from the area including: 70 on 5th, 42 on 18th, 14 on 20th and 30 on 31st. At Baggy Pt, 23 flew S on 5 Jan and 11 on 28 Jan. On the S coast, Dawlish Warren hosted the largest counts, with a peak of 26 on 16 Feb, 20 on 3rd, 11 on 13th and 12 on 13 Mar. The only other double-figure count was of 12 from Hope's Nose on 1 Jan, of which 11 passed S 08.45-10.45h. Records of smaller numbers from Torbay, Prawle Pt, Westward Ho!, Baggy Pt, Plymouth, Kingsbridge Est, Cornborough and Start Bay show that these birds can appear anywhere off either coastline. Very few reports gave details of passage movement; in addition to those already mentioned, the exceptions included four W past Prawle on 1 Jan, and at Hope's Nose, nine S on 1 Apr and four S on 22 Apr.

Summer. Only two records: one from Baggy Pt on 1 Jun, and one in bp around Start Bay on 6th.

Autumn/second winter-period. An early record was from Dawlish Warren on 2 Sep, with the next bird also there on 27 Sep. Hope's Nose produced the only other Sep bird, one heading S on 30th. Only three Oct records at Dawlish Warren, but regularly recorded there in Nov, with a max of six on 12 & 18 Nov. Oct records elsewhere were: one at Hartland Pt on 9 Oct; one at Broadsands on 23rd; two past Berry Hd on 26th; and four past Baggy Pt & one in Start Bay on 29th. In Nov, records from many sites mostly involved singles, the exceptions being: three at Hartland Pt on 11 Nov; two from Berry Hd on 16th; and two at Prawle on 18th. In Dec, records ranged from Prawle Pt to Seaton in the S and Hartland to Westward Ho! in the N, but large counts remained scarce. From Hartland Pt ten were seen on 8 Dec, yet the following day *c.*120 were counted from nearby Shipload Bay (MD,AR); unfortunately records from this area are infrequent, and all other counts were in single figures. Torbay peak counts included nine on 19 Dec and 15+ on 23 Dec, whilst four birds were seen from Buck's Mills on 27 Dec and Brandy Hd on 30 Dec.

BLACK-THROATED DIVER *Gavia arctica*

Very scarce passage and winter visitor to estuaries and offshore; rare inland. SPEC(3), BoCC(Amber).

Generally lower numbers than in 2000, particularly in Torbay and Start Bay.

Table 2: Monthly maxima at main/well-watched sites:

	J	F	M	A	M	J	J	A	S	O	N	D
Dawlish Warren / Bay	-	2+(1*)	1	-	-	1	-	1	-	-	3+	1
Hartland Point	-	-	-	-	-	-	-	-	-	-	-	-
Otter Est / Ladram Bay	-	-	-	-	-	-	-	-	-	-	1	1
Start Bay	-	1*	-	-	1*	-	-	-	-	-	1*	1
Torbay	2	2	-	-	-	-	-	-	-	-	1+	1

** refers to passage birds*

First winter-period/spring. A lean period and, unless otherwise stated, records concern single birds. The first was from Brixham with two on 1 Jan (PMM), and a single also noted between 10-13 Jan (ML,MRAB,CG). The only other records around *Torbay* were from Brixham-Broadsands area 6-17 Feb (PBo,ML,MD,AR), with two on 7th. From Dawlish Warren, the first for the year was on 3 Feb (JRD), with a bird also seen heading E on 8 Feb (IL), plus another seen from Exmouth on 18th (DRCx), and one on 14-15 Mar (IL,KRy). Elsewhere on the S coast, only two other records: one W past Prawle Pt on 10 Feb (PMM), and a late bp bird E past Thurlestone Bay on 10 May (ML). From N Devon: one from Green Cliff, Abbotsham on 19 Jan (JEW) and one from Baggy Pt on 13 Feb (RJ).

Summer. Records suggest three birds over this period, two 1s and the other not specified: Dawlish Warren played host to both 1s birds, the first 5-9 Jun (IL), plus another (possibly the same?) 1-3 Aug (IL,THS); and an unexpected find at Batson Creek, Kingsbridge Est. on 1 Jul (JBB).

Autumn/second winter-period. During Nov & Dec, 25 records were received and all concerned single birds except three at Dawlish Warren on 12 Nov (JEF). The earliest was from the Skern, on 2 Nov (IM). Dawlish Warren had numerous records but possibly just one bird; the first was 10-16 Nov (KRy *et al.*), then on 30 Nov, 6-7 & 16 Dec (IL). Nearby Budleigh Salterton had records on 26 Nov and 18 Dec that may well relate to the same bird (DWH,RJO). Torbay had birds off Elberry on 10 Nov (MD,AR), Sharkham Pt on 12 & 24 Nov (ML,RBe) and Broadsands on 1 Dec (MKn). Other records were from: Staddon Pt, Bovisand on 11 Nov (SGe); another W past Prawle Pt on 24 Nov (JCN); Jennycliff Bay, Plymouth on 14 Dec (BG); and Start Bay on 30 Dec (BW).

GREAT NORTHERN DIVER *Gavia immer*

Scarce passage and winter visitor to estuaries and offshore; rare inland. BoCC(Amber).

More in Torbay and Plymouth Sound than in 2000, but fewer in Start Bay and past the headlands.

Table: 3 Monthly maxima at main/well-watched sites:

	J	F	M	A	M	J	J	A	S	O	N	D
Berry Head	-	1	2	1*	-	-	-	-	-	2	-	1
Dawlish Warren / Bay	16	5	5	5	2	2	-	1	-	1	5	3
Hope's Nose	4	3	-	-	-	-	-	-	-	-	-	1
Plymouth Sound	9	6	2	1	-	-	-	-	-	-	-	-
Prawle Point	1	-	-	-	1	3	-	-	-	-	2	2
Start Bay	1	4	1	2	1	-	-	-	-	-	1	2
Torbay	11	16	2	-	2	4	-	-	1	-	10	2

* *refers to passage birds*

First winter-period. The 187 submitted records for this period were mostly from Plymouth to Dawlish Bay, with only one record further E at Ladram and a handful of sightings from N Devon. The largest counts included 11 at Broadsands and 16 in Dawlish Bay on 2 Jan, nine from Plymouth on 28 Jan and 16 in Torbay on 11 Feb.

Spring/summer. All single figures, with max of five at Dawlish Warren on 14 Mar and 24 Apr; elsewhere, ones and twos. Passage difficult to evaluate, but appeared low compared to 2000 with the only records: four past Dawlish Warren on 13 Mar; singles at Berry Hd on 24 & 30 Apr; two off Dawlish Warren on 17 May; one past Slapton on 28 May; and one on 2 Jun and three on 3rd past Prawle. Other Jun sightings were all from Dawlish Warren; initially seen on 5 Jun, there were two on 6-7th, with a first summer lingering until 18th. The only Lundy record was one imm 5-8 May.

Autumn/second winter-period. The earliest was one in bp off Exmouth on 9 Aug. The next were on 2 Sep, with singles at Appledore, Broadsands and Dawlish Warren. Another lull ensued until early Oct, when birds started to appear in Torbay and Dawlish Warren. Both these sites then held the majority of all submissions, with peaks of five from Dawlish on 14 Nov and from Torbay, four off Three Beaches on 2 Nov. Elsewhere in S Devon: Start Bay had its first on 3 Nov, with only three other records on 10 & 22 Nov and 21 Dec; and Prawle and Thurlestone accounted for the remaining sightings. In N Devon, singles were at Bull Pt, Skern, Hartland Pt, Yelland and Bideford Quay; and on Lundy, up to two, 28 Oct – 5 Nov.

DIVER species *Gavia spp.*

Several records, mostly of distant birds in flight, in ones or twos from the usual well watched S Devon sites, particularly in the first winter-period.

PIED-BILLED GREBE *Podilymbus podiceps*

Rare vagrant.

The first record for the county. This long overdue county addition has now appeared 36 times in Great Britain, with neighbouring Cornwall accounting for five.

One at Upper Tamar L on the Cornwall/Devon border, 8 Jan – 24 Feb (S.M.Christophers,B.M.Phi lips *et al.*). (Accepted by BBRC, this was the only one in Britain in 2001.)

LITTLE GREBE *Tachybaptus ruficollis*

Scarce winter visitor, rare or very scarce resident and migrant breeder.

Number of breeding pairs down on 2000; WeBS county winter totals broadly similar, though slightly lower in autumn and higher in Dec.

Table 4: Monthly maxima at main/well-watched sites and WeBS county totals:

	J	F	M	A	M	J	J	A	S	O	N	D
Avon Est	6	5	4	1	-	-	-	2	-	2	4	4
Axe Est	3	2	-	-	-	-	2	2	3	2	2	5
Bicton L	2	2	-	-	-	-	-	-	1	3	1	10
Burlescombe	-	2	-	-	5	6	9	9	5	3	3	-
Erme Est	4	2	-	-	-	-	-	-	3	2	6	2
Exe Est	3	5	8	3	-	2	1	7	2	7	6	3
Grand Western Canal	2	-	-	-	2	-	-	-	-	-	-	1
Kingsbridge Est	16	21	-	-	-	-	-	1	-	5	16	15
Otter Est	3	4	4	-	-	-	1	-	1	2	3	7
Plym Est	6	8	6	-	2	-	2	-	4	3	6	4
Radford L	*6*	*10*	*8*	-	*2*	*4*	*4*	*4*	*3*	-	*6*	*3*
Roadford Res	29	27	-	-	-	-	2	16	-	20	20	11
Shobrooke L	-	-	-	-	-	-	2	1	1	1	1	-
Slapton Ley	3	4	2	-	-	2	-	5	3	2	-	1
Thurlestone area	1	1	2	-	-	-	-	-	-	-	-	1
Tamar Complex WeBS	11	11	-	-	-	-	-	-	2	12	18	17
Tavy Est	-	-	-	-	-	-	-	-	-	*3*	*7*	*4*
Taw / Torridge Est	14	16	5	8	-	2	3	4	6	3	9	15
Teign Est	5	3	5	4	-	-	-	-	-	1	1	8
Yealm Est	12	4	-	-	-	-	-	-	2	3	12	16
Kitley Pond	*5*	*1*	-	-	-	-	-	-	*9*	*4*	-	-
WeBS county total '00	106	107	73	7	3	11	15	52	59	80	96	79
WeBS county total '01	107	100	4	-	-	-	2	12	29	61	76	103

Breeding. Records confirm breeding at ten different sites with 12 prs counted (*cf.* 13 sites with 24 prs in 2000, and counts in parentheses show 2000 results for comparison): Radford Lake, 2 prs noted trilling early, yet only 1 pr bred, rearing 2 yg *(4prs with 5 yg)*; Uffculme, 2 prs reared 3 yg *(pr with 2 broods)*; Roadford Res, a juv noted on 5 Jul *(8-10 prs)*; Higher Metcombe, pr bred, but no other details supplied *(pr with 4 yg)*; Horsham Pond, Manaton, pr with 2 yg *(pr with 3 yg)*; Huntsham Lake, pr reared single yg *(pr noted)*; Yelland Pond, pr with 2 yg *(pr with 1 yg)*; Taw Est, pr reared 2 broods and one juv from each brood noted *(no details last year)*; Velator, pr reared 1 yg *(no details last year)*. Also of note: a bird trilling in Jun at Wrafton Pond; a pr displaying in Feb at Kitley Pond; and two juvs on Exminster Marsh on 8 Aug. Sites with only summer sightings (but breeding confirmed in 2000) included Burlescombe and Bowling Green Marsh, whilst other regular summer sightings were from Isley Marsh and Shobrooke Lake. No breeding records or summer sightings from Dawlish Warren, Buckfast, Borras pit nr Seaton Marshes and Lopwell Dam (all with confirmed breeding in 2000).

First winter-period. The highest count was 29 at Roadford Res in Jan, and other double-figure counts occurred on the Kingsbridge, Tamar, Taw/Torridge and Yealm Ests. The highest count from an untabulated site was eight on R Dart at Dartington in Feb.

Second winter-period. Roadford Res again produced the max count with 20 in both Oct and Nov, and double-figure counts came from Bicton L and the above four estuaries.

GREAT CRESTED GREBE *Podiceps cristatus*

Scarce winter visitor and very scarce or scarce resident breeder.

Fewer breeding prs and some lower WeBS county totals than in 2000. No new breeding sites.

Table 5: Monthly maxima at main/well-watched sites and WeBS county totals:

	J	F	M	A	M	J	J	A	S	O	N	D
Beesands Ley	3	2	1	2	-	-	-	-	-	-	-	1
Exe Est / Dawlish Bay	6	21	20	6	3	3	4	2	2	5	10	13
Fernworthy Res	-	1	-	-	-	-	4	1	2	-	-	-
Hennock Res	-	1	-	2	2	1	-	-	-	-	-	1
Kingsbridge Est	15	6	-	-	-	2	-	1	-	10	5	7
Roadford Res	2	7	-	-	-	-	1	-	-	3	4	-
Shobrooke L	2	4	-	-	-	-	3	3	3	3	3	-
Slapton Ley	11	24	27	35	34	32	61	59	58	45	32	21
Start Bay	6	5	-	-	2	-	-	-	-	-	-	-
Tamar Lakes	4	7	-	-	-	-	3	4	2	3	4	2
Tamar Complex WeBS	25	15	-	-	-	-	-	-	3	48	32	24
Upper / mid Tamar	*7*	-	-	-	-	-	-	-	-	-	-	*32*
Taw / Torridge Est	1	1	-	-	-	-	-	-	-	1	1	2
Torbay	71	26	14	1	-	-	1	-	3	2	16	19
Babbacombe	-	*120*	-	-	-	-	-	-	-	-	-	-
Hope's Nose	*49*	*73*	-	*1*	-	-	-	-	-	*1*	*1*	*1*
WeBS county total '00	53	53	60	37	29	23	21	20	22	52	23	34
WeBS county total '01	32	37	2	2	2	4	4	5	10	63	55	33

Breeding. Confirmed at six sites and involved only 15 prs (*cf* 27 prs in 2000, 21 in 1999 & 14 in 1998): Slapton, with 8 prs hatching 22 yg, 10+ fledging (down on 2000, when 17 prs had 36 yg); Stover CP, two broods of 2 yg fledged (more success than 2000); New Cross Pond, one juv noted (similar to 2000); Shobrooke L, one pr with a late brood of 4 hatched and all still present in Sep (2prs on 2000); Lower Tamar L, one juv noted (similar to 2000); and Fernworthy Res, possibly two broods; Hennock, one on nest at Trenchford Res, and an ad and one juv at nearby Kennick Res on 14 Aug. No records from Bishop's Brook Pond (new site in 2000), Rackerhayes or Beesands.

First winter-period. Counts were in line with previous years, with max numbers around the usual wintering sites: 120 from Babbacombe on 4 Feb (JWn), once again exceeding the criterion for *national importance* (100); 24 at Slapton on 17 Feb; 25 on the Tamar Est in Feb; 21 at Dawlish Warren on 21 Feb; and 15 on the Kingsbridge Est on 21 Jan.

Second winter-period. With no particularly large counts from Torbay, max counts came from the Tamar Est and Slapton Ley, though Dawlish Warren, Kingsbridge Est and Torbay also produced counts ≥10.

RED-NECKED GREBE *Podiceps grisegena*

Scarce but regular winter and passage visitor; rare in summer and inland. BoCC(Amber).

Fifty-three records received (cf. 76 in 2000).

Table 6: Monthly maxima at main sites:

	J	F	M	A	M	J	J	A	S	O	N	D
Dawlish Warren	1	-	-	-	-	-	-	-	-	-	4	1
Plymouth Sound	1	-	-	-	-	-	-	-	-	-	-	-
Start Bay	2	2	-	1	1	-	-	-	-	-	1	-
Torbay	1	-	1	-	-	-	-	-	-	1	2	1

First winter-period/spring. The first was on 1 Jan, with a single bird seen at Plymouth, around Cattewater/Mt Batten area for one day (VRT,SRT,RWBu) though the same or another was seen off Drake's Island on 21 Jan (SGe). On 2 Jan, one also appeared at Broadsands and remained in the area until 23 Jan (ML), although a sighting on 27 Jan from Anstey Lea may have been another bird (PBo). On the Exe Est. another single was found on 5 Jan (MRAB), but one on 20 Jan from Dawlish Warren may have been the same (IL). The only *inland* record was from Roadford Res 20 Jan - 2 Feb (GAV). Off Slapton there were two on 14 Jan and 18 Feb (GEC) with one then lingering until 27 Feb (RWBu,KEM). A more unusual record was of a single bird at Dittisham on 3 Feb (JMW). A bird in bp was noted on 22 Mar from Three Beaches (ML), whilst another in bp was at Slapton on 20 Apr (PHA), and again 4-5 May (MKn).

Autumn/second winter-period. The earliest bird was at Broadsands on 8-13 Oct (RBe,ML), and there were two at nearby Preston on 11 Nov (MRAB). Dawlish Warren also had two on 11 Nov (LC,KRy), but on 12th there were four - an exceptional count and a site record (KRy,JEF); thereafter, singles were noted regularly until 29 Nov, with two also seen on 17th & 19th. Few other birds were noted, all singles: Goodrington on 24 Nov (MRAB); Prawle Pt on 30 Nov (MD); R Clyst on 18 Dec (MSW); and finally Torbay on 23 Dec (MD).

SLAVONIAN GREBE *Podiceps auritus*

Scarce winter and passage visitor. BoCC(Amber).

Two hundred and twenty six records received (cf. 102 in 2000).

Table 7: Monthly maxima at main sites:

	J	F	M	A	M	J	J	A	S	O	N	D
Exe Est / Dawlish Bay	18	17	17	3	-	-	-	-	-	-	3	5
Kingsbridge Est	6	1	-	-	-	-	-	-	-	-	-	-
Plymouth Sound	-	-	1	-	-	-	-	-	-	-	-	-
Start Bay	4	7	4	3	-	-	-	-	-	-	4	1
Taw / Torridge Est	-	-	-	-	-	-	-	-	-	-	-	-
Torbay	2	3	1	-	-	-	-	-	-	-	-	-

First winter-period/spring. Dawlish Warren remains the Devon stronghold, with peak counts over the first three months of 18, 17 & 17 similar to those of recent years. Elsewhere: at Slapton, a max of seven on 23 Feb, and counts of four on 20 Jan, 9 Feb and 19-21 Feb; on the Kingsbridge Est, four 7-28 Jan, five on 15 Jan and six on 27 Jan; in Torbay, many records of singles 25 Jan - 23 Feb, with three on 14 Feb and two on 17 Feb at Three Beaches; at Beesands, three on 29 Jan; at Prawle, two on 16 Feb; at Seaton Hole, two on 28 Feb; and finally the only Plymouth record, one from the Mt Batten area on 24 Mar.

Autumn/second winter-period. The first were two off Dawlish Warren on 2 Nov, followed by mostly singles but with max counts of five on 2 & 11 Dec. Elsewhere: Slapton had four on 3 Nov and one the following day; in Torbay, singles were at Elberry Cove on 25 Nov and Three Beaches on 2 Dec; and in Thurlestone Bay there was one18-19 Dec.

BLACK-NECKED GREBE *Podiceps nigricollis*

Very scarce winter and passage visitor. BoCC(Amber)

One hundred and fifty eight records received (cf. 137 in 2000).

Table 8: Monthly maxima at main sites:

	J	F	M	A	M	J	J	A	S	O	N	D
Exe Est / Dawlish Bay	1	1	3	-	-	-	-	-	1	-	1	-
Roadford Res	-	-	-	-	-	-	-	-	-	-	-	-
Start Bay	1	1	-	1	-	-	-	1	-	1	-	2
Torbay	3	38	11	-	-	-	-	-	-	-	12	17

First winter-period/spring. In *Torbay*, over the first five weeks of the year, counts never exceeded three from any site; yet following an influx in early Feb, there were max counts for the bay of 35 on 11 Feb and 38 on 21 Feb (both ML), the latter being a Torbay record count. *Elsewhere*: at Slapton, singles on 14 & 19 Jan (ML,PHA,BW), 5th & 17-21 Feb (DIJ,BW) and one on 11 Apr (EM); at Dawlish Warren, singles on 2 Jan (JEF), 24 Feb (IL) and 25 Mar (TF), with three on 29-30 Mar (DC,JEW).

Autumn/second winter-period. In *Torbay*, the earliest were five on 4 Nov at Preston seafront (MRAB); numerous records followed, but the only double-figure counts were: 12 on 10 Nov (MD,AR), ten on 26 Nov (CJP) & 3 Dec (MD,AR), 17 on 29 Dec (MD,AR) and 12 on 30 Dec (MRAB). *Elsewhere*, early birds appeared from mid-August: one at Slapton Ley 19-31 Aug (BW,PSa); a moulting bird on the

Plym Est on 24 Aug (BG); one at Dawlish Warren on 22 Sep (JRD,GV); and two at Dartmouth on 24 Sep (AF). Subsequently, singles were noted: at Start Bay on 27-29 Oct (GEC); at Dawlish Warren on 2 Nov (JEF); on the Kingsbridge Est. on 3 Nov (PSa); and at Slapton Ley 15-24 Dec (BW). Counts of more than one included three past Prawle Pt on 10 Nov (JLt), two on the Kingsbridge Est on 18 Nov (JHB) and two at Slapton Ley on 30 Dec (BW).

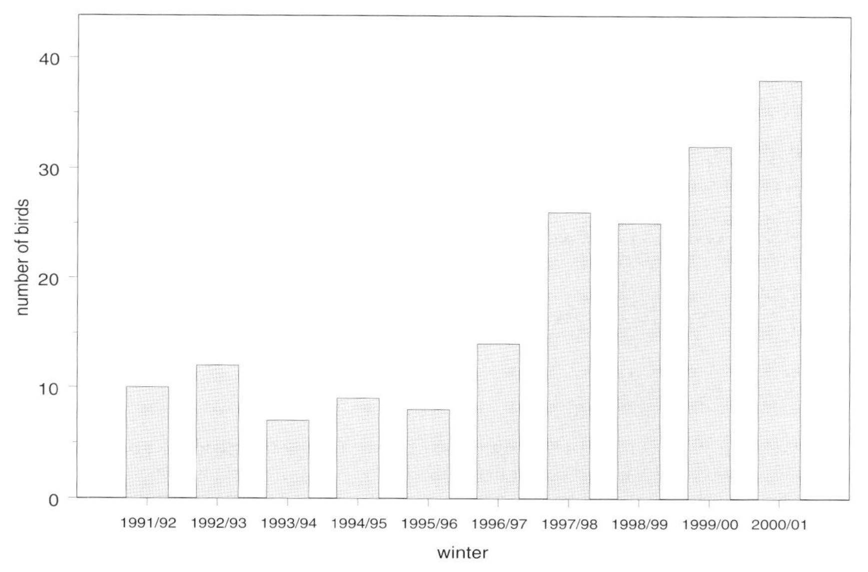

Fig 15: Black-necked Grebe **-** Maximum counts in Torbay for 1991/92 - 2000/01 winters (Source: DBRs)

FULMAR *Fulmarus glacialis* — Northern Fulmar

Not scarce, both as passage visitor offshore and coastal migrant breeder. BoCC (Amber).

Fewer and smaller large counts outside the breeding season compared to 2000, but at last some definitive breeding counts via the Seabird 2000 census (even though it shows a population decline – see separate report).

Table: Monthly maxima for coastal areas:

	J	F	M	A	M	J	J	A	S	O	N	D
N coast (incl Lundy)	100	64	-	-	-	10	-	20	3	-	15	10
S coast (W of Dart)	1	20	1	4	22	10	28	100	1	-	6	13
S coast (E of Dart)	36	10	10	100	20	20	30	15	6	2	30	5

Breeding. The only direct counts received for 2001 were from: Jennycliff (two prs) & Bovisand (11 prs) in Plymouth Sound; Berry Head (seven nests); and Burgh Island (five nests). However, the *Seabird 2000* census (see separate report) gave an estimated total of 473 prs for the 1999-2001 period, of which 190 were on Lundy. On the mainland, breeding occurred on all coastal sectors, the most important being Braunton-Foreland Pt with 94 of the 283 mainland prs, and the mainland total was about equally divided between N and S coasts. The most important individual sites included Clovelly (29 prs), Hallsands (28 prs), and Cow & Calf and Wringcliff (each with 22 prs).

Non-breeding. Many healthy counts from both coastlines included: 40 at Wringcliff Bay on 8 Jan; 36 from Otterton Cliffs on 20 Jan; 100 from Wringapeak-Highveer coast on 24 Jan; 64 at Highveer Pt on 9 Feb; *c*.50 at Hartland Pt (09.00-10.00h) on 23 Feb; from Hope's Nose, 30 on 14 Apr and *c*.100 on 22 Apr; 91 at Dawlish Warren on 24 Apr; and from Prawle Pt, *c*.100 on 7 Aug and 44 on 12 Aug. One blue-phase bird off Prawle on 22 Jul (TM).

CORY'S SHEARWATER *Calonectris diomedea*

Rare passage visitor offshore between spring and autumn. SPEC (2).

With no records in 2000, this year was only slightly better, with just one accepted record.

One at Hope's Nose on 12 Aug (CJP).

Record request *(from ML). This is a tricky species to identify, especially as it is normally seen at distance, often in poor weather conditions. Full descriptions are essential, and should enable the Records Committee to rule out Great Shearwater – an easy confusion species at distance.*

GREAT SHEARWATER *Puffinus gravis*

Rare passage visitor offshore between spring and autumn.

Five birds (cf. two in 2000 and a record 440 in 1999), all in Aug, and from both coasts.

The five records all concerned single birds: the first from Saunton Sands on 1 Aug (RJ), on the same day as a Cory's Shearwater there; Berry Hd on 2 Aug (ML); Sharkham Pt (DH) and Prawle Pt (TM) on 7 Aug; and, finally, Westward Ho! on 30 Aug (RGM)

CORY'S/GREAT SHEARWATER *Calonectris diomedea/Puffinus gravis*

Large shearwaters can be difficult to identify when seen at long range and/or in conditions of poor visibility.

Records this year included: one at Saunton Sands on 1 Aug (RJ); four at Prawle Pt on 3 Aug (JLb); and a bird heading S at 08.45h on 18 Aug from Berry Hd (MD,AR).

SOOTY SHEARWATER *Puffinus griseus*

Scarce autumn passage visitor off S coast, very scarce off N coast in recent years, although probably under-recorded.

Seventy-two submitted records (cf. 38 in 2000), but small numbers accounted for a rather low minimum of 77 birds (cf. 122 in 2000 and 139 in 99). Birds appeared from mid-Jul to mid-Oct with a late record in early Nov. Peak passage 5-8 Oct, with the only double figure count on 7th.

Table 10: Monthly totals (birds):

	J	F	M	A	M	J	J	A	S	O	N	D
Berry Head	-	-	-	-	-	-	1	4	1	19	-	-
Dawlish Warren / Bay	-	-	-	-	-	-	-	-	1	2	-	-
Hartland Point	-	-	-	-	-	-	-	6	9	-	1	-
Hope's Nose	-	-	-	-	-	-	1	-	-	2	-	-
Prawle Point	-	-	-	-	-	-	1	8	4	25	-	-
Start Bay	-	-	-	-	-	-	-	-	-	2	-	-
Other sites	-	-	-	-	-	-	-	-	-	1	-	-
Total 2000	-	-	-	-	-	-	-	7	71	63	4	2
Total 2001	-	-	-	-	-	-	3	18	15	51	1	-

Jul-Sep. Two *Jul* records on 17th from Hose's Nose (CJP,MRAB) and Berry Hd (ML,MD) may have referred to the same bird; another was from Prawle Pt on 22nd (JCN). *Aug* passage started from 11th, with multiple counts: three at Prawle Pt on 12 Aug (MD,AR); three from Hartland Pt on 16 Aug (DC,RD); two at Berry Hd on 18 Aug (MD *et al.*); two on 22 Aug at Prawle Pt (JCN); and two from Hartland on 30 Aug (MSS,DC). *Sep* started with regular small counts until 13th, with the highest: five from Hartland Pt on 7 Sep (DC,MSS) and three there on 8th (MD,AR).

Oct-Nov. Passage resumed in early *Oct*: six on 5 Oct and seven on 6th at Berry Hd (MD,ML); and four at Prawle Pt also on 6th (JLt,PMM,TWW); on 7 Oct, five off Berry Hd (MD *et al.*) and 12 past Prawle Pt, 07.00-10.30h, (PMM *et al.*); and four past Prawle on 8th in 4.5 hours (PMM). Records then became scarce, with only single counts noted on 10, 11 & 15 Oct, all from Prawle Pt. Only one *Nov* record, a single past Hartland Pt on 7th (DC,MSS).

MANX SHEARWATER *Puffinus puffinus*

Not scarce summer and passage visitor offshore; not scarce / fairly numerous (but poorly recorded) breeder on Lundy. SPEC(3), BoCC(Amber).

As usual the highest counts emanate from N Devon, presumably relating to the Lundy and/or the S Wales breeding populations. Max counts on both coasts higher than in 2000.

Table 11: ***Monthly maxima for coastal areas:***

	J	F	M	A	M	J	J	A	S	O	N	D
Lundy Island Crossing	-	-	-	-	166	-	-	-	2	-	-	-
N coast (excl Lundy)	-	-	-	-	20	300	1127	2000	432	1	-	-
S coast (W of Dart)	-	-	-	-	119	565	591	225	20	6	-	-
S coast (E of Dart)	-	-	-	82	15	150	847	20	2	-	-	-

Breeding. Using tape-recorded calls, a *complete* survey of *c.*7,000 holes on Lundy 19-26 May found 166 occupied, considerably less than the 300-400 extrapolated from the survey along sample transects undertaken in Jun 2000. A population of 166 prs should be regarded as the best estimate and is the official total for the *Seabird 2000* census. (All information from DJP).

North coast. The *first record* was of a lone bird on the 12 April, yet the next sighting was not until 13 May with 13 birds noted on the Lundy crossing; the only other May record was of 20 on 16th from Bideford Bay. *Hartland Pt* was once again the focus of most sightings, the first on 23 Jun with a count of three, then some much larger counts, including: 300 on 29 Jun; 420 on 7 Aug; 2,000 (four hour period) on 9 Aug (MSS); 1,400 (90 minute period) on 16 Aug (RD,DC); 200 on 30 Aug; 432 on 8 Sep; and 343 on 13 Sep. *Elsewhere* 1,127 were counted from Westward Ho! on 10 Jul & 150 past Baggy Pt on 8 Sep.

South coast. The *first sighting* was of three on 13 April from Hope's Nose, with 12 there the next day; other *Apr* records show movement on 22nd, with 82 from Hope's Nose and 33 from Berry Hd, the only other sighting being 15 from Dawlish Warren on 24th. *May* had 12 records, all single figures except: 30 on 5th, 17 on 12th & 119 (in 90 minutes) on 27th, all from Prawle Pt; and 80 on 27th from Start Pt. *Jun* had 21 records, with peak counts: 565 W on 3rd from Prawle Pt; 103 on 5th from Dawlish Warren; 150 from Shoalstone on 14th; and 60 at Berry Hd on 15th. *Jul* produced 47 records, 19 of which were counts >100, the largest being: 580 from Prawle Pt on 7th; 847 (over four hours) from Berry Hd on 21st; 591 (in two hours) from Start Pt on 21st; and 520 from Prawle Pt on 28th. In *Aug,* out of 47 records, the only >100 counts were from Prawle Pt, with 125 on 4th and 225 on 9th. *Sep* onwards produced only a handful of records, the highest being: 20 from Bolt Tail and 12 from Prawle Pt on 4 Sep; and 15 from Budleigh Salterton on 5 Oct. The last was a single late bird from Prawle Pt on 30 Nov.

BALEARIC SHEARWATER *Puffinus mauretanicus*

(Formerly known as P. yelkouan* 'Mediterranean Shearwater'*). Scarce passage visitor off S coast, mainly late summer and autumn; rare or very scarce off N coast. SPEC(4).

A record-breaking year, with 260 submitted records (cf. 49 in 2000) and an estimated 772 birds (cf. 131 in 2000 and 88 in 1999). (More information for records of this and other seabird species, including start/end times of observations and comments on whether passage or lingering bird, would greatly improve the accuracy of estimating numbers.) Lyme Bay, offshore from Portland, Dorset, to Berry Head, is thought to be internationally important for this species, which gathers in post-breeding flocks in Jul and Aug to moult (RSPB, 1997).

Table 12: Monthly totals (birds):

	J	F	M	A	M	J	J	A	S	O	N	D
Berry Head	-	-	-	-	-	-	118	129	48	46	-	-
Dawlish Warren / Bay	-	-	-	-	-	6	78	32	30	12	-	-
Hartland Point	-	-	-	-	-	-	-	39	134	-	-	-
Hope's Nose	-	-	-	-	-	-	21	35	18	29	-	-
Prawle Point	-	-	-	-	-	-	74	111	16	10	-	-
Start Bay	-	-	-	-	-	-	12	3	1	6	-	-
Other sites	-	-	-	-	-	-	46	10	3	4	-	-
Total 2000	-	-	-	-	-	1	22	49	60	6	-	-
Total 2001	-	-	-	-	-	6	349	369	250	107	-	-

With so many records, mostly only those ≥20 have been listed.

Jun-Jul. The first records were in the first week of Jun, with small counts focused around Dawlish Bay, and birds were seen almost daily from 4 Jul. The first large counts were from Prawle Pt, with 24 on 21 Jul (BG) and 25 on 28th (JCN,TM); there were also up to 27, a site record, at Dawlish Warren on 30 Jul (KRy,IL,JEF) and *c.*70 at Berry Hd on 31st (BHNNR). Away from the main sites, there was one on 28 Jul and 11 on 29 Jul from Littlecombe Shoot (DWH).

Aug-Sep. More high counts from Berry Hd included 31 on 1 Aug (ML,BMc) and 28 on 2nd (ML); and 32 were seen from Prawle Pt on 12 Aug (MD,AR). Hartland Pt experienced three high counts, with 29 on 30 Aug (DC,MSS), 66 on 3 Sep (MD,AR) and 41 on 8 Sep (MD,AR). At Berry Hd there were eight submitted records for 30 Sep, the highest count being 45 (07.05 - 16.30h) (MD,AR *et al.*). Away from the main sites, singles from: Sidmouth on 1 Aug (THS); Seaton on 7 & 11 Aug (BG); Stoke Pt on 13 Aug (BG); Bolt Hd (two) on 4 Sep (PAS); and Baggy Pt on 8 Sep (BG).

Oct. Records continued from the main sites with a max of 21 on 1 Oct at Berry Hd (MD,AR). Elsewhere: three past Teignmouth on 6 Oct (LC); and, the *last* of the year, one W past Wembury Beach on 24 Oct (SGe), proving that seeing seabirds need not be confined to the well-watched sites.

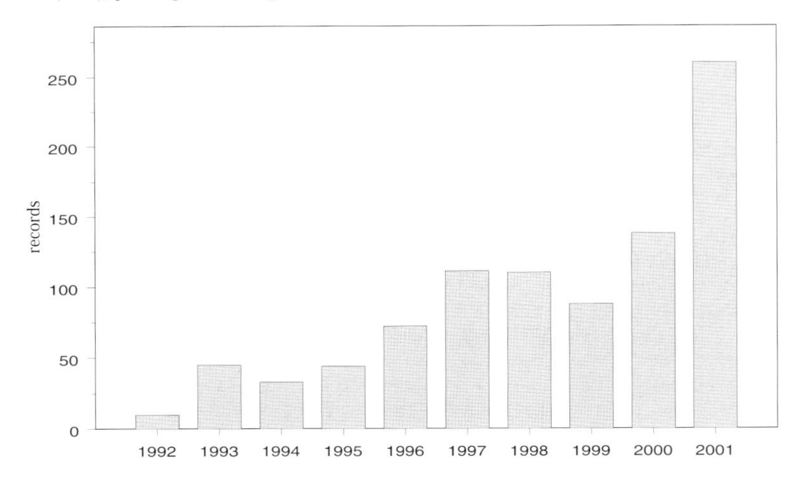

Fig. 16: Balearic Shearwater - Annual number of records submitted 1992-2001 (Source: DBRs)

FEA'S/ZINO'S PETREL *Pterodroma feae/Madeira* CapeVerde/Madeira Petrel

Rare vagrant

With the first recorded as recently as 1999, this much wanted ocean seabird has again been seen off the Devon coast. To date there are only 23 accepted British records.

Devon's second bird passed S early on 17 Jul, first through Hope's Nose (CJP) and then Berry Hd (ML,MD,BMc). The seasoned stalwarts from the latter site gained excellent views and, accompanied with some meticulous note-taking, were able to identify this bird (*cf. DBR* 1999, including report on p.167). (Still under consideration by BBRC, this was one of six reported in Britain in 2001)

STORM PETREL *Hydrobates pelagicus* European Storm-petrel

Scarce passage and summer visitor offshore (bred 1874 and 1950); rare in winter. SPEC(2), BoCC(Amber).

More in Jul & Aug than in 2000, but no large autumn influx.

Table 13: Monthly totals (birds):

	J	F	M	A	M	J	J	A	S	O	N	D
Berry Head	-	-	-	-	-	-	15	31	-	11	-	-
Hope's Nose	-	-	-	-	-	-	8	-	1	-	-	-
Prawle Point	-	-	-	-	-	-	55	62	-	-	-	-
N coast	-	-	-	-	-	-	-	-	-	-	-	-
Other sites	-	-	-	-	5	-	23	15	-	-	-	-
Total 2000	-	-	-	-	-	-	-	7	71	63	4	2
Total 2001	-	-	-	-	5	-	101	108	1	11	-	-

Spring. The only record concerned five birds from Start Pt on 27 May (ML,BMc).

Summer. For the period 1 Jul - 14 Aug, there were 41 submitted records covering 20 different dates, most of which focused around the regular S Devon headlands. Sightings started on 1 Jul with 23 heading W at Prawle Pt (JCN), then birds seen regularly up to 22 Jul, with 12 on 10 Jul from Berry Hd (ML) and 22 on 21 Jul from Start Pt (ML) being the highest counts. Birds reappeared on 3 Aug with three from Prawle Pt (JLb), then occurring almost daily until 14 Aug, with the largest counts: 12 from Berry Hd (ML,MD,AR) and 14 from Prawle Pt (TM) on 7 Aug; 28 form Prawle Pt on 12 Aug (MD,AR); 16 at Berry Hd (ML) and 14 from Stoke Pt (BG) on 14 Aug. Away from the usual headlands, one on 7-8 Aug at Dawlish Warren (KRy,JEF,IL). No records from N Devon.

Autumn. Only two records received, both from Berry Hd: one on 9 Sep (BG) and 11 on 8 Oct (ML,BBH).

LEACH'S STORM-PETREL *Oceanodroma leucorhoa*

Rare passage/storm-blown visitor, very scarce in some years. SPEC(3), BoCC(Amber).

One record (*cf.* three in 2000); a bird on 17 Oct from Hope's Nose at 08.28h (WJD).

GANNET *Morus bassanus* — Northern Gannet

Not scarce/fairly numerous passage visitor offshore throughout year. SPEC(2), BoCC (Amber).

Similar pattern to 2000, but max monthly counts on S coast mostly lower, particularly in winter. Highest counts & hourly passage rates, and most feeding activity, in Sep and Oct. On N coast, numbers higher and the season longer than in 2000.

Table 14: Monthly maxima for coastal areas, and overall maximum passage rates:

	J	F	M	A	M	J	J	A	S	O	N	D
N coast (inc Lundy)	2	1	-	5	15	60	15	200	100	10	20	-
S coast (W of Dart)	1	5	-	40	68	124	260	500	1200	1830	111	138
S coast (E of Dart)	4	4	21	121	45	66	400	750	2500	400	40	3
Max hourly passage rate	-	-	-	40	45	53	130	69	1200	915	51	69

Maximum counts. On the *S coast*, the 19 counts ≥500 were from: Prawle Pt (6), Exmouth/Dawlish Bay (6), Start Bay/Pt (3), Berry Hd (2), Hope's Nose and Bolt Tail, and seasonally, they occurred in Aug (2), Sep (13) and Oct (4). The highest were: 2,500 N past Hope's Nose on 4 Sep; 1,830 W past Prawle Pt in 2h on 15 Oct; 1,200 past Prawle Pt in 1h on 2 Sep; >1,000 passing Berry Hd in steady stream on 15 Sep; 865 W past Prawle Pt in 3.5h on 7 Oct; and 800 off Dawlish on 4 Sep. On the *N coast*, the highest counts were off Hartland Pt: 200 on 30 Aug and *c.*100 on 7 Aug & 13 Sep.

Passage rates and weather effects. The highest recorded *hourly passage rates* are included in the table for the first time; the highest were at Prawle Pt with 1,200 on 2 Sep and 915 W on 15 Oct; in *c.*40h seawatching there May-Dec, PMM recorded a total of 3,232 going W and 975 going E. No major weather effects, though about 100 inshore at Plymouth on 23 Oct and 23 at Wembury on 24th coincided with SW gales, and 'many' were in Thurlestone Bay on 17 Feb after strong NE wind.

Feeding flocks. Reported from May to Sep (*cf.* from Jul to Dec in 2000) along both N and S coasts, the largest counts being 500 off Start Pt on 16 Aug and 500 off Prawle Pt on 2 Sep, with several other large feeding flocks along the S coast in early Sep coinciding with an abundance of Sprat and Mackerel. Presence inshore was probably dictated more by food than weather in 2000.

BOX – SEABIRD PASSAGE RATES

Most of the records of seabirds in Devon are of birds passing the headlands on their migrations, and most species are, at least in part, passage visitors. The records may be the result of brief, casual visits to the coast, but many are from a dedicated core of 'sea-watchers' who spend many hours, mostly May to November at the favoured watch points - Hope's Nose, Berry Head, Prawle Point and Hartland Point – carefully logging every passing bird.

In the *DBR*s, these records are presented in a variety of ways. For the rarest species (the extreme example being **Fea's/Zino's Petrel**) every record is described. For scarce species like most shearwaters and skuas, an attempt is made to estimate the numbers passing each headland each month and present these in tables. For the commonest species such as **Gannet**, the maximum daily count for each month is usually given because the total numbers recorded each month would often run into several thousand, and unlike the other species, it is often only the largest numbers that are recorded anyway.

For all species, the number of birds seen will depend partly on the number of hours spent observing. Since the key observers actually send in these times as part of their records, another way of expressing the abundance of passing seabirds is to calculate ***hourly passage rates***. Although these would not be very helpful for the rare and scarce species, because the numbers would be extremely small and we have reasonable alternatives as described above, it is perhaps instructive to know that the 12 **Puffins** that were recorded passing Berry Hd in 2001 did so during a total of 19 hours of observation, and that the *maximum rate* was four in two hours (i.e. two per hour) on 21 July. Note however, that the 19 hours does not include all those days when no Puffins were recorded, so the *mean passage rate* would be far less than one per hour. For common species of seabird, however, passage rates are much more useful. Thus, the *mean passage rates* for **Gannet** at Prawle Pt for the May-November period can be calculated as 80 per hour going west and 24 per hour going east, based on the totals given by Pat Mayer for 40 hours of observations, but it is *maximum passage rates* that most observers record and which we now use for this species in the *DBR*. The overall maxima in 2001 in fact came from Prawle Point with 1,200 per hour on 2 September and 915 on 15 October, both rather less than the 1,800 maximum recorded in 2000. A larger count, of 2,500 streaming north past Hope's Nose on 4 September 2001 may have reflected a higher passage rate, but no information on the period of observation was given. This helps to emphasise the importance of including the period of observation with records of all passage birds, whether at sea or on land.

Record request*. Records of seabirds should indicate whether the birds are passing, or at rest or feeding. If passing, direction of travel and period of observation should be given, as well as more specific times for rare/scarce species and large groups. This helps to identify individuals and minimise double-counting from different headlands.*

CORMORANT *Phalacrocorax carbo* **Great Cormorant**

Not scarce resident breeder and winter visitor. BoCC (Amber).

Evidence for a continuing increase in numbers from a record overall WeBS maxima, some monthly maxima and record counts from the major estuaries. Numbers in excess of the threshold for national importance occurred on the Kingsbridge Est for the first time, as well as on the Exe.

Table 15: Monthly maxima at main/well-watched sites and WeBS county totals:

	J	F	M	A	M	J	J	A	S	O	N	D
Avon Est	5	6	6	6	4	4	4	6	6	6	7	6
Axe Est	13	11	-	-	-	-	4	5	4	5	7	10
Burrator Res	4	3	-	-	-	1	4	8	6	10	4	6
Exe Est	127	103	16	20	20	31	54	**193**	**205**	**150**	87	85
Kingsbridge Est	7	22	-	-	-	-	-	21	24	**163**	**160**	84
Plym Est	-	7	-	-	3	6	6	-	14	8	8	21
Roadford Res	11	12	-	-	-	-	-	-	-	11	7	14
Slapton Ley	5	10	10	12	12	6	-	-	15	20	19	13
Tamar Complex WeBS	49	42	-	-	-	-	-	-	89	104	80	40
Upper / mid Tamar	*19*	*30*	-	-	-	-	-	-	*36*	*56*	*55*	*19*
Tavy Est	*7*	*4*	-	-	-	-	-	-	*12*	*6*	*5*	*7*
Taw / Torridge Est	65	35	-	-	-	17	-	-	80	54	75	52
Teign Est	1	-	10	-	12	-	9	10	-	9	3	-
Wembury	-	6	-	-	26	36	38	39	6	1	-	3
WeBS county total '00	243	229	180	198	118	222	295	262	348	291	312	336
WeBS county total '01	344	278	4	6	1	7	11	224	193	432	460	329

Breeding. The only nest counts were 19 on Burgh Island and a pr on nest at Seaton Hole on 18 Feb. The *Seabird 2000* census (see separate report), however, reveals a current Devon population of 194 prs, all on the mainland and all but three (Cow & Calf and Woody Bay) on the S coast: Wembury Gt Mewstone (84), Dartmouth Mewstone (30), Burgh Island (17), Blackaterry Pt (13) and Thatcher Rock and Dawlish (both 12) were the main sites: others included Brandy Hd - Chiselbury Bay (9), Wadham Rocks – St Anchorite's Bay (8), Eastern Blackstone (3), Hope's Nose (2) and Hope Cove (1).

Maximum counts. The *Exe Est* continues to be the most important site, and, as in most recent years held *nationally important* numbers (>130) with a peak of 205, sitting on sandbanks between Lympstone and Exmouth on 30 Sep (DJG). Other individual counts there included a max of 101 on 13 Jan for the pylon roosters at Countess Wear, 41 on 12 Sep at Starcross and 35 at Exeter Riverside CP on 15 Feb. On the *Kingsbridge Est*, in addition to Oct & Nov WeBS counts of 163 & 160, 117 were counted from Wareham Pt on 31 Oct and a single flock of 100 between Saltstone and Kingsbridge was seen on 18 Nov; these unprecedented numbers coincided with movements of Sprat and Mackerel into the estuary. High counts on the *Taw/Torridge Est* included 41 at Braunton Burrows on 3 Sep, and in *Plymouth Cattewater* there were 56 on 22 Jan resting on buoys and boats. *Inland*, where odd singles can occur almost anywhere, Roadford Res continues to be the main truly inland site, though recorded numbers were down on 2000. *Elsewhere*, the only unlisted sites with maxima >five were: Saunton Sands (32 on 31 Jan); Straight Pt (24 on 18 Feb); Radford L (21 on 20 Dec); Three Beaches, Torbay (18 on 14 Oct); New Cross Pond (15 on 27 Jan); Upper Tamar L (10 in Sep); and Lower Tamar L, Huntsham L and Shobrooke L (all with six).

Passage and behaviour. The only records of sea movements were of a group of 32 moving SW off Hartland Pt on 8 Sep, and 25 E off Prawle Pt on 12 Oct. Co-operative fishing was observed in Radford L on four dates in Jun & Dec, with a max of 21 on 20 Dec (SRT,VRT); shoals of Grey Mullet were seen being driven to the bank side and, on one occasion, Little Egrets also took advantage of the increased fish availability.

Ringing recoveries. From birds ringed on the Gt Mewstone, Wembury, one was recovered only 4km away after five years and another 358km away after one year, reflecting the fact that many young birds may move across the Channel but later return to breed; for more details, see Ringing Report.

P.c.sinensis* (‘Continental’ Cormorant): *probably very scarce passage and winter visitor.

This race is currently under review by the Devon Records Committee. Full descriptions are required for all sightings and must include the angle formed by the rear edge of the gular patch.

Two were reported during the year.

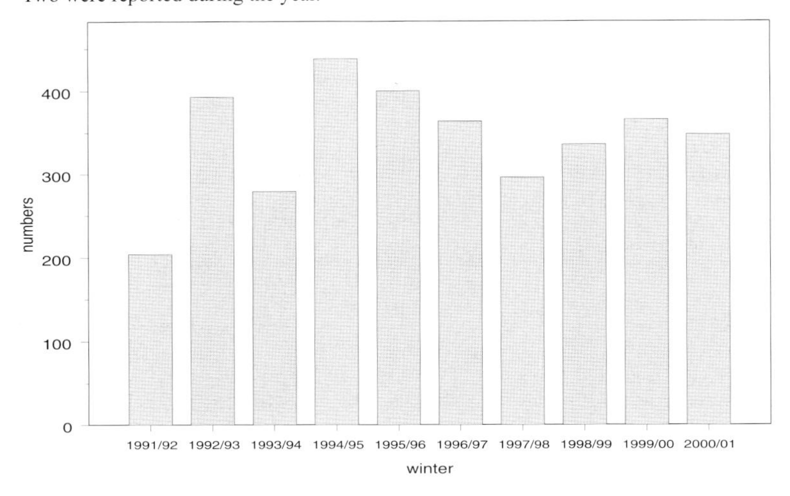

Fig. 17: Cormorant - Maximum WeBS County Totals for the 1991/92 - 2000/01 winters (Source: WeBS data in DBR's)

SHAG *Phalacrocorax aristotelis* — European Shag

Not scarce resident breeder and winter visitor. SPEC(4), BoCC (Amber).

Numbers and occurrences mostly similar to 2000, but with a lower max count on the Exe Est. Almost all records were from the S coast.

Table 16: Monthly maxima at main sites:

	J	F	M	A	M	J	J	A	S	O	N	D
Exe Est / Dawlish Bay	130	106	50	100	30	82	200	70	200	50	3	50
Kingsbridge Est	-	10	-	-	-	-	-	-	60	-	-	4
Wembury	-	12	-	-	17	37	71	80	15	17	-	13

Breeding. Again poorly recorded, with nest counts from just five sites: Berry Hd (seven, briefly eight); Burgh Island (12); Baggy Pt, Gammon Hd and Prawle Pt. The *Seabird 2000* census, however (see separate report), reveals a current Devon population of 260 prs, of which 55 are on Lundy, two on the N coast (Elwill Bay and Woody Bay) and 203 on the S coast, mostly in the Torbay and Plymouth-Prawle sectors: Wembury Gt Mewstone (47), Dartmouth Mewstone (also 47), Thatcher Rock (30), Ore Stone (15) and Otterton Ledge – Brandy Hd (12) were the main sites, with others at: Shag Rock (9); Burgh Island (7); Dawlish & Berry Hd (both 6); Bolt Tail, Dancing Beggars & Lead Stone (all 5); Hope's Nose (4), Gammon East & Eastern Blackstone (both 2); and Leonard's Cove (1).

Maximum counts. The highest counts as usual came from the *Exe Est/Dawlish Bay* area where, although the max count was less than half that in 2000, more monthly records were broken, notably Feb. The max storm-driven incursions to Starcross involved only 70 birds in early Jan and 56 on 2 Oct (*cf.* 242 in 2000). Upriver there was a max of eight at Topsham on 22 Feb. On the *Kingsbridge Est* and at *Wembury* there were fewer records than in 2000, but the max counts were similar. *Elsewhere*, high counts included: 165 on sea off Otterton on 4 Sep; 60 off South Sands, Salcombe on 20 Nov; 47 resting on Netton Island, Noss Mayo on 19 Sep; 45 off Beer Hd on 21 Sep; 36 in Start Bay in Jan and 30 there in Sep; 34 off Bolt Tail on 25 Aug; 14 in the Avon Est on 31 Dec; and ten in Plymouth Sound on 1 Apr. No *inland* and very few *N coast* records received.

Corpses and pale birds. Two found dead, Elberry Cove on 13 Jan. One pale bird (resembling *P.a. desmarestii*, the Mediterranean Shag) reported, in Torbay on 12 Nov.

BITTERN *Botaurus stellaris* — Great Bittern

Rare/very scarce winter visitor, usually associated with cold weather; casual breeder (bred 1996 and 1997). SPEC(3), BoCC (Red), UK BAP.

Compared to 2000, a similar number of records from Slapton, but additional records from two other sites, including a booming bird in residence for over two weeks.

Slapton Ley area. Sightings came from Ireland Bay, Slapton Bridge and the Higher Ley on 2 Feb (BW), 21-22 Aug (DJa,BW), 19 & 21 Oct (PBo) and 30 Oct (DJSu). Nearby, one flew past the hide at Beesands on 2 Dec at 15.53h (KEM).

Sherpa Marsh. A booming bird was present 27 May - 12 Jun (ArSt,DC *et al.*); on I Jun it was heard to make 'whump' calls at 10-15 min intervals from 20.50 to 22.00h (DC,MAS) and the booming was described as a lot quieter than that of a normal booming male, leading to speculation that it may have been a 1y (IM).

NIGHT HERON *Nycticorax nycticorax* — Black-crowned Night Heron

Rare vagrant. SPEC(3).

The first since 1999 was a 1s at Stover CP 12-15 April, though found dead on the last day of its stay (JDA,RCh). (One of four accepted by BBRC in Britain in 2001, all in spring/summer).(*see col plates*)

CATTLE EGRET *Bulbulcus ibis*

Rare vagrant/rare escape

Only two previous records, the last in 1986, but the 2001 bird was considered to be an escape.

One inhabited fields near Colyton 28-30 Dec (IM). It carried a metal ring with a number untraceable to any wild ringing scheme, and was very tame, walking to within feet of the observer; it also failed to interact with livestock in nearby fields.

LITTLE EGRET *Egretta garzetta*

Formerly rare, now not scarce, passage and winter visitor; breeding attempted 1996. BoCC (Amber).

The increase continues. Most estuarine sites experienced record counts, seven exceeded 50, including the Dart and Yealm Ests for the first time, and thirteen had maxima of at least ten (the threshold for national importance). A record WeBS county total of 405 was reached in Sep.

Table 17: Monthly maxima at main/well-watched sites and WeBS county totals:

	J	F	M	A	M	J	J	A	S	O	N	D
Avon Est	15	22	**12**	**11**	5	4	**19**	**24**	**33**	**31**	**17**	**18**
Axe Est	7	**10**	-	4	-	3	8	**14**	4	6	2	**27**
Dart Est	2	6	**17**	9	-	-	-	**50**	**53**	2	**11**	3
Erme Est	8	**13**	5	-	1	-	-	**25**	**14**	**11**		8
Exe Est	**60**	**13**	**56**	**26**	**18**	7	**41**	**149**	**81**	**35**	126	59
Kingsbridge Est	**30**	**39**	**27**	-	-	-	**23**	**95**	**100**	**65**	**55**	**35**
Otter Est	2	**12**	6	7	1	-	1	7	**15**	4	9	3
Plym Est	**13**	8	8	5	7	1	5	**20**	**23**	**22**	**26**	5
Tamar Complex WeBS	61	71	-	-	-	-	-	-	144	127	208	60
Upper / mid Tamar	***12***	***26***	-	-	-	-	-	-	***49***	***32***	***25***	***13***
Tavy Est	***42***	***42***	-	-	-	-	-	-	***19***	***18***	***21***	***13***
Taw / Torridge Est	**18**	**25**	**25**	4	4	5	**25**	**38**	**64**	**36**	**40**	**40**
Teign Est	5	**42**	**12**	**20**	7	-	35	65	34	46	16	21
Yealm Est	2	2	-	-	-	-	-	**53**	**16**	**12**	5	2
WeBS county total '00	126	176	179	89	43	26	113	285	355	272	152	133
WeBS county total '01	125	169	1	5	-	4	21	268	405	254	268	136

Breeding season. Still no breeding records, and generally scarce in May and Jun, with birds recorded from only seven sites from mid-May to mid-Jun: eight at Matford Marsh, Exeter on 24 May; one at Budleigh Salterton on 25 May; singles at Stoke Canon and Plym Est on 1 Jun; four on the Avon Est 1-4 Jun; five on the Taw/Torridge Est on 12 Jun; and three on the Dart Est on 16 Jun. Juveniles were first not-

MAP 1: Little Egret.
Distribution of records1997-2001 (usually 2001). The main estuarine sites are grouped into three abundance catagories based on max count; the other sites, inland and coastal are where egrets have been recorded in the same period - with one exception (Exwick in 2001) always less than 10 and usually singles.
(Data: Based upon a five year period 1997-2001. Source: DBRs) (see Overview on page 25)

ed on 10 Jul, at Dawlish Warren.

Maximum counts. Most sites showed increases from 2000, but there were decreases on Upper/mid Tamar and Taw/Torridge Ests. The Yealm and Dart Ests showed particularly large increases, the latter resulting from counts made from Totnes - Dartmouth boat cruises on 24 Aug and 7 Sep. As usual, peak counts were mostly Aug-Oct arising from the influx in late summer and autumn, with an additional small influx in early spring. Peak *roost or pre-roost counts* at the main sites were: on the Exe Est, 126 on 23 Nov (GS), 60 at Dawlish on 8 Jan, 56 at Powderham Pk on Mar 12, 59 at Bowling Green Marsh on 24 Oct and 42 at Exminster Marshes on 10 Sep; on the Kingsbridge Est, 95 at Lincombe on 24 Aug (MBk); on the Teign Est, 65 on 31 Aug (MKn); on the Dart Est, 53 on 7 Sep (MD,AR); on the Yealm Est, 53 on 24 Aug (LHH *et al.*); and on the Taw/Torridge Est, 40 at Cleave, Bideford on 15 Dec (AF). *Other large counts* included: a flock of 42 at Starcross flying to roost at Cockwood on 9 Jan; a group of 22 together on creek mud and then a ploughed field at Kingsbridge on 21 Mar; and 20 around a small drying pool at Exwick on 2 Aug.

Coastal and inland records [see map on previous page]. Around the *coast*, singles were recorded past or over Berry Hd (Aug & Sep), Croyde Bay (Jul), Lynmouth (eight W on 24 Jun), Prawle Pt (Oct), Staddon Pt (Nov) and Start Pt (May & Oct, including one half a mile out to sea on 12 Oct), and six tired birds at Thurlestone on 22 Sep may have recently arrived. Records of *feeding on rocky shores* along the open coast involved mostly singles, but there were maxima of seven at Wembury in Oct, five at Prawle Pt in Nov and five at Mt Batten, Plymouth Sound in Oct. Three on Lundy on 12 Aug, with one to 16th, was only the island's eighth record. *Inland,* there were records from almost 30 sites mostly involving one or two birds around the tidal limit of the estuaries, but one at Fernworthy Res on 25 Jul and three at Torrington on 14 Dec were more noteworthy. At Slapton, four on lily pads in Ireland Bay on 9 Jul was the highest count.

GREY HERON *Ardea cinerea*

Not scarce resident breeder; passage status unknown.

Incomplete breeding data, but no evidence of any overall change in breeding numbers. Estuary counts and WeBS totals were similar to 2000, though the autumn peak and counts on some estuaries, notably the Taw/Torridge, were down on the previous year.

Table 18: Best estimates' of numbers of occupied nests in each heronry, grouped by river catchment:
(2000 results in parentheses: nc = not counted (sites shaded))

		Rivers / Estuaries			
AVON	Cockleridge	6 (6)	TAVY	Double Waters	nc (9)
AXE	Axe Farm	3 (1)		New Barn Farm	nc (20)
DART	Beardown	nc (6)	TAW	Molland	nc (2)
	Buckfast	2 (4)		Arlington	nc (18)
	Fleet	2 (0)		Kings Nympton	nc (6)
	Maypool	5 (5)		Pedley Wood	3 (2)
	Sharpham	2 (1)	TEIGN	Wanford Wood	5 (6)
ERME	Orcheton	3 (4)		Netherton	nc (7)
EXE	Ashilford	nc (10)	TORRIDGE	Hele Bridge	nc (8)
	Brampford Speke	nc (2)		Landcross	7 (9)
	Powderham	38 (36)		Okehampton	nc (1)
	Shobrooke	6 (6)	YEALM	Beechwood	nc (25)
KINGSBRIDGE	Gullet Plantation	2 (0)		Puslinch	nc (0)
	Lincombe	nc (5)		West Woods	nc (3)
	Wall Park	nc (3)			
OTTER	Tracey	17 (23)			
PLYM	Hoo Meavy	nc (5)			
TAMAR	Dunsdon	nc (4)			
	Lifton Park	3 (4)			
	Milford	1 (1)			
		Lakes / Reservoirs			
GARA	Slapton Ley	4 (2)	TEIGN	Fernworthy	1 (1)
TAMAR	Roadford	nc (3)			

Table 19: Monthly maxima** at **main/well watched sites** and **WeBS county totals:

	J	F	M	A	M	J	J	A	S	O	N	D
Avon Est	8	11	10	-	9	7	6	8	9	7	7	6
Axe Est	3	5	-	-	-	-	6	4	3	8	3	4
Burrator Res	3	3	-	-	-	3	5	6	5	5	2	4
Dart Est	13	12	-	-	-	-	-	10	3	2	5	2
Erme Est	1	1	-	-	-	-	-	-	7	3	2	1
Exe Est	6	6	4	3	5	6	10	10	39	26	16	6
Kingsbridge Est	19	25	-	-	-	-	-	17	22	26	24	18
Otter Est	5		-	2	3	1	3	3	2	3	2	2
Tamar Complex WeBS	15	19	-	-	-	-	-	-	44	40	39	32
Upper / mid Tamar	*4*	*6*	-	-	-	-	-	-	*10*	*12*	*9*	*13*
Tavy Est	*6*	*5*	-	-	-	-	-	-	*16*	*9*	*11*	*11*
Taw / Torridge Est	15	14	8	2	2	2	6	19	16	20	15	15
Teign Est	2		10	3	-	-	5	5	-	2	8	-
WeBS county total '00	91	92	98	70	68	116	114	181	165	163	62	87
WeBS county total '01	93	87	-	2	2	10	16	44	154	135	111	93

Breeding. The Heronry Census Devon Organiser, DR, reports that, in spite of the access restrictions, returns were received from 18 of the 37 active Devon heronries. Obviously no total figures for occupied sites or nests can be calculated, but the 18 counted colonies comprised 109 occupied nests – very similar to the 2000 total of 111, tentatively suggesting that the Devon population has remained more or less constant since last year. Seven sites showed an increase, six a decrease and five no change. Powderham, the largest colony, rose from 36 to 38, while Tracey dropped from 23 to 17. Full results are listed in Table 18, with the heronries grouped by river catchment.

Maximum counts. The WeBS total peaked at 154 in Sep (*cf.* 181 in Aug 2000), and, with lower counts from the Taw/Torridge Est, none exceeded the qualifying level for *national importance*, provisionally set at 50. Peak counts on most estuaries were similar to those in 2000, but on the Dart, Kingsbridge, Taw/Torridge and Teign there were substantial decreases. The highest counts of birds seen together were: 26 at Exminster Marshes on 10 Sep; 21 at Honiton on 16 Feb; 19 at Ashford (Taw Est) on 26 Aug; 15 at Penhill Marsh on 25 Aug and at Powderham Pk on 10 Aug; and 12 at Isley Marsh on 14 Aug. The only unlisted sites with counts >five were: Stafford's Bridge (8), Shobrooke L (6) and Upper Tamar L (6).

Coastal records. Only small numbers were recorded from coastal sites, and included up to three in both Torbay and the Prawle Pt area, and singles on Lundy on 14 dates Jun-Nov. None seen at sea, but an intriguing record was off a group of seven flying W high over Plymstock and Plymouth Sound at *c*.0830h on 22 Sep.

Aberrant plumage. A very dark (melanistic) juv on Matford Marshes, Exeter on 1 Jun.

Grey Heron - *John Walters*

PURPLE HERON *Ardea purpurea*

Rare vagrant. SPEC(3).

One briefly at Bradiford NR, Taw/Torridge Est on 20 Jun (MHC), was the first since 1999.

BLACK STORK *Ciconia nigra*

Rare vagrant. SPEC(2).

One at Marley Hd, South Brent 26 Aug – 20 Sep (SHo,SSk,SBr), was the first since 1995, and only the seventh since 1958. (Accepted by BBRC, along with one on Scilly on 27 Aug, the only other record in Britain in 2001).

WHITE STORK *Ciconia ciconia*

Rare vagrant. SPEC(2).

Three records of at least two birds, the first since 1998.

One reported from fields nr Bideford 18-19 Mar (R Shaw & G Tucker); one, possibly the same, in a roadside field at Exebridge 21-22 Mar (RGr,JEW); and one soaring over Bampton on 25 Jun (Mr Trew).

SPOONBILL *Platalea leucorodia* **Eurasian Spoonbill**

Rare passage and winter visitor. SPEC(2). BoCC (Amber).

Although a regular winterer since 1987/88, especially on the Taw/Torridge Est, this was the best year to date. Similar max numbers to 2000, and for the second year running a good series of records from the Exe Est, but longer periods of residence on both estuaries resulted in a big increase in bird-days from the previous year. At least eight were present in the county in both winter-periods, and involved both ads and imms.

Table 20: Monthly maxima at main sites:

	J	F	M	A	M	J	J	A	S	O	N	D
Exe Est	6	4	4	6	4	4	4	1	-	4	4	3
Taw / Torridge Est	4	4	2	2	4	-	-	1	1	4	5	5

Bird-days. Bird-days (see Box on p.74 of *DBR* 2000) provide a valid unit of abundance for this conspicuous, well-recorded species. Based on weekly maxima for the main sites, it is estimated that the county total in 2001 was 1,333 bird-days, more than double that of 2000 (650) in spite of similarities in max counts. The total comprised 593 from the Taw/Torridge, 738 from the Exe and two from elsewhere. Thus usage of Devon estuaries has increased, and the Exe has overtaken the Taw/Torridge.

Taw/Torridge Est. Up to four and five present during first and second winter-periods respectively, and recorded from Ashford, Barnstaple, Bideford Bay, Braunton Marsh, Caen Est, Isley Marsh, Penhill Marsh, Pottington, Velator and Yelland (mo). No more than two recorded during Mar & Apr, and the only records between 14 Apr and 6 Oct were four on 1 May, and singles on 1,29 & 30 Aug and 3 Sep. Wintering group started with two on 6 Oct, but the max of five not seen until 10 Nov.

Exe Est. In contrast to the Taw/Torridge, stayed later in spring and arrived later in Oct (mo). The only records between 5 Jun and 29 Oct were four on 2-4 Jul, one on 1 Aug (see also Taw/Torridge) and four on 2 Oct. In the first winter-period, four were consistently recorded together, six only being noted on 15 Jan and 15-16 Apr at Bowling Green Marsh. Four may have been present throughout the second winter-period, but were only recorded together on 2 Oct and 5 Nov, and almost all records referred to ones and twos. Recorded throughout the estuary from Dawlish to Bowling Green and Exminster Marshes, and including Powderham Pk and the golf course at Starcross. One juv flew in off the sea at Dawlish at 07.52h on 1 Nov (JEF).

Ages. In the first winter-period, the Exe birds were all imms, but on the Taw/Torridge the group consisted of two ads and two imms. In the second winter-period, the birds on the Exe were mostly described as imms and occasionally as ads, but both may have been present even if not seen together. The five on the Taw/Torridge consisted of two ads and three imms, and JEW recorded the build-up of the group from the single ad and imm in early Oct. Unseasonal ads appeared on the Exe on 4 Jun and on the Taw/Torridge in Apr and late Aug – early Sep.

Elsewhere. One on Stoke Gabriel Mill Pond on 28 Nov (EJL) and one E at Prawle Pt on 10 Nov (JLt) were the only other records.

MUTE SWAN *Cygnus olor*

Scarce resident breeder . BoCC (Amber).

Table 21: Monthly maxima at main sites (recording >10) and WeBS county totals:

	J	F	M	A	M	J	J	A	S	O	N	D
Avon Est	30	30	22	15	27	33	**20**	49	34	58	58	65
Axe Est	30	6	52	20	10	10		44	4	2	4	10
Exe Est	33	25	22	4	12	28	52	112	80	45	29	37
Kingsbridge Est	15	20	-	-	-	-	-	19	6	13	7	7
Slapton Ley	27	6	6	15	7	6	6	35	40	29	37	46
Stoke Canon	20	-	-	4	-	-	-	6	-	-	-	-
Tamar Complex WeBS	28	35	-	-	-	-	-	-	34	49	29	46
Upper / mid Tamar	*12*	*10*	-	-	-	-	-	-	*12*	*18*	*10*	*31*
Taw / Torridge Est	49	44	2	-	7	22	9	-	26	37	56	40
Teign Est	56	-	14	2	2	-	-	-	-	5	-	-
Yealm Est	-	-	-	-	-	-	-	-	17	7	4	-
WeBS county total '00	196	220	179	197	207	247	299	255	260	249	176	172
WeBS county total '01	154	161	4	6	4	36	31	167	210	242	255	240

Breeding. The number of breeding prs remained similar to recent years, but cygnet survival rate was a great improvement on 2000. Breeding confirmed at 23 sites, with 37 prs, though the total of 71 fledged yg came only from 28 prs for which adequate data were provided. The breeding sites (showing prs/yg/fledged, and ? = no data) were: Avon Est (3/6/?); Bowling Green Marsh (1/5/3); Braunton (1/4/4); Clennon Valley (1/1/0); Countess Wear (1/3/3); Exminster Marshes (5/7/?); Grand Western Canal (3/?/11); Kenwith NR (1/6/6); Killerton (3/0/0; all nests lost due to trampling by cattle); Matford (1/?/2); Passage House, Teign Est (1/?/?); Plymouth Hooe Lake (1/4/3; Powderham (1/?/2); Radford Lake (1/8/8); Slapton Ley (3/?/2); Stafford's Bridge, Upton Pyne (2/?/5); Stoke Canon (2/8/8); Stover CP (1/6/3; the cob raised the young after the death of the pen); Taw Est (1/5/3); Tiverton (1/3/3); West Charleton Marsh (1/6/6; the first breeding record at this locality for 30 years); and finally Wrafton Pond (1/?/3).

Maximum counts. The highest WeBS count was again from the Exe Est with a max of 112 during Aug (*cf.*189 in Jul 2000). On the Avon Est, the local population was augmented by RSPCA releases from West Hatch, Taunton, of ten cygnets on 21 Oct and six on 6 Dec (RWBo)

Ringing recoveries. Although recoveries in 2001 have been up to 139km away (nr Bath), five were of Slapton birds recovered at Shaldon, nr Teignmouth, a move associated with the lack of submerged vegetation in the Ley; for more details see Ringing Report.

BEWICK'S SWAN *Cygnus columbianus* — **Tundra Swan**

Very scarce winter visitor. SPEC (3^w), BoCC (Amber).

First winter-period. One ad at Braunton Great Field on 27 Feb (RJ).

Second winter period. Four flew in off the sea and landed briefly near the Axe Est on 28 Oct; their stay was brief, as five minutes later a fifth bird flew over calling, and the four took off and joined it (BG).

WHOOPER SWAN *Cygnus cygnus*

Very scarce winter visitor. SPEC(4^w), BoCC (Amber).

First winter-period. A juv on the Axe Est 1-17 Jan (PA *et al.*), and another on various Taw/Torridge Est sites 13 Jan - 4 Apr (WHT *et al.*).

Second winter-period. A single at Roadford Res on 1 Nov (KEM) and four (two ads & two juvs) on Lundy on 2-3 Nov (LFS) followed by three individuals on the Taw/Torridge: an ad at Bideford on 13 Nov (NCW); a presumed different ad 15–29 Dec (DC *et al.*); and a juv at Weare Giffard 19–27 Dec (WHT,MAS,MSS).

PINK-FOOTED GOOSE *Anser brachyrhynchus*

Rare winter visitor; most records probably refer to feral birds/escapes. BoCC (Amber), SPEC(4).

All birds assumed to be feral: two at Bowling Green Marsh on 12 Jul (DFr); one at Trenchford Res on 30 Jul (PMM); one around the Exe Est 26 Aug – 16 Sep (DJG,RSPB); and one at Totnes on 1 Nov (ML).

WHITE-FRONTED GOOSE *Anser albifrons* Greater W-f Goose

A. a. albifrons* 'European White-fronted Goose': *very scarce winter visitor. BoCC (Amber).

An ad of unknown origin was noted at Exminster Marshes on 19 May (BDW,DS), remained around the upper Exe Est (mo), and was last seen at Bowling Green Marsh on 14 Jul (CC). Seven at Upper Tamar L on 16 Dec (GPS), and perhaps the same as four ads and three juvs the next day at S Huish Marsh (VRT,SRT); two remained until 18 Dec (AJL), and one ad with three juvs on 30 Dec were also probably members of the original family party (AJL,VRT).

A.a. flavirostris* 'Greenland White-fronted Goose': *very scarce winter visitor. BoCC (Amber).

One juv on Lundy on 6 Nov (JRD).

GREYLAG GOOSE *Anser anser*

Very scarce or scarce feral resident (naturalised establishment); genuinely wild birds are unlikely to occur. BoCC (Amber).

Table 22: Monthly county maxima

	J	F	M	A	M	J	J	A	S	O	N	D
Total 2000	8	2	2	3	5	4	4	40	4	3	37	33
Total 2001	4	25	4	4	4	4	33	33	28	6	5	7

Breeding. Two prs nested at Exminster Marshes (RSPB), though success/productivity unknown.

Non-breeding. Max count was a flock of 33 at Bowling Green Marsh on 9 Aug. Other records included: singles at Upper Tamar L 7–14 Feb and 26 Nov, four at Slapton on 8 Jun, and two on the Taw/Torridge Est on 26 Feb, with singles there on 25 Oct and 1 Dec.

Presumed GREYLAG x CANADA GOOSE hybrids.

Six at Dawlish Warren on 23 Jan, and three there on 22 Aug.

UNIDENTIFIED GREY GEESE

A flock of 12 geese flew N high over broadsands on 14 Dec (ML). From bill and belly colour, the most likely species was Bean Goose, and there was a national influx at about this time; for example, there were five at Colliford Res in SE Cornwall 15-17 Dec (Madge 2002).

SNOW GOOSE *Anser caerulescens*

Rare feral visitor/escape.

One was at Bowling Green Marsh on 6 Nov and also at Exminster Marshes on 27 Nov (RSPB).

CANADA GOOSE *Branta canadensis*

Not scarce resident breeder (naturalised introduction).

Table 23: Monthly maxima at main/well-counted sites and WeBS county totals:

	J	F	M	A	M	J	J	A	S	O	N	D
Avon Est	2	6	17	2	-	-	-	-	22	200	15	-
Axe Est	-	-	-	-	-	-	-	42	-	-	-	-
Beesands Ley	8	26	7	2	-	-	14	61	-	-	9	25
Bicton Lake	53	55	-	-	-	-	-	-	9	14	27	-
Burrator Res	-	-	-	-	-	35	38	25	1	49	13	10
Elfordleigh	214	62	17	21	43	51	72	136	63	135	85	94

continued.........	J	F	M	A	M	J	J	A	S	O	N	D
Exe Est	369	156	21	18	14	65	140	500	**800**	170	180	390
Fernworthy Res	-	5	-	-	-	-	39	8	32	38	17	-
Hennock Res	-	14	-	-	-	-	-	-	-	-	-	-
Kingsbridge Est	12	8	2	-	-	-	29	144	239	1	4	-
Meldon Res	-	-	3	-	-	-	15	-	-	-	-	-
Otter Est	-	3	-	2	2	9	29	460	300	150	-	-
Portworthy Dam	25	8	-	-	-	12	59	105	119	22	128	-
Roadford Res	579	100	-	-	-	240	-	-	-	134	507	425
Shobrooke Lake	340	80	-	-	-		110	332	290	346	226	154
Slapton Ley	54	30	-	22	9	12		321	286	346	77	160
South Huish area	240	81	-	6	6	8	57	4	104	285	262	143
Tamar Complex WeBS	147	50	-	-	-	-	-	-	309	259	126	189
Upper/Mid-Tamar	*12*	*10*	-	-	-	-	-	-	*12*	*18*	*10*	*31*
Tavy Est	*43*	*27*	-	-	-	-	-	-	*260*	*145*	-	*197*
Taw / Torridge Est	500	262	48	36	2	270	283	400	396	**723**	**888**	548
Upper Tamar Lake	166	153	-	-	-	-	127	**1048**	295	**777**	184	510
Yealm Est	-	2	-	-	-	-	-	-	84	62	-	-
WeBS county total '00	1257	1268	463	285	224	749	839	2228	2597	2053	2063	1837
WeBS county total '01	2478	1065	17	23	45	72	175	1142	2985	2272	2541	2304

Breeding. Breeding success was on a par with the previous year; 19 prs at 15 sites produced a minimum of 53 yg.

Maximum counts. Three sites achieved the threshold for *national importance* (600): 800 at Bowling Green Marsh on 17 Sep, 675 in the Taw/Torridge area on 12 Oct and a probable site record of 1,048 at Upper Tamar L on 29 Aug (RGM). Some counts >50 from untabulated sites include: 110 at Burlescombe on 1 Dec, 192 at Grand Western Canal on 31 Aug and 97 at Huntsham L on 7 Sep. Two flying N over Lundy on 31 Oct was only the seventh record for the island.

BARNACLE GOOSE *Branta leucopsis*

Rare winter visitor; very scarce feral visitor/escape. SPEC(4/2), BoCC (Amber).

None of the 66 records are thought to refer to genuine wild birds.

Table 24: Monthly county maxima

	J	F	M	A	M	J	J	A	S	O	N	D
Total 2000	2	-	1	1	1	3	1	1	7	13	6	4
Total 2001	5	1	-	4	5	-	1	2	5	3	4	4

First winter-period/spring. Three at S Huish Marsh in Jan (perhaps the Torbay birds from the previous year), and the same or another three on the Taw/Torridge Est on 14 Apr and 13 May. Singles at: Beesands Ley on 29 Apr; Bowling Green Marsh Apr-May; Bradiford NR 1-13 Jan; Countess Wear on 13 Jan; Passage House Inn on 1 Jan; Shobrooke L 1 Jan – 28 Feb; and Slapton Ley 5 Apr – 6 May.

Autumn/second-winter period. One at Bowling Green Marsh 1 Aug – 6 Nov, with two there on 17 Sep, and two at Bradiford NR on 1 Dec. Singles at: Broadnymet Pool on 20 Sep; Burlescombe 14 Nov – 31 Dec; Exminster Marshes 6 Nov – 4 Dec; Fernworthy Res 21 Sep – 3 Nov; Huntsham Lake on 8 Sep; Otter Est on 28 Aug; Shobrooke L 1 Jul – 31 Oct; and S Huish area 4 - 18 Dec.

Presumed BARNACLE x BRENT GOOSE hybrid

One at Starcross on 21 Oct, and again on 11 Nov (DS).

BRENT GOOSE *Branta bernicla*

B. b. bernicla* 'Dark-bellied Brent Goose': *fairly numerous winter visitor; very scarce in summer. SPEC(3), BoCC (Amber).

An improvement on last year's numbers with a max of 2,000 at Exminster Marshes during Dec. The Exe Est once again exceeded the qualifying level (1,000) for national importance during both winter periods.

Table 25: Monthly maxima at main/well-counted sites and WeBS county totals:

	J	F	M	A	M	J	J	A	S	O	N	D
Exe Est	**1400**	960	30	7	2	-	6	2	240	**1096**	**1179**	**2000**
Kingsbridge Est	80	68	-	-	-	-	-	-	-	1	39	64
Otter Est	9	15	-	-	-	-	-	-	1	-	-	-
Taw / Torridge Est	184	175	-	6	11	-	-	-	9	73	118	151
WeBS county total '00	1558	1641	852	45	13	-	-	-	48	1411	1334	1464
WeBS county total '01	978	1173	-	-	-	-	-	1	53	1102	1314	871

Maximum counts. In addition to the count of 2,000 at Exminster Marshes, other *Exe Est* sub-site counts included: 547 at Bowling Green Marsh on 31 Dec; 550 at Clyst; 372 at Cockwood on 15 Jan; 200 at Dawlish Warren on 24 Oct; 680 at Exmouth (Mudbank Lane) on 24 Oct; and 970 at Turf on 15 Jan. On the *Taw/Torridge*: 130 on 22 Feb at Isley Marsh and 151 at Skern on 2 Dec. On the *Kingsbridge Est*: 80 on 28 Jan and 64 on 19 Dec. *Elsewhere*: ten flew S past Hope's Nose on 19 Oct, four past Prawle Pt on 11 Nov and nine at Thurlestone on 17 Dec.

Breeding success. Another poor breeding year, indicated by only two reports of young: eight juvs on the Exe Est on 21 Oct, and four at Skern on 23 Oct.

B.b. hrota* 'Light-bellied Brent Goose'** ***(very scarce passage visitor; rarely winters).

Table 26: Monthly maxima at main sites:

	J	F	M	A	M	J	J	A	S	O	N	D
Exe Est	1	2	5	1	1	-	-	-	8	2	5	1
Taw / Torridge Est	13	4	-	-	-	-	-	-	-	-	-	-

First winter-period. Most on the Taw/Torridge Est, with 12 at Penhill Marshes 13-25 Jan and two there on 22 Feb (JEW). On the Exe Est, a single bird seen at various sites from 27 Jan (MKn) through to 2 Jul (THS), though five reported from Dawlish Warren on 27 Mar (BG).

Second winter-period. On the Exe Est, two at Dawlish Warren on 18 Sep (BG), eight on 23rd (DS) and two again on 19 Oct (MKn); also up to five on four dates in Nov. The only Taw/Torridge record was one at Yelland on 10 Nov (RJJ), and the only Dec record was one on the Kingsbridge Est on 28th (PSa).

RED-BREASTED GOOSE *Branta ruficollis*

Rare escape/vagrant.

Two records of presumed escapes: one, with a narrow green ring on right leg, on 5 Jan at Skern on the Taw/Torridge Est (DC,RD,MSS); and two on 19 Jun at S Huish Marsh (RBu).

EGYPTIAN GOOSE *Alopochen aegyptiacus*

Very scarce feral resident/escape (naturalised introduction in E England). SPEC(3).

Long-staying residents throughout the year at: Slapton Ley, where first recorded on 20 Feb 1999 (mo); Goodrington (ML *et al.*); and Totnes, where two remained from the previous year with the final sighting of one on 1 Feb (ML,JLb).

Elsewhere: four on the Teign Est on 25 Feb (ML); one at Stover CP 13–31 Mar (JDA); two at Rackerhayes on 16 Jul (EM); four at Bowling Green Marsh on 4 Aug, with singles present there on 31 Aug and 6 Dec (RSPB); and finally two at Sharpham Marsh on 2 Nov (PMM).

RUDDY SHELDUCK *Tadorna ferruginea*

Rare vagrant or feral visitor/escape. SPEC(3).

Up to four free-flying birds from Paignton Zoo commuted regularly to Clennon Valley Lakes throughout the year (ML *et al.*).

SHELDUCK *Tadorna tadorna* — Common Shelduck

Not scarce migrant breeder and winter visitor. BoCC (Amber).

Table 27: Monthly maxima at main/well-counted sites and WeBS county totals:

	J	F	M	A	M	J	J	A	S	O	N	D
Avon Est	9	14	16	-	-	9	-	-	1	1	1	-
Axe Est	60	73	-	20	30	-	46	-	9	-	11	50
Dart Est	44	38	-	-	-	60	-	-	-	-	10	44
Erme Est	2	12	-	-	-	19	-	31	-	-	-	-
Exe Est	192	185	100	34	-	32	21	30	73	48	96	168
Kingsbridge Est	312	271	-	-	-	-	43	16	6	-	131	211
Otter Est	40	27	-	32	33	16	5	5	-	-	2	-
Plym Est	18	95	20	39	17	20	19	-	-	-	3	37
Tamar Complex WeBS	126	103	-	-	-	-	-	-	11	32	78	144
Upper / mid Tamar	*126*	*103*	-	-	-	-	-	-	*11*	*32*	*78*	*144*
Tavy Est	*21*	*32*	-	-	-	-	-	-	-	*1*	*25*	*36*
Taw / Torridge Est	282	297	-	2	10	25	-	-	49	34	131	208
Teign Est	34	10	15	20	5	60	-	-	-	1	2	-
Yealm Est	22	50	-	-	-	-	-	-	-	-	10	39
WeBS county total '00	1291	1154	1321	1030	643	540	175	69	56	128	425	945
WeBS county total '01	1083	1188	-	32	33	18	54	51	126	123	478	879

Breeding. A much better year, with a minimum of 26 prs producing 184 yg. (*cf.* 2000 when 11 prs raised 59 yg), but fledging success poor in some areas. Prs/yg were recorded from the following sites: Avon Est (3 or 4/39) but none fledged (RWBo,PJR); Exe Est (5/35); Dart Est (3/36); Erme Est (4/19); Kingsbridge Est (3/12) but only one survived (GECW); Otter Est (1/2); Plym Est (3/12); Taw/Torridge Est (2/9); Teign Est (1/14); and Westward Ho! (1/6).

Inland and island records. Three at Roadford Res on 22 Jan, with one there on 14 Feb, two at Velator on 25 Feb and five at Portworthy Dam on 23 Apr were the only inland records. A pr on Lundy on 7 May was only the 11th record there (LFS).

MANDARIN *Aix galericulata* — Mandarin Duck

Very scarce resident breeder (naturalised introduction)/escape.

Breeding. There were many sightings on both the Dart and Plym, but the only confirmed breeding records were: a ♀ with four yg at Cann Wood, R Plym, on 16 May (CC); a ♀ with three yg on the Plym Est on 16 May (CC); and a ♀ with a juv on the R Dart (Hembury Woods) on 19 Jun (PMM).

Non-breeding. Maximum numbers were 14 on the Plym Est on 30 July and nine at Hannaford Pond (R Dart) on 14 Oct (SMRY). Elsewhere: a ♂ at Radford L on 15 Feb (PFG) and two ♂♂ there on 15 - 17 Mar (PFG,JCN); two prs at Chudleigh Knighton on 31 Mar (RK); a pr at Totnes on 29 Dec (JLt), and two prs noted at Dawlish Water were assumed to be part of the permanent wildfowl collection (EM,AFo).

WIGEON *Anas penelope* — Eurasian Wigeon

Fairly numerous winter visitor. BoCC (Amber)

Table 28: Monthly maxima at main/well-counted sites and WeBS county totals:

	J	F	M	A	M	J	J	A	S	O	N	D
Avon Est	25	27	10	-	-	-	-	-	7	86	180	160
Axe Est	175	177	-	-	-	-	-	-	-	93	80	220
Erme Est	5	138	-	-	-	-	-	-	19	112	47	73
Exe Est	1173	686	360	17	1	1	3	140	1344	**3355**	2728	**3703**
Kingsbridge Est	1269	1035	-	-	-	-	-	10	70	604	620	615
Otter Est	383	237	-	2	-	-	-	1	70	135	312	290
Roadford Res	116	247	-	-	-	-	-	-	-	18	15	230
Slapton / Beesands Ley	-	-	-	-	-	-	-	-	13	152	37	10
South Huish area	180	116	-	-	-	-	-	-	-	-	26	127
Tamar Complex WeBS	446	245	-	-	-	-	-	-	-	118	262	315
Upper / mid Tamar	*65*	*18*	-	-	-	-	-	-	-	*24*	*32*	*69*
Tavy Est	*22*	*23*	-	-	-	-	-	-	-	-	*6*	*51*
Taw / Torridge Est	920	855	21	-	-	-	-	-	62	160	176	319
Yealm Est	50	37	-	-	-	-	-	-	-	10	30	74
WeBS county total '00	4297	3395	1807	107	1	3	4	4	1545	3163	2879	3190
WeBS county total '01	4315	3548	-	2	-	-	-	141	1533	4551	4230	5929

Maximum numbers. WeBS County Totals showed an improvement over the previous year in both winter-periods, particularly during the second. The Exe Est easily managed to regain the qualifying level (2,800) for *national importance* after failing to do so in 2000. In addition to the tabulated sites there were: 40 at Goosemoor NR on 9 Feb; two on Lundy 6-13 Oct; a ♂ at Passage House (Teign Est) on 22 Dec; one E past Prawle on 9 Dec; four at Dartington on 23 Dec; one at Squabmoor Res on 13 Jan; a ♀ at Stover CP on 25 Sep; and two at Wistlandpound Res on 19 Feb, one 29 Sep, two 25 Sep and one on 9 Nov.

Departures and arrivals. A ♂ summered on Exminster Marshes. The first returns of autumn were two at Bowling Green Marsh on 20 Jul.

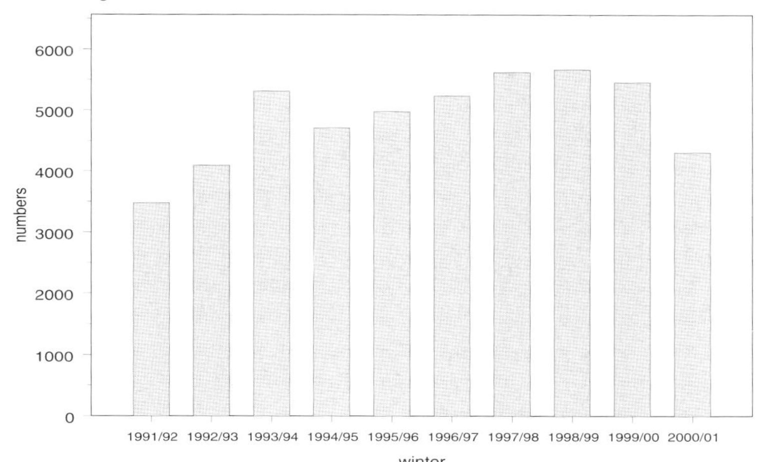

Fig. 18: Wigeon Maximum WeBS County Totals for the *1991/92 - 2000/01* winters (Source: DBRs)

GADWALL *Anas strepera*

Scarce passage and winter visitor; very scarce in summer; rarely breeds. SPEC(3), BoCC (Amber).

Table 29: Monthly maxima at main/well-counted sites and WeBS county totals:

	J	F	M	A	M	J	J	A	S	O	N	D
Exe Est	14	3	-	5	3	2	1	-	4	2	3	4
Roadford Res	3	2	-	-	-	-	-	-	2	2	2	2
Beesands Ley	-	5	8	4	2	10	-	6	21	6	13	4
Slapton Ley	5	4		3	5	2	-	10	**109**	70	**87**	**108**
WeBS county total '00	93	47	51	2	4	2	-	2	49	19	31	30
WeBS county total '01	24	8	-	2	-	-	-	-	20	74	87	121

Breeding. No breeding confirmed, although noted in May-Jun from Beesands Ley (< 5prs), Slapton Ley (<5), Exminster Marshes (<3), Bowling Green Marsh (<2), South Huish area (pr) and Kingsbridge Est (1).

Maximum counts. The important focus for this species continues to be centred on the Torcross area. Elsewhere, recorded from seven sites, all with counts under five.

EURASIAN TEAL *Anas crecca* **Common Teal**

A. c. crecca: not scarce passage and winter visitor; last bred in 1999. BoCC (Amber).

Table 30: Monthly maxima at main/well-counted sites and WeBS county totals:

	J	F	M	A	M	J	J	A	S	O	N	D
Arlington Lake	35	6	-	-	-	-	-	-	-	-	-	17
Avon Est	60	64	60	-	-	-	-	-	1	4	32	140
Axe Est	5	21	-	-	-	-	-	15	14	8	31	60
Burrator Res	19	6	-	-	-	-	-	-	-	-	8	13

continued	J	F	M	A	M	J	J	A	S	O	N	D
Exe Est	702	400	75	50	9	2	5	74	330	443	325	711
Dawlish Warren	*280*	*248*	*32*	*1*	-	-	*4*	*75*	*528*	*551*	*470*	*565*
Fernworthy Res	22	19	-	-	-	-	4	4	12	8	12	13
Huntsham Lake	12	3	3	-	-	-	-	-	7	6	12	11
Kenwith NR	6	7	-	-	-	-	-	2	52	25	30	37
Kingsbridge Est	161	84	-	4	1	-	2	21	50	7	36	294
Lopwell Dam	-	-	-	-	-	-	-	-	-	-	-	6
Otter Est	171	128	-	6	-	-	-	30	257	475	397	341
Roadford Res	111	114	-	-	-	-	-	-	20	69	54	147
Shobrooke Park	18	12	-	-	-	-	-	-	2	2	20	62
Slapton Ley	-	-	-	-	-	-	-	-	-	9	8	53
South Huish area	112	89	47	-	-	-	-	-	23	33	68	75
Tamar Complex WeBS	275	63	-	-	-	-	-	-	-	13	154	364
Upper / mid Tamar	*62*	*10*	-	-	-	-	-	-	-	-	*20*	*58*
Tavy Est	*19*	*25*	-	-	-	-	-	-	-	-	*4*	*56*
Taw / Torridge Est	400	100	13	-	-	-	-	-	33	110	103	950
Yealm Est & Kitley Pond	26	55	-	-	-	-	-	-	11	-	42	142
WeBS county total '00	1394	867	438	96	5	12	15	104	696	899	736	1363
WeBS county total '01	1682	1053	-	6	-	-	-	126	628	1125	1147	2534

Breeding. No confirmed breeding, though up to four birds at Fernworthy Res (a recent breeding site) in Jul and Aug may have attempted to breed, and some summering birds were noted around the Exe Est.

Maximum counts. The highest count of the year was 950 on Taw/Torridge on 28 Dec, a great improvement on the previous year's max of 548. The highest count on the Exe Est was 711 during Dec.

GREEN-WINGED TEAL *Anas carolinensis*

(Formerly ***A. crecca carolinensis***). ***Rare vagrant.***

A ♂ at Isley Marsh on 13 Jan (PMM) was reported there again on 20-21 Feb (IT,IM,JEW).

MALLARD *Anas platyrhynchos*

Fairly numerous resident breeder and winter visitor.

Table 31: Monthly maxima at main/well-counted sites and WeBS county totals:

	J	F	M	A	M	J	J	A	S	O	N	D
Arlington Lake	24	5	-	-	-	-	-	-	-	-	-	5
Avon Est	92	75	50	50	39	40	48	82	178	255	139	120
Axe Est	99	96	-	-	-	-	81	95	120	75	71	60
Burrator Res	42	29	-	-	-	31	19	26	24	43	39	52
Dart Est	120	40	-	-	-	-	-	-	123	13	149	75
Exe Est	415	110	12	11	16	46	84	592	187	582	372	493
Fernworthy Res	12	14	-	-	-	-	33	26	14	16	9	-
Grand Western Canal	25	32	-	-	16	-	7	-	19	44	46	43
Huntsham Lake	27	14	6	5	15	-	68	90	80	44	45	74
Kenwith Valley	40	60	-	1	-	-	59	-	48	60	65	40
Kingsbridge Est	135	130	-	-	-	52	-	140	189	313	340	215
Otter Est	100	55	-	12	25	87	50	108	143	170	94	88
Plym Est	34	22	37	10	45	59	50	25	12	150	127	-
Roadford Res	117	78	-	-	-	-	-	-	-	198	135	85
Shobrooke Lake	65	52	-	-	-	-	104	76	108	86	68	73
Slapton Ley	5	47	-	25	16	30	-	50	-	17	74	52
South Huish area	72	55	-	10	-	-	-	-	40	13	47	39
Stover CP	98	48	20	40	54	60	20	20	40	40	40	20
Tamar Complex WeBS	317	157	-	-	-	-	-	-	231	356	335	361
Upper / mid Tamar	*129*	*57*	-	-	-	-	-	-	*103*	*106*	*95*	*213*
Tavy Est	*50*	*48*	-	-	-	-	-	-	*63*	*51*	*113*	*80*
Taw / Torridge Est	247	186	8	10	8	8	44	60	451	406	406	377
Yealm Est & Kitley Pond	68	15	-	-	-	-	-	-	44	52	100	6
WeBS county total '00	1980	1383	1060	742	784	1109	1501	2049	2560	2591	2418	2876
WeBS county total '01	2279	1447	55	69	97	193	225	1058	2210	2879	2628	2427

Breeding. In spite of the FMD access restrictions, a much better year with 31 sites, 61 prs and a minimum of 292 yg reared (*cf.* 18/48/98 in 2000).

Maximum counts. With 592 on Exe Est during Aug and 451 on Taw/Torridge in Sep, numbers were similar to those of 2000. Some maxima from untabulated sites were: 25 at Stover CP on 6 Jan; 72 at Radford L on 20 Jan; 18 at Elfordleigh in May; 24 at Wembury on 24 May; 89 at Umborne Bridge (Colyton) on 10 Jun; 33 at Fernworthy Res on 15 Jul; 49 at Wistlandpound Res on 27 Jul; ten at Beesands in Aug; 18 at Hound Tor on 17 Sep; 30 at Broadnymet Pond on 20 Sep; 193 on R Dart at Dartington on 1 Oct; 40 at Lynmouth on 14 Oct; 42 at Landcross on 23 Oct and 22 at Newcross Pond, Kingsteignton on 21 Dec.

AMERICAN BLACK DUCK *Anas rubripes* — Black Duck

Rare vagrant

The long-staying ♂ was noted from its familiar haunts in the South Hams at the following localities (ASCB,PMM *et al.*): Slapton Ley, 1 Jan - 8 Jun and again on 26 Sep, 20 Oct and from 2 Nov to year end; W Charleton Marsh, 30 July - 24 Aug; and Bowcombe Creek, 4 Oct - 2 Nov. (One of four accepted by BBRC in 2001, but two had persisted from 2000). *(see col plate)*

PINTAIL *Anas acuta* — Northern Pintail

Scarce winter visitor. SPEC(3). BoCC (Amber).

Table 32: Monthly maxima at main sites and WeBS county totals:

	J	F	M	A	M	J	J	A	S	O	N	D
Exe Est	81	19	11	1	1	1	-	-	3	28	91	76
Kingsbridge Est	8	4	-	-	-	-	-	-	-	-	10	20
Taw / Torridge Est	54	43	-	-	-	-	-	-	-	9	40	20
WeBS county total '00	27	41	3	4	-	-	-	-	27	61	60	66
WeBS county total '01	100	23	-	-	-	-	-	-	3	30	113	95

Maximum counts. As usual, the highest counts came from the Exe Est with a max of 91 in Dec (*cf.* 72 in 2000, and 133 in 1999). There was also a small increase from 2000 on the Taw/Torridge, with a max of 54 in Jan (*cf.* 40 in 2000, and 56 in 1999). Away from the tabulated sites: two on the Teign Est on 6 Jan; a ♂ on the Otter Est 12–28 Jan; a ♂ at Burlescombe on 18 Jan; five at Stover CP on 5 Feb; a ♀ on Upper Tamar L on 24 Sep; and five on the Avon Est on 1 Dec. An interesting mid-summer record was of a ♂ at Beesands Ley on 21 Jun (PBo) and a ♂ & two ♀♀ the next day (H&JH).

Arrivals. The first of the autumn, a single, arrived in off the sea with Wigeon at Dawlish Warren on 17 Sep, and on 4 Oct five came in with Gadwall (IL).

GARGANEY *Anas querquedula*

Very scarce passage visitor; casual breeder (last bred 1998). SPEC(3). BoCC (Amber).

Spring/summer. The first spring arrival was of a ♂ at Exminster Marshes on 5 Apr (THS); thereafter, one or two ♂♂s at this site 11 Apr – 20 May (mo) with a max of two ♂♂s and a ♀ there on 14 Apr (MKn). Elsewhere: a ♀ at Exeter Riverside CP on 8 Apr (BG); a ♂ at Bowling Green Marsh on 14 Apr (BBH,MSW,MRAB); and a ♂ at S Huish Marsh 13–21 May (PBo,PSa,RBu).

Autumn. The first autumn return was a ♀/imm at Velator on 22 Aug (RJ), followed by three at Bowling Green Marsh on 31 Aug (PBO,RSPB), with a single there on 16 Sep (BG) and an imm at Slapton Ley 8–26 Sep (SL,PBO,GV *et al.*)

SHOVELER *Anas clypeata* — Northern Shoveler

Scarce winter visitor; scarce/very scarce in summer; casual breeder (eg. 2000). BoCC (Amber).

Table 33: Monthly maxima at main/well-counted sites and WeBS county totals:

	J	F	M	A	M	J	J	A	S	O	N	D
Avon Est	8	-	-	-	-	-	-	-	-	-	-	5
Beesands Ley	-	-	1	-	-	-	-	3	-	-	2	-
Exe Est	81	60	24	17	3	2	10	8	15	25	45	83

continued	J	F	M	A	M	J	J	A	S	O	N	D
Slapton Ley	-	-	-	-	-	-	-	10	-	17	22	52
South Huish area	40	32	-	-	-	-	-	-	-	-	-	4
Taw / Torridge Est	22	12	12	-	1	-	-	-	-	2	12	2
WeBS county total '00	27	41	3	4	-	-	-	-	27	61	60	66
WeBS county total '01	121	83	-	-	-	-	-	9	17	36	59	123

Breeding. Two or three birds summered on the Exe Est but no confirmed report of breeding.

Maximum counts. Much improved WeBS county totals with Jan, Feb & Dec totals about double those in 2000. Most were on the Exe Est with maxima of 81 in Jan and 83 in Dec, but also high Jan/Feb counts in the S Huish area (inc. Thurlestone Marsh) and at Slapton in Dec. Elsewhere: seven on Axe Est during Dec; two Kenwith NR on 14 Dec; one at Kingsbridge Est on 7 Jan; 25 at Matford Park on 2 Jan; five at Newcross Pond (Kingsteignton) on 21 Dec; three on Otter Est on 27 Jan; four at Roadford Res on 16 Jan; 17 at Seaton Marshes on 17 Dec; one at Shobrooke L on 1 Dec; and two at Stover CP during Sep/Oct. Ten at S Huish on 23 Dec were on the sea (AJL).

RED-CRESTED POCHARD *Netta rufina*

Rare vagrant or escape. SPEC(3).

All records may be of escape/feral origin: a ♀ at Wistlandpound Res on 24 Jan (NWCo) was seen again at this location on 25 Feb (MAP); a ♀ at Slapton Ley on 23 Aug (NPR); and a ♂ at Isley Marsh on 2 Dec (MAS, DC).

POCHARD *Aythya ferina* — Common Pochard

Not scarce winter visitor; very scarce in summer; first bred in 2000. SPEC(4). BoCC (Amber)

Table 34: Monthly maxima at main/well-counted sites and WeBS county totals:

	J	F	M	A	M	J	J	A	S	O	N	D
Beesands Ley	11	10	6	-	-	-	-	-	-	-	10	54
Burlescombe	-	8	-	-	-	-	1	-	-	1	5	11
Exe Est	22	14	13	2	-	-	1	-	1	-	18	16
Hennock Res	5	3	-	-	-	-	-	-	-	-	-	18
Huntsham Lake	-	-	-	-	-	-	-	-	-	-	-	-
Roadford Res	43	8	-	-	1	1	3	-	2	-	44	24
Slapton Ley	78	37	-	3	4	3	-	15	-	44	82	154
Stover CP	21	6	5	1	1	1	1	1	1	4	10	4
Yealm Est & Kitley Pond	7	-	-	-	-	-	-	-	-	1	5	-
WeBS county total '00	158	73	26	5	2	-	1	4	5	37	77	62
WeBS county total '01	112	34	5	1	1	1	1	1	2	50	100	199

Maximum counts. WeBS county totals were lower in the first winter-period than in recent years, but rallied well during the second winter-period, with 199 in Dec (*cf.* 62 in 2000). Slapton Ley once again had the highest counts with a max of 154 in Dec, (*cf.* 77 in 2000, and 136 in 1999). Away from tabulated sites, recorded from ten other locations, but the only count >10 was 24 at Newcross Pond, (Kingsteignton) on 21 Dec (MKn).

RING-NECKED DUCK *Aythya collaris*

Rare vagrant; most records during autumn and winter.

The ♂ from 1999 continued its prolonged association with Burrator Res, remaining until 3 Mar (mo). It returned on 6 Dec (PMPh) and was present at Burrator until 15th before being re-located on 18th at nearby Lopwell Dam (RHi) where it remained until at least the year end (mo).

FERRUGINOUS DUCK *Aythya nyroca*

Rare vagrant or escape. SPEC(1).

The escaped ♂ from the previous year remained in the Totnes area and was regularly reported throughout the year (mo), also being noted at Buckfast on 24 Jul (BrN).

TUFTED DUCK *Aythya fuligula*

Not scarce winter visitor; scarce in summer; rare breeder

Table 35: Monthly maxima at main/well-counted sites and WeBS county totals:

	J	F	M	A	M	J	J	A	S	O	N	D
Beesands Ley	16	21	12	14	-	-	-	1	4	2	12	26
Bicton Lake	8	9	10	3	-	-	5	-	-	-	-	1
Burrator Res	24	21	-	-	-	-	-	-	-	1	14	4
Exe Est	7	7	5	10	4	5	2	3	3	-	6	8
Elfordleigh Ponds	20	4	-	-	-	-	-	-	-	-	4	13
Exe Est	7	7	5	10	4	5	2	3	3	-	6	8
New Cross Pond	24	11	-	12	-	-	-	-	-	-	-	44
Hennock Res	-	6	-	-	-	-	-	-	-	-	-	12
Huntsham Lake	2	-	-	-	-	1	-	1	2	1	-	-
Portworthy Dam	-	8	-	-	-	-	-	-	-	-	-	-
Rackerhayes	16	23	-	-	-	-	-	-	-	-	-	17
Roadford Res	92	128	-	-	-	1	27	-	32	50	64	129
Shobrooke Lake	5	1	-	-	-	-	-	-	1	-	3	2
Slapton Ley	60	49	5	35	20	20		30	1	82	95	108
Stover CP	33	20	1	1	1	1	2	5	5	8	12	4
Wistlandpound Res	10	12	-	-	-	-	1	2	1	3	5	8
Yealm Est & Kitley Pond	18	28	-	-	-	-	-	-	-	-	2	-
WeBS county total '00	361	251	117	41	29	14	54	73	71	97	192	204
WeBS county total '01	224	213	4	1	4	4	3	6	16	124	170	276

Breeding. Only two breeding records: at Clennon Valley, an ad ♀ with three yg on 4 Aug and an ad with eight yg on 11th; and a ♀ with three half-grown yg at Beesands Ley on 11 Aug.

Maximum counts. WeBS county totals, although significantly lower in Jan, were on a par with those in 2000. Roadford Res continues to produce the highest counts, with 129 in Dec, though there were also 108 at Slapton Ley in Dec. Counts additional to those tabulated included: a max of 17 at Goodrington Seafront on 24 Feb; a pr at Lee Moor China Clay works on 25 Apr, a pr on Lundy 18-23 May; one at Venford Res on 22 Sep; eight at Lopwell Dam on 18 Dec; and eight on the Taw/Torridge Est on 17 Feb.

SCAUP *Aythya marila* — Greater Scaup

Very scarce/scarce passage/winter visitor to coast; rare/very scarce inland. SPEC(3^{W}), BoCC (Amber).

First winter-period/spring. Five birds: the ♀ from the previous year remained at Wistlandpound Res until 25 Feb (mo); at Roadford Res, a ♂ on 5 Jan (GAV) and a ♀ 13 Jan – 19 Feb (DC,MAS *et al.*); and a ♀/imm on the Tamar Est (DIJ) and a ♀ at Dawlish on 12 Feb (JEF).

Summer. A ♀ at Radford L on 4 June (VRT,SRT).

Autumn/second winter-period. Eight birds: one at Wistlandpound Res on 1 Oct (JEW); a ♀ at Slapton Ley on 13 Oct, followed by two 1w ♂♂ 16-28 Nov (PSa,BW); a 1w # on the Torridge Est on 20-21 Nov (RG), and perhaps the same at Wrafton Park on 10 Dec (JEW) and at Sherpa Marsh on 12th (IM); and a ♂ at Sharkham Pt on 29 Dec (MD,AR).

EIDER *Somateria mollissima* — Common Eider

Scarce winter/passage visitor to coast; very scarce/scarce in summer; rare inland. BoCC(Amber)

Table 36: Monthly maxima at main/well-counted sites and WeBS county totals:

	J	F	M	A	M	J	J	A	S	O	N	D
Dawlish Warren / Bay	17	20	19	19	20	8	12	-	2	14	19	20
Hope's Nose	-	-	-	-	-	4	2	2	2	1	1	-
Start Bay	1	1	-	-	-	-	-	-	-	2	-	-
Torbay	-	-	-	-	-	-	-	-	-	2	1	4
Taw / Torridge Est	20	36	6	20	16	4	13	17	-	7	6	13
WeBS county total '00	83	34	4	8	51	14	15	47	-	-	8	-
WeBS county total '01	20	36	-	-	-	4	-	2	2	3	1	9

Away from tabulated sites: an imm ♂ in Plymouth Sound 14-21 Jan; and ten flying E past Prawle Pt on 18 Oct.

LONG-TAILED DUCK *Clangula hyemalis*

Very scarce winter visitor. BoCC (Amber).

First winter-period/spring. Seven birds, three from 2000: the 1w from 2000 remained at Roadford Res until 19 Feb (mo) and was joined by a ♀ on 14 Feb (GAV *et al.*) remaining until 1 Apr; an ad ♂ at Hope's Nose from 2000 until at least 17 Jan (PMM,CJP *et al.*); a ♀ at Kingsbridge from 2000 until 24 Mar (PBo); a ♂ on the Exe Est 20 Jan – 18 Feb (RWBu *et al.*); a ♀ at Wistlandpound Res on 26 Jan (MAP); and a ♀ in Start Bay on 29 Apr (BBH).

Autumn/second winter-period. Seven birds: a ♀/imm at Torbay from 4 Nov (MRAB), a different ♀/imm from 18 Nov, with both seen on 20 Nov, one remaining until 24th (BMc,PAS,ML *et al.*); a juv flew S past Dawlish Warren on 15 Nov (JEF); and a 1w ♂ and ♀ at Topsham on 23 Nov (RJ,MSW), sometimes joined by an ad ♂ from 30 Nov, all remaining on the Exe Est until 31 Dec (mo).

COMMON SCOTER *Melanitta nigra* — Black Scoter

Not scarce migrant, scarce at other times; very scarce inland. BoCC (Red).

Table 37: Monthly maxima at main/well-watched sites:

	J	F	M	A	M	J	J	A	S	O	N	D
Budleigh/Sidmouth	30	4	-	-	52	1	24	-	-	7	30	70
Dawlish Warren / Bay	55	70	31	40	100	36	120	60	164	160	97	60
Prawle Point	-	-	-	-	17	14	169	46	25	16	22	8
Start Bay	50	45	20	16	2	-	25	4	-	-	50	125
Torbay Area	10	1	13	2	15	13	118	31	55	41	16	24
Westwd Ho!/Hartland	1	1	1	-	-	50	4	14	1	1	27	21

Maximum counts. Additional to tabulated counts: 15 at Bolt Tail on 9 Feb; 53 off Teignmouth sea-front on 7 Feb, and 43 there on 7 Jul; 25 W past Thurlestone on 19 Mar; 18 at Seaton on 7 Oct; 50 at Man Sands on 21 Oct; and 15 off Lynmouth on 16 Dec;

Autumn passage. Migrants were noted moving mainly W past regular sea-watching locations, peaking in mid-Jul with 70 at Orcombe Pt and 169 at Prawle Pt on 16th, and 118 at Berry Hd and 90 at Hope's Nose on 17th. Fifty past Westward Ho! on 10 Jun was the only significant N coast record.

SURF SCOTER *Melanitta perspicillata*

Rare vagrant.

Recorded for the fourth year in succession at Dawlish Warren; an ad ♀ on 16 Dec (LC) became the tenth individual recorded at this site, seven of them since 1998.

VELVET SCOTER *Melanitta fusca*

Very scarce winter visitor to coast. SPEC(3^W), BoCC (Amber).

First winter-period/spring. Two singles from Dawlish Warren: an imm on 29 Jan and a ♀/imm on 16 Feb (JEF).

Autumn/second winter-period. *Dawlish Warren*: eight on 4 Nov at 09.30h and again in the afternoon having visited Torbay in the interim (JEF); two on 5 Nov (JEF); two (one being a ♂) on 11 Nov (JBa,LC); a 1w on 26 Nov (JEF); an ad ♂ and ♀ on 7 Dec (JEF); and a ♀ on 16 Dec (LC). *Torbay*: a ♀ on 10 Nov (RBe,MD,AR); two imms from 2 Dec (BMc,ML), joined by a ♀ and remained until 31st (mo). *Sharkham Pt*: one on 12-13 Nov (ML,PMM), three on 24 Nov (CRB), five – different to nearby Torbay birds - on 19 Dec (ML); and six on 29 Dec (MD,AR). *Prawle Pt*: two flying W on 1 Dec (PMM); two W and two E on 13 Dec (JLt); and one E on 29 Dec (JCN,JLt).

GOLDENEYE *Bucephala clangula* — Common Goldeneye

Scarce winter visitor; rare in summer. BoCC (Amber).

Table 38: Monthly maxima at main/well-counted sites and WeBS county totals:

	J	F	M	A	M	J	J	A	S	O	N	D
Exe Est	14	12	3	1	-	-	-	-	-	2	3	6
Kingsbridge Est	14	11	3	-	-	-	-	-	-	-	2	7
Roadford Res	36	35	-	-	-	-	-	-	-	1	7	30
Slapton Ley	4	10	-	-	-	-	-	-	-	-	18	8
Taw / Torridge Est	3	2	2	-	-	-	-	-	-	-	-	1
Teign Est	3	-	3	-	-	-	-	-	-	2	-	-
WeBS county total '00	94	102	66	9	-	-	-	-	-	1	21	35
WeBS county total '01	67	56	-	-	-	-	-	-	-	-	2	51

First winter-period. WeBS total and site counts were lower than in 2000, with Slapton Ley showing the largest drop in numbers. In addition to tabulated sites, records came from (singles unless stated): Beesands Ley (2), Burrator Res, Fernworthy Res, Mill Leat (Exeter), Shobrooke L (2), Stoke Gabriel (2) and Upper Tamar L.

Autumn/second winter-period. Numbers were up on 2000/01 winter at both of the main sites, Roadford Res, with 30 in Dec, and Slapton Ley, with 18 in Nov. Additional records away from tabulated sites (singles unless stated) from: Beesands Ley, Burrator Res, Paignton (2) and Shobrooke L.

SMEW *Mergellus albellus*

Rare winter visitor. SPEC(3).

The only record was of a ♀ at Bowling Green Marsh 14 Jan – 11 Mar (BBH,MSW *et al.*).

RED-BREASTED MERGANSER *Mergus serrator*

Scarce winter visitor; rare in summer; casual breeder (first bred 1993).

Table 39: Monthly maxima at main/well-counted sites and WeBS county totals:

	J	F	M	A	M	J	J	A	S	O	N	D
Exe Est	101	135	25	37	2	2	1	1	4	29	94	69
Dawlish Warren	*60*	*67*	*35*	*79*	*2*	*2*	*2*	*1*	*3*	*25*	*91*	*135*
Kingsbridge Est	46	24	7	-	-	-	-	-	-	-	35	41
Plym Est	2	20	-	9	-	-	-	-	-	-	-	-30
Tamar Complex WeBS	30	21	-	-	-	-	-	-	-	-	4	24
Torbay area	10	9	15	3	-	-	-	-	-	3	2	5
Taw / Torridge Est	6	1	-	3	-	-	-	-	-	1	3	1
Teign Est	9	5	8	42	15	-	-	-	-	-	18	-
WeBS county total '00	152	199	160	32	5	3	-	1	1	41	69	97
WeBS county total '01	161	193	-	11	-	-	-	-	4	29	131	127

Breeding season. No records of breeding, but a pr summered at Dawlish Warren until 30 Jul, thereafter one until late Sep (IL *et al.*).

Maximum counts. The *Exe Est* continues to be the primary site, with numbers in both winter-periods exceeding the qualifying level (100) for *national significance*, with a max of 135 in Dec. Late first winter-period flocks included 42 on the Teign Est on 14 Mar and 79 at Dawlish Warren on 20 Apr. The first returns were of three at Dawlish Warren on 26 Sep. In addition to the tabulated sites: three at Budleigh Salterton on 13 Jan; four at Chiselbury Bay on 27 Feb; two on the Otter Est 14 Jan – 31 Mar; and one at W Charleton Marsh on 14 Feb; and during the second winter-period at Prawle Point, one on 2 Dec and two on 8th.

GOOSANDER *Mergus merganser*

Scarce winter visitor; scarce breeder (first bred 1980).

Table 40: Monthly maxima at main/well-counted sites and WeBS county totals:

	J	F	M	A	M	J	J	A	S	O	N	D
Burrator Res	10	6	-	-	-	5	-	-	-	-	2	9
Fernworthy Res	2	5	-	-	-	-	-	-	-	-	-	16
Roadford Res	15	9	-	2	-	1	-	-	-	-	4	14
Shobrooke Lake	-	-	-	-	-	-	-	-	-	-	-	-

continued	J	F	M	A	M	J	J	A	S	O	N	D
Lower Tamar Lake	8	13	-	-	-	-	-	-	-	-	-	6
Taw / Torridge Est	1	4	-	2	-	-	-	-	-	-	1	10
Venford Res	28	16	-	-	-	-	-	-	16	20	26	27
WeBS county total '00	37	66	60	28	14	2	-	-	-	17	22	21
WeBS county total '01	44	38	-	-	-	-	-	-	16	-	21	37

Breeding. Confirmed records: at Buckfastleigh (♀ and nine ducklings on 6 Jun) (JMW); Buckfast (two ♀♀ with 17 yg on 14 Jun) (BrN); at Staverton (a ♀ with ten yg on 5 May) (RCh); and at Huccaby (a ♀ raised three yg) (RHi).

Maximum counts. WeBS totals probably do not accurately reflect county numbers, nor do dusk roost counts unless synchronised. Roosts on *Dartmoor* reservoirs as usual provided the highest counts, with a max of 28 at Venford during Jan and 27 in Dec. The first returning ad ♂ after post-breeding absence from area was noted here on 27 Oct, with numbers of ♂♂ gradually rising to seven on 19 Dec, fewer than in first winter-period when ♂♂ outnumbered brownheads (RHi). Max roost count elsewhere on *Dartmoor* reservoirs was ten at Burrator Res on 21 Jan (RS). *Away from Dartmoor*, three at Aveton Gifford on 21 Jan, and two on the Avon Est during Apr; singles on the Axe Est on 12 Nov and 24 Dec; four at Bampton on 21 Jan; six at Bickleigh on 10 Dec; a ♂ at Cockwood on 20 Mar; a ♀ at Countess Wear on 10 Nov; six at Cove on 21 Jan; a ♀ at Grenofen on 16 Jan; a ♀ at Otter Est during Feb; a ♂ at Slapton on 21 Feb; up to three present at Stover CP during Jan/Feb with a ♂ present there between 14 – 21 Dec; one at Lynmouth on 14 Oct and 4 Nov; five in the Tiverton area on 31 Dec; and a ♂ at Totnes 1-11 Feb.

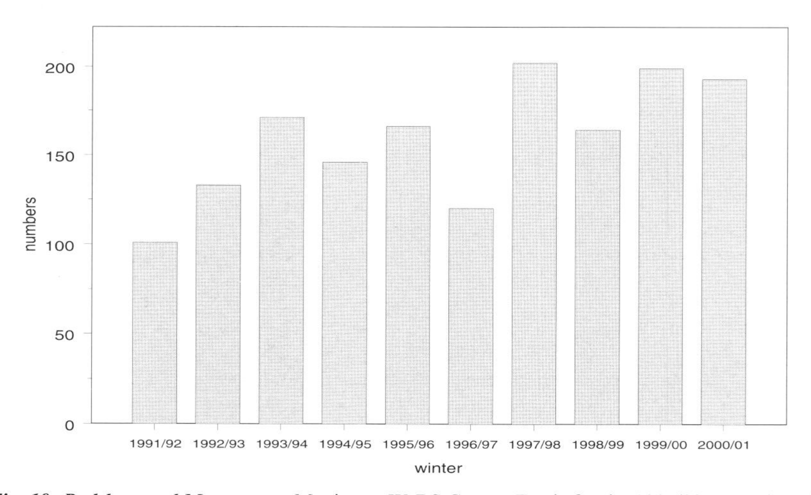

Fig. 19: Red-breasted Merganser - Maximum WeBS County Totals for the 1991/92 - 2000/01 winters (Source: WeBS data in DBRs)

RUDDY DUCK *Oxyura jamaicensis*

Very scarce winter visitor (naturalised introduction).

Breeding. A ♀ with four yg reported from Slapton Ley on 21 Sep, reducing to two by 1 Oct (PMM *et al.*) represents *the first Devon breeding record.*

First winter-period/spring. Up to three at Slapton Ley from the beginning of the year until 22 Apr, and one on the Kingsbridge Est was the first for this locality (PSa).

Autumn/second winter period. Again, mostly noted from Slapton Ley, with maxima of eight in Sep, 16 in Oct, 12 in Nov and 16 in Dec. In addition: one at Powderham on the Exe Est on 24 Nov, one at Exminster Marshes 19-22 Dec; and two at Roadford Res on 10 Sep, eight on 17 Sep and one 4-20 Nov.

HONEY BUZZARD *Pernis apivorus* **European Honey-buzzard**

Very scarce passage visitor; rare migrant breeder (bred 1979–1996). SPEC(4), BoCC (Amber).

An uneventful year, only four records (May-Jul) and again no breeding.

The first of the year, an imm/♀, was a relatively long stayer in the Haldon area 15 May-15 Jul (RK), followed by one over Starcross on 17 May (DJG) (possibly relating to Haldon records); then a ♂ at Haldon (BOP Viewpoint) on 21 May (RK) and finally, one over Ashclyst Forest on 29 May (A Stevenson).

RED KITE *Milvus milvus*

Rare or very scarce passage and winter visitor (some records refer to released birds from reintroduction programme). SPEC(4), BoCC (Amber).

The recent trend of wandering individuals, mostly emanating from the release sites in England and Scotland, continues; records in Feb, Mar, May, Jun and Nov.

The individual at Stoney Bridge, North Molton, present at the end of 2000, remained until 28 Jan (DC,MAS,JEW *et al.*). This (now 2cy bird) was wing-tagged (L blue, R pink with the number five in black) as a nestling in Inverness, Scotland, where it fledged in Jun 2000 (third generation from there); amazingly, it got back to Inverness from Devon in just two days, being back at the Scottish roost site on 30 Jan. *Elsewhere* (all singles unless otherwise stated; some records could refer to the same bird): at Huntsham on 9 Feb (RGr); near Braunton 23-27 Mar (RJ); South Brent (MRG) and Burrator (BG) on 5 May (probably the same bird); Lydford on 10 May (*per* DSGp); Jennett Pt NR on 12 May, and two over Seaton on the same day (BG); Exminster Marshes 23-24 May (MD,RSPB); Ide 1 Jun (SWa); nr Westcott 2 Jun (AMR); nr Fernworthy Res during May (DSGp); Sampford Peverell on 3 Jun (GMc); Okehampton on 4 Jun (MGM); Bampton on 12 Jun (CCh); Barnstaple on 22 Jun (KGe); and the last, at Chudleigh on 5 Nov (SBg).

2000 addition. One at North Molton on 23 Dec (A Lansdell, JEW).

MARSH HARRIER *Circus aeruginosus* **Eurasian Marsh Harrier**

Very scarce passage visitor. BoCC (Amber).

A good year, with around ten individuals recorded on spring passage, while just two during summer/ autumn. Again Exminster Marshes proved to be the top site.

Spring. First was a ♂ at *Exminster Marshes* on 8 Apr (DS), followed by other individuals there including: an imm on 16 Apr (DS), an imm ♂ on 2 May (RK), with one on Lundy the same day (LFS); one (nearby over the Exe Est) on 9 May (RSPB); and a ♀/imm 12-13 May (RSPB,MKn,DJSu). *Elsewhere*: at Slapton Ley, a ♂ on 10 Apr (BW) and a 3cy ♂ on 25 Apr (DEk); and a 2cy ♀ over Holne Moor on 11 May (RHi).

Autumn. First return was a ♀ hunting Braunton Marshes on 23 Aug (RJ), followed by a ♀/imm at Slapton Ley 18-23 Oct (PBo,PSa,DR,BW).

2000 corrections. Delete one at Kenwith NR 13-14 May; and the ♀ at Berry Head on 22 Apr should be 2cy, not 1cy.

HEN HARRIER *Circus cyaneus*

Scarce passage and winter visitor. SPEC(3), BoCC (Red).

Numbers appear similar to last year despite FMD precautions restricting access to Dartmoor (affecting the national roost counts) and other areas in the spring. Peaks from the regular roosts suggest around five ## and two ringtails during the first winter-period/spring period, and around five ## and three ringtails in the autumn/second-winter period.

First winter-period/spring. At the regular *roosts*: at *Bursdon Moor* up to three, one ♂ and two ringtails, used the roost 1 Jan – 2 Mar, although never all three seen on any one occasion, then just one ringtail on 1 Apr (FHCK); and in the *Warren House area of Dartmoor*, the roost was only used intermittently, with no roosting birds seen until 25 Jan when four present (three ♂♂ and one ringtail ad ♀), then three ♂♂ on 12 Feb, then just one ♂ on 18 Feb (DSGp). *Other records from Dartmoor* included single ♂♂ at East Bovey Head on 28 Jan, Huccaby on 29 Jan, Halshanger Common and Trendlebere Down on 24 Feb. There was a migrant on Lundy on 1 May.

Autumn/second-winter period. The first returning birds at the *Bursdon Moor roost* were two on 1 Oct, then four (two ♂♂ and two ringtails) by 5 Nov and five (three ♂♂ and two ringtails) 9-31 Dec (FHCK). On *Dartmoor*, the first return was a ringtail at Huccaby on 4 Oct, and one also there on 3 Nov, with other ranging ♂♂ at Skir Hill on 3 Dec; Cut Hill on 8 Dec, Holne Moor on 15 Dec, Holne on 18 Dec and over Buckfastleigh on 28 Dec. At the *Warren House area roost*, the first return was a ♂ on 21 Oct, with two (one ♂, one ringtail ad ♀) by 16 Nov and three (an ad ♂, a 2cy ♂ and ad ♀ ringtail) on 28 Dec (DSGp). *Elsewhere* a ♂ was at Trinity Hill NR on 10 Dec, a ♂ at Isley Marsh 12 Dec, a ringtail at Moorhouse Ridge on 19 Dec and two ♂♂ at Upper Tamar L on 22 Dec.

MONTAGU'S HARRIER *Circus pygargus*

Rare or very scarce passage visitor (last bred in 1976). BoCC (Amber),

In contrast to 2000 (seven records), just one: a ♂ on Little Haldon on 12 May (RK).

2000 additions. Additional records, in a very good year: ad ♀♀ at Roborough Down on 12 May (RS) and South Molton on 15 May (FHCK,IK).

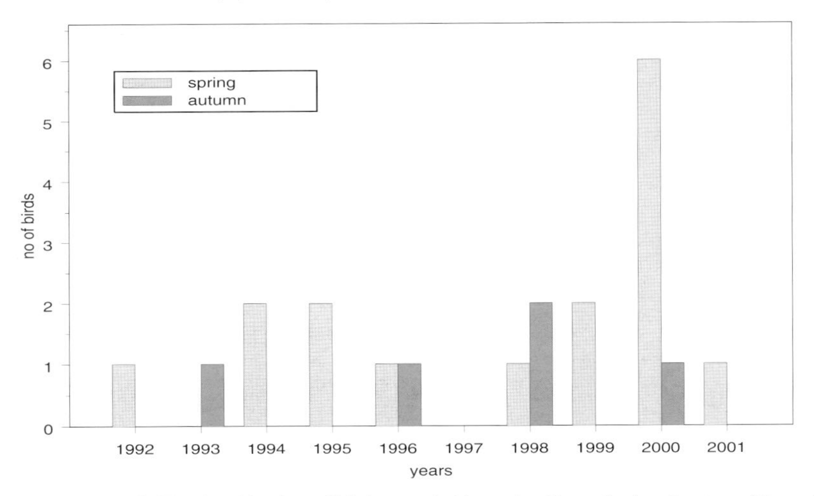

Fig. 20: Montagu's Harrier - Number of birds recorded in spring (Apr - Jun) and autumn (Aug - Sep) 1992-2001. (Source: DBRs)

Unidentified HARRIER species *Circus spp.*

A single ringtail over Clennon Valley on 5 May was seen too briefly for positive identification.

GOSHAWK *Accipiter gentilis* — Northern Goshawk

Very scarce resident breeder and passage visitor.

Limited monitoring of breeding sites could take place, due to FMD access restrictions.

Breeding. Four prs known to be successful and one pr failed (mean 2.3 yg fledged per pr) (MD,RK,AR).

Non-breeding. Away from known breeding areas, accepted records of singles included: a ♀ over Exeter (Middlemoor) on 8 Apr (NCW); a ♀ at Bolt Hd on 29 Aug (DJa); a ♀ at Halwell on 4 Nov (MD,AR); and birds at Sourton Cross on 5 Nov and Dartmeet on 22 Nov (both RHi).

SPARROWHAWK *Accipiter nisus* — Eurasian Sparrowhawk

Fairly numerous resident breeder; passage status uncertain.

Well-recorded, with a total 344 records from 146 widespread locations across the county, and well up on last year (similar to 1997 and 1999) despite FMD access restrictions.

Breeding. Displaying birds seen, or successful breeding noted (prs and yg produced if known) at: Bowling Green Marsh (pr bred), Bridford (pr), Buck's Mills (pr), Clennon Valley (pr), Dawlish Warren (pr), Decoy L (pr), East Budleigh Common, Exminster Marshes (pr), Exminster Marshes (2 prs displaying from local woods), Haldon Forest (7-8 prs, 4 successful, but with only 7yg (RK)), Lethole Plantation (pr, 2yg), Pickwell Down (pr), East Prawle (pr), Roadford Res (pr), Soar area (pr bred), Stover CP (pr bred), Tiverton (pr) and Weare Giffard.

Prey species. Only Woodpigeon, Blackbird, Starling and Great Tit recorded.

BUZZARD *Buteo buteo* — Common Buzzard

Fairly numerous resident breeder; passage status uncertain.

An increase in reported sightings, with c.430 records from 175 sites widespread throughout the county; although substantially higher than the 134 sites in 2000, not up to the 1999 level (c.200 sites) of recording. Flocks of up to 37 and some visible migration reported.

Breeding. Successful breeding under-recorded, but noted at 38 sites.

Jan-Jun (*large gatherings of ten or more*). In *Jan*, ten feeding on the ground at Huxham on 14th. In *Feb*, soaring over Bridford were 11 on 5th, 22 on 13th, 26 on 14th and 12 on the 18th; and 15 over Heatree Down on 9th; ten over Haldon on 13th; and 11 at Barton Pines on 18th. During *Apr*, 18 soaring at Huntsham on 18th, ten in a field at Hendom Cross and 15 at Sandford on 21st, while on Haldon (Underdown) *visible migration* was noted on 29th, with 71 birds passing in a NE direction in three dispersed groups of 13, 21, and 37 (JGa). During spring and summer, when large gatherings are less frequent, 20 were at Prawle on 7 *May* and ten were at Huntsham on 17th; and 21 in a field by the Grand Western Canal on 8 *Jun.*

Jul-Dec (*large gatherings of ten or more*). The first gatherings were in *Sep*, with: 32 feeding in a ploughed field at Zeal Monachorum on 6th; 17 soaring over Wembury on 20th; *c*.20 near Berry Pomeroy Castle on 21st; 11 in newly planted kale at Cator Gate and 15 in ploughed field at East Charleton on 22th; and 20 at Totnes on 24th. In *Oct*, 15 at Blackborough on 8th, 13 at Ugborough and 14 in a field at South Brent on 26th. In *Nov*, ten at Holcombe on 1st and 11 at South Huish on 11th. And in *Dec*, a series of gatherings at Kerswell Cross of 11 on 6th, ten on 11th and 18th, 11 on 19th and 18 on 20th.

Mortalities and a ringing recovery. Jul 20 was not a good day for Buzzards, with one hit by a car on the A38 at Kennford and one electrocuted on a powerline at Hexworthy. A luckier individual was one rescued from the sea in 1990, ringed and released at Dartmouth and recovered there, albeit dead, over ten years later; further details in the Ringing Report. No poisonings relating to pesticide abuse were confirmed by DEFRA this year; however, one corpse did contain levels of second generation rodenticide. Observers should report any possible pesticide (abuse) poisoning incidents to the *Wildlife Incidents Investigation Scheme* (freephone 0800 321600).

OSPREY *Pandion haliaetus*

Very scarce passage visitor. SPEC(3), BoCC (Amber).

A very good year; recorded every month Apr-Nov without the usual break in records, indicating summering bird(s) in the county. As usual, migrating birds dropping in on the county's estuaries from Aug onwards provided the majority of records, most departing by Oct (apart from one injured juv which stayed into Nov).

Apr-Jul. *Passage* started with singles at Bowling Green Marsh on 12 Apr (RSPB), then Buckfastleigh on 14 April (MD,AR), Penhill Marshes on 15 April (JEW), Dawlish Warren (IL) and Torquay (MRAB) on 5 May, Teign Est on 6 May (MKn), Turf on 8 May (JGa) and Bowling Green Marsh on 11 May (RSPB). Records continued through the *summer* with singles at Countess Wear (MKn) and Dawlish Warren (IL) on 1 Jun, Bowling Green Marsh (RSPB,MKn) and R Dart on 7 Jun (AF), Torridge Est 16 Jul (IM) and again at Skern on 31 Jul (AF), and another on the Plym Est the same day (ASCB).

Aug-Nov (S coast estuaries). The more usual *return passage* began on 9 Aug. On the *Dart Est*: two at Dittisham on 9 Aug (*per* RHi), and one there on 28th (PDd); singles at Stoke Gabriel on 25 Aug and 2 Sep (EJL); one at Sharpham on 1 Sep (A&PAM); and a juv at Duncannon on 7 Sep (MD,AR). On the *Kingsbridge Est*; a juv at Charleton Bay on 26 Aug (DJa,NPR); and singles at Wall Park Wood

29 Sep-6 Oct (MBk) and Frogmore Creek 14 Oct (JHB). On the *Yealm Est*: one on 24 Aug (SWor), followed by at least two, possibly three, including an ad and juv (with red ring on left leg) from 9-13 Sep (MD,LHH,RHH,CJR,DRe). On the *Exe Est*, a juv at Dawlish Warren (PBo,LC,IL,TM,GV) 22-23 Sep. On the *Tavy Est*, a juv at Lopwell Dam on 25 Sep (RS).

Aug-Nov (elsewhere). Singles were at: Roadford Res on 16 Aug (FHCK); SE over Haldon on 21 Aug (RK); Prawle area on 29 Sep (TM) and 16 Oct (PMM); and Lundy on 27 Oct (LFS). Finally, a long-staying juv was present at Hatchlands fishery, Rattery from at least mid Sep-20 Nov (MD,PMM,ML *et al.*), and was also seen over nearby Harbourneford on 15 Nov (LBr). The long stay appeared to be due to an injured left leg; although the bird could still perch and fish, it seems that the small pools packed with Trout were an easier option than attempting to fish elsewhere, hence the delayed migration.

KESTREL *Falco tinnunculus* — Common Kestrel

Not scarce/fairly numerous resident breeder and passage visitor. SPEC(3), BoCC (Amber).

Recorded from 151 widespread sites (468 records), similar to 1999, and higher than 1997–98 and 2000, although breeding poorly recorded perhaps due to FMD restictions.

Breeding. Only 17 prs reported in the breeding season, of which 11 prs were known to breed (average 2.8 yg per successful pr). Breeding was recorded or suspected at: Bowcombe Creek, Bowling Green Marsh, Braunton, Broadsands (2 yg), Exton, Grand Western Canal, Haldon Forest area (2 prs reared 5 yg), Start area, Sharkham Point, Sourton Cross, Stoke Pt, Jennycliff (2 yg), Prawle area, Sherpa Marsh (4 yg), Thurlestone (4 yg), Wembury and Wild Tor area.

Non-breeding. The highest count of the year outside the breeding season was ten in the Start area on 3-4 Oct (PJR,JMW). Interesting sightings included a ♂ walking amongst 100 unconcerned Pied Wagtails at Sourton Cross on 3 Oct (RHi), and birds noted regularly at Virginstow for the first time in 12 years (HM).

Mortality. Unfortunate individuals included: a carcass (presumed raptor kill) found at Broad Falls, R Avon on 10 Aug (DSGp); one pulled out of the sea by fishermen at Berry Hd on 27 Oct, which, although taken by MD and kept warm, quickly died; and one hit by a golf ball at Thurlestone during the autumn which also died (H&JH).

***Record request**. Please continue to submit all breeding records, including sightings of prs and number of yg if known.*

MERLIN *Falco columbarius*

Very scarce passage and winter visitor; rarely breeds. BoCC (Amber).

A total of 173 mainland records from 73 sites (similar to the last three years); with the majority of records occurring during autumn migration (50% during Sep-Nov, with 27% in Oct alone) and over-wintering birds (37% during Dec-Feb), particularly at Bursdon Moor, on Dartmoor and around the Exe Est. No confirmed breeding.

First winter-period/spring (all single ♀/imm unless stated otherwise). At *Bursdon Moor*: two regularly through Jan-Feb including a ♂; and then one on 1 Apr, also one seen at nearby Welsford Moor on 13 Jan. On *Dartmoor* at: Hennock Res on 4 Jan; Soussons Farm, a ♂ on 14 Jan; Ugborough Moor, a ♂ on 7 Feb, and a ♀/imm there on 14 Feb; Meldon Res on 14 Feb; Hooten Wheals on 16 Feb; Ditsworthy Warren on 17 Feb; Fice's Well on 18 Feb; and Trowlesworthy Tors, a ♀ on 19 Feb. *Elsewhere* individuals were at: Kenton on 1 Jan; a ♂ at Woodbury Castle on 2 Jan; Huntsham Lake on 3 Jan; at Dawlish Warren a ♀ on 5th, 7th, 15th, 28th & 31st Jan, 27 Feb and 7th, 10th, 11th & 12th Mar; a ♂ at Exminster Marshes on 5th, 7th & 11th Jan and 23rd & 27th Mar; Exwick on 10th; Turf (Exe Est) on 13 Jan; East Allington on 16 Jan; Haldon on 23 Jan; a ♂ at Berrynarbor on 26 Jan; Kennford on 26 Jan; a ♀ at Kenn on 28 Jan; a ♂ at Witheridge Moor on 11 Feb; a ♀ at Aylesbeare Common on 12 Feb; Wembury on 20 Feb; West Putford on 11 Mar; a ♀ Hallsands on 22 Mar; Braunton Burrows on 31 Mar; Taw Est on 9 Apr, a ♀ at Stumpy Post Cross on 13 Apr and Velator on 14 Apr.

Autumn/ second winter-period. Early returning birds were a ♂ at Bowling Green Marsh on 4 Aug, a ♀/imm at Routrundle on 27 Aug and one at Upper Tamar L on 29 Aug. Then the more usual arrivals were (all single ♀/imm unless stated otherwise): at Prawle Point (one over the sea during seawatch) on 12 Sep and in the area 24-30 Sep, 10-27 Oct, 11-24 Nov; a ♂ at Skern on 15 Sep and nearby Northam

Burrows on 18 Sep; Soar area on 22-23 Sep, 9-20 Oct, 3 Nov; Staddon Point 1 Oct; Bigbury area on 5th & 28 Oct; a ♀ at Bowling Green Marsh on 5 Oct; Bowcombe Creek on 10 Oct and on the Kingsbridge Est on 16 Oct; Lundy Island on 10 Oct; Dawlish Warren 11 & 24 Oct; Seaton Marshes on 11-12 Oct; Upper Tamar Lake on 11 Oct; Wembury on 16 Oct; Slapton on 16 Oct; Start area on 16-21 Oct; Axe Est 21 Oct; Staddon Point on 23 Oct; a ♂ at Blackborough on 29 Oct; at Exminster Marshes a ♀ on 29 Oct, a ♂ 10-11 Nov, then up to three (♂ & 2♀♀) 1-29 Dec; two (♂ & ♀) Braunton Burrows on 1 Nov and one 9 Dec; a ♂ Haldon (Underdown) on 5 Nov and 10 Dec; Babbacombe on 6 Nov; Isley Marsh and Skern on 12 Nov; Stoke Point on 13 Nov; Slapton Ley on 14 Nov; two at Burlescombe and one at Darracott on 22 Nov; Malborough on 4 Nov; Kenton on 19 Dec; two (♂ & ♀) at Huxham on 26 Dec; and Otter Est on 30 Dec. On *Dartmoor* at: Moretonhampstead on 11 Oct; Venford Res on 17 Oct; Warren House area on 21 Oct, 2 Nov, 16 Nov and 28 Dec; Huccaby on 24 Oct; East Okement Farm on 28 Oct; a ♂ at Rippon Tor on 4 Nov; Green Combe on 9 Nov; Brent Moor on 22 Nov; Burrator Res on 22 Nov, and a ♂ on 18 Dec; a ♀ at Gutter Tor on 26 Nov; Ryder's Hill on 22 Dec and Bridford on 31 Dec. On *Bursdon Moor* the first return was one on 1 Oct, and from then on seen regularly, with two (♂ & ♀) by 2 Nov and up to three including a ♂ from 11 Nov until year end, often seen to roost in willow scrub.

Hunting behaviour. A ♀/imm was seen (following) using a ♂ Hen Harrier to flush prey in the Warren House area on 16 Dec, waiting on Peregrine-style above the Harrier and even hovering twice (MD,AR); and again seen using similar tactics in the same area on 28 Dec (RS).

HOBBY *Falco subbuteo* — Eurasian Hobby

Scarce passage visitor and migrant breeder.

Well recorded throughout the county with c.170 records received (similar to 2000), many relating to migrating birds.

Hobby - (eating dragonfly) - *Mike Langman*

Breeding. In the breeding season, prs were present in at least 17 areas producing a minimum of 26yg; interestingly, at one site three ads were observed bringing food into the yg (HAWo *et al.*).

Apr-Jul. The first *spring arrival* was in *Apr* at Exminster Marshes on 5th (THS) followed by singles (unless otherwise stated): on Lundy on 18th; at Ugborough Moor on 21st; Exminster Marshes on 26th; Blackborough on 28th; and Dawlish Warren on 29th. Records then picked up in *May* with numbers building to five at Exminster Marshes (seen hawking insects) by 15th, and new arrivals mainly at S coastal locations such as: Exmouth on 4th-5th; Avon Est on 6th; Prawle on 7th; Bigbury on 9th; Dawlish Warren on 11th; Bowling Green Marsh on 18th; and Slapton Ley on 20th. *Non-breeders* were also noted at coastal locations during the breeding season with singles at Berry Hd on 10 Jul and Prawle Point on 28 Jul

Aug-Oct. In *Aug*, some early dispersal (or just lingering birds) at coastal locations involved singles at Starcross on 1st, Avon Est on 2nd, Bowling Green Marsh and Prawle on 5th, Dawlish Warren on 6th, Northam Burrows on 19th, Prawle on 19th and Exminster Marshes on 20th. The usual *autumn migration* then picked up into *Sep*, with individuals seen at Topsham on 2nd; Dawlish Warren on 3rd; Slapton Ley 4th-5th, 8th, 14th & 28th (but apparently no substantial hirundine roost this year to attract birds); Prawle on 9th, 22nd-25th; at Noss Mayo, a juv on 15 Sep; at Salcombe Regis a juv on 17th; at Ugborough Moor a 2cy ♂ on 22nd-23rd (see Box); Lundy on 25 Sep; Wembury, a juv on 28th; and Brixham on 30th. Numbers tailed off into *Oct* with the last departing birds at: Slapton Ley, with one on 1st, three on 3rd and a juv on 18-19th; Prawle with two on 6th and one on 13th; and Beesands Ley, with one on 9th and three (one ad and two juvs) on 10th. The *last* were seen inland at Uplyme on 21 Oct and at West Putford on 22nd (AMJ).

Record request*. Please pass on any breeding season sightings (other than in the Haldon area) as they occur to HAWo, the Species Co-ordinator.*

HOBBY RESCUE

An interesting success story started with an unfortunate second year male Hobby seen rowing himself towards the bank of a small privately owned lake in south Devon in June. Obviously the bird should not have been in the lake, but neither perhaps should it have been in Britain since it is believed that most young Hobbies remain in Africa until mature. The bird, which had evidently sustained a soft tissue wing injury, was rescued and taken into care, but initially showed no inclination to fly even though kept in a spacious enclosure. It wasn't until 23 August that Leonard Hurrell became involved in trying to confirm recovery and ensure the bird's fitness for the autumn migration. Although Leonard is long experienced in the rehabilitation of birds of prey, Hobbies do have the reputation that it is exceedingly difficult to induce them to pursue their natural prey species. This one eventually responded to a lure, and by 16 September, weighing 170g, joined a trained Merlin in pursuit of Skylarks on Dartmoor. The Merlin was not keen to chase those larks that ringed up rapidly, and soon abandoned them. In marked contrast, the Hobby declined to chase those that remained low, probably knowing that they would be likely to drop into cover easily if pressed. He would wait to see if they were going to mount up, and if so, would set off in pursuit. Consequently, he always gave them a most sporting start, and had long testing pursuits as well as a great deal of exercise. On 22 September he caught his first Skylark and though seen the following day, stayed free thereafter. At the time, there were large numbers of migrant hirundines in the area, and it is hoped that having achieved full fitness, this bird made a successful migration to Africa for the winter. A more detailed account has been submitted to *Devon Birds*; see also photograph on the inside front cover.

PEREGRINE *Falco peregrinus* — Peregrine Falcon

Scarce resident breeder and winter visitor. SPEC(3), BoCC (Amber).

Recorded throughout year with many widespread records, relating to breeding, migrating and wintering birds. During the breeding season FMD precautions restricted access to the countryside; as most of S coast path was later re-opened, breeding success there could be gauged, but for most of the N coast and some inland sites few data were available. FMD also forced the postponement of the national Peregrine survey (10 year intervals, last in 1991).

Breeding. The average numbers of yg fledged per pr were 1.3 on the N coast (only 43% site coverage), 1.7yg inland (66% site coverage) and 1.8 on the S Coast (100% coverage; PJo). This was the most successful year on record for the S coast in terms of yg fledged, possibly helped by the lack of public access due to FMD restrictions during early part of the breeding season, or pure coincidence! Other success stories were the pr on the Exeter church rearing three yg (successful for the fifth consecutive year), with a Webcam installed here later in the year by Eco-watch, which should provide thrilling live footage of the eyrie during the 2002 breeding season. Positive action by the National Trust and RSPB set up wardening at Cann Quarry, Plym Valley, where the birds had been deliberately poisoned last year (see *DBR 2000*); this resulted in successful breeding with three yg reared.

Prey species. Species recorded (or seen taken) included Ringed Plover, Lapwing, Dunlin, Meadow Pipit, Feral Pigeon, Stock Dove and Starling; unsuccessful attempts were seen on Teal, Wigeon, Redshank, tern sp., Fieldfare and Swallow. An interesting observation involved an ad ♀ seen to strike and bind to a Woodpigeon over the sea at Wembury on 24 Oct, but it was dropped and she failed to retrieve the prey which eventually drowned in the surf (SGe); an imm bird at Hallsands also failed, after trying to retrieve a wounded Feral Pigeon from the sea at Hallsands in Aug (PE). See also under Rock Dove, and a more detailed analysis of urban Peregrine diet in a separate report.

Record request. *No deliberate poisonings were confirmed by DEFRA (cf. incidents in 1992, 1994 and 2000). However, observers are urged to report any possible pesticide (abuse) poisoning incidents to the Wildlife Incidents Investigation Scheme (freephone 0800 321600). Please also report any suspicious activities regarding theft of eggs/yg or disturbance, especially 'feral' pigeons released or tethered near known eyries (if possible get vehicle registration numbers), to Inspector Nevin Hunter (Police Wildlife Liaison Officer) - direct 01884 232702 or through 'crime stoppers' 0800 555111.*

RED GROUSE *Lagopus lagopus* — Willow Ptarmigan

L. l. scoticus: scarce resident breeder (naturalised (re-)establishment), confined to Dartmoor and Exmoor. BoCC (Amber).

Only 14 records from 11 Dartmoor sites, an understandable reduction from 19 records at 14 sites in 2000. Again there were no records from Exmoor. There was a single record showing evidence of breeding. All except three records came Jun-Nov.

Records (involving one-two birds unless stated otherwise) came from:

N Dartmoor: Black Hill, Dinger Tor, 'NW Dartmoor', Ockerton Court (eight on 8 Feb) (MD), Okement Hill, Yes Tor (five on 29 Sep) (DHWT).

E Dartmoor: Warren House area.

S Dartmoor: Burrator area, Green Hill, Ryder's Hill, Three Barrows (pr with at least three yg on 9 Jun) (MLHS).

RED-LEGGED PARTRIDGE *Alectoris rufa*

Scarce resident breeder (naturalised introduction); many artificially reared for release. SPEC(2).

A total of 45 records from 43 sites, both considerably up on any year since 1997, with clusters in N Devon around the Taw/Torridge Est (13 records) and SE Devon around Exeter (eight records). There were six records from Dartmoor, three from both NE and E Devon and one from Exmoor.

Breeding. Unsurprisingly for such a widely neglected species, only one record appeared to refer to breeding: an ad with two juvs at Morchard Bishop on 8 Jul.

Maximum counts. The only records of >ten birds were of 30 at Kenn on 10 Nov (MRAB), 20 at Roborough Down on 2 Jun, 13 at Barton Pines on 9 Dec and 12 near Otterton on 2 Dec.

GREY PARTRIDGE *Perdix perdix*

Resident breeder, declining and now probably scarce; some records refer to local releases. SPEC(3), BoCC (Red), UK BAP.

Fourteen records of birds thought genuinely wild, again from 11 sites (cf. 15 sites in 1999 and 13 in 1998), largely from areas typical of recent years, though a single at Dawlish Warren was unexpected and a breeding record from Hittisleigh encouraging. A breeding season record from Chudleigh Knighton was

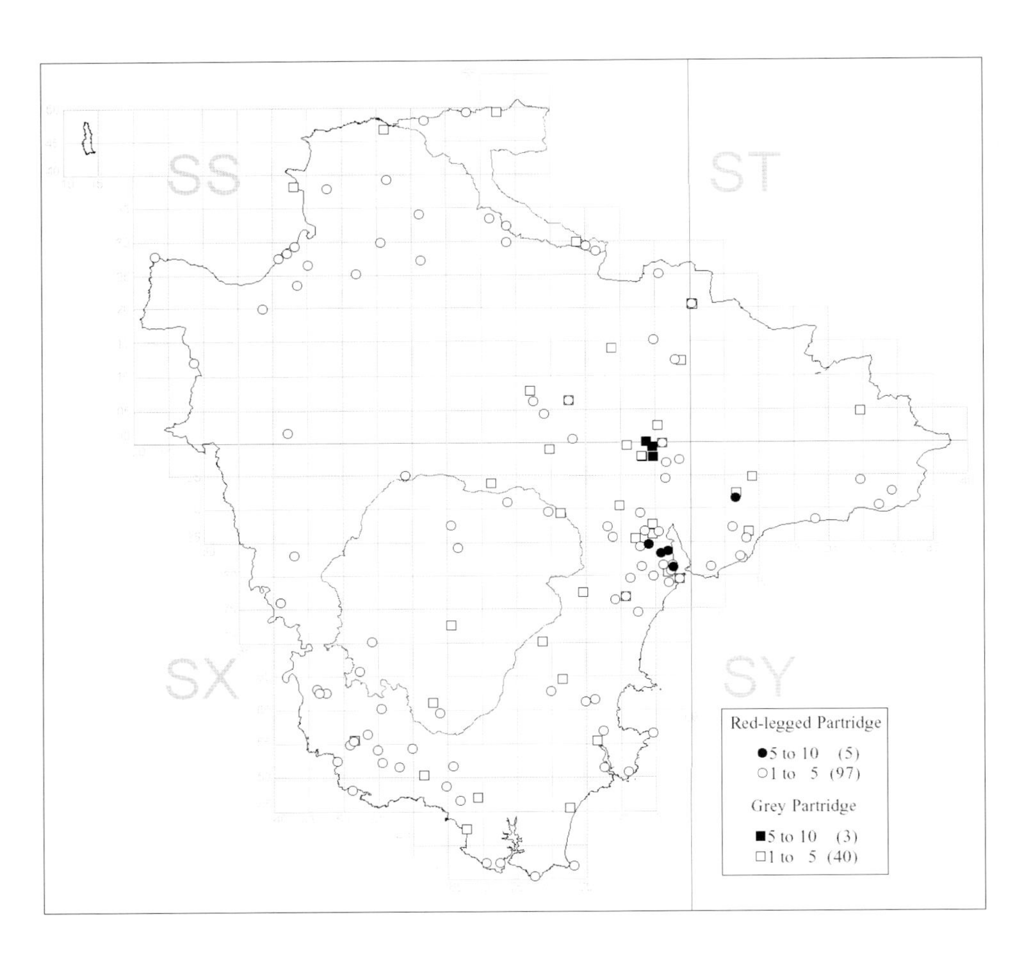

Map 2: Red-legged Partridge & Grey Partridge.
A comparative distribution. Maximum count of pairs (Apr-Aug) at each location. Records may include released birds
Data: Based upon a ten year period 1992-2001. Source: DBRs) (see Overview on page 25)

the first southern record from west of the Haldon Ridge since 1999. Records came from a wider spread of areas than last year, with one from N Devon and another from Exmoor.

Breeding. Again only one proven record - ♀ with 11 yg between Hittisleigh and Whiddon Down on 30 Aug (AMR). It is not thought that birds are released for shooting in this area. Other records suggestive of breeding: three prs present at Huxham in Jan and Feb and a group of 16 going to roost there on 27 Dec "suggests some breeding success" (DJJ); three/four birds seen regularly during period at Chudleigh Knighton (RK); two at Columbjohn, Killerton on 19 Apr (MSW); two at Huntsham on 1-2 May and 27 Aug (RGr).

Other records. All other records: two at Winkleigh on 19 Feb (MRG); two at Kenton on 8 Mar and seven on 8 Dec (HWa); one at Seven Crosses, Tiverton on 29 Mar (RJJ); one at Dawlish Warren on 21 Apr (JEF,KRy) – the first here since 1974; one at Roborough Down on 2 Jun (bird released for shooting) (DSGp); one at Berrynarbor on 1 Aug (MAP); three at Anstey Gate on 22 Aug (MAP).

QUAIL *Coturnix coturnix* — Common Quail

Scarce but highly variable summer visitor; breeding rarely proved (last in 1995). SPEC(3), BoCC (Red).

The best year since 1997, with at least six birds at the same number of sites.

One resting under a tractor at Hollacombe, Holsworthy on 25 Apr (NBri); one calling W of Bridford in the second half of Jun (DJP); one in the Soar area 26-29 Jun and 10-13 Jul (probably one bird) (PSa,PMM,RJW); one at Zeal Monachorum throughout Jul (SGM); one at Whitcombe, Kenn on 8 Jul (KRG) and one at Frogmore on 24 Jul (JHB).

PHEASANT *Phasianus colchicus* — Common Pheasant

Fairly numerous resident breeder (naturalised introduction), with large numbers reared artificially for release.

With 172 records from 37 sites (cf. 48 records from 30 sites in 2000), probably the most ever records for a species perhaps made suddenly fascinating to observers frustrated by FMD restrictions.

Breeding. Despite the plethora of records, confirmed breeding (not including released poults) was only reported from Dawlish Warren (first breeding record for site), Higher Metcombe, Huccaby, Prawle area and Watertown Farm.

Maximum count. The highest count away from shooting estates was 27 at Exminster Marshes on 6 Nov.

GOLDEN PHEASANT *Chrysolophus pictus*

Rare escape from captivity (naturalised introduction in E England)

Although no feral populations of this 'Category C' species are known in the county, many are kept in captivity and escapes/releases are occasionally recorded.

Only one record: a ♂ in a garden in the Gara valley, Slapton 20 Nov-3 Dec (PSpt).

WATER RAIL *Rallus aquaticus*

Scarce winter visitor; very scarce/rare breeder. BoCC (Amber).

Reported from 47 sites, typical of most recent years. No definite breeding records.

Table 41: Sum of maxima at all sites and WeBS county totals:

	J	F	M	A	M	J	J	A	S	O	N	D
Sum max from all sites	48	41	12	3	13	3	1	18	12	26	86	43
WeBS county total '00	14	6	6	-	-	-	-	-	2	6	8	6
WeBS county total '01	8	8	-	1	-	-	-	1	2	3	7	5

Breeding season. Yg suspected present at Slapton Ley on 23 May (BW), and up to three juvs with ads at Topsham Quay during Aug (MKn,BBH) were likely to have been locally reared. Other records Apr-Jul came from Bere Ferrers, Exeter Canal, Exminster Marshes, Lower Otter Valley/Otter Est, Pottington, West Charleton Marsh and Yelland.

First winter-period/spring. Most records of one or two birds, with the exception of these site maxima: ten at Dawlish Warren on 2 Jan, eight at Slapton Ley on 5 Jan, six at Budleigh Salterton on 16 Feb and four at South Milton Ley on 25 Feb. Two at Berry Hd on 25 Mar were almost certainly migrants.

Autumn/second winter-period. *Autumn passage* evinced by singles at a pond at Halwell, Brixton on 14 Sep; in urban Barnstaple on 28 Sep (see below); at Chudleigh Knighton on 1 Nov and Berry Hd on 25 Nov (see below). Lundy had a max of four on 4 Nov. *Winter maxima*: 11 at Exminster Marshes on 21 Nov, followed by seven at Slapton Ley on 12 Dec and five at Clennon Valley on 29 Oct and 14 Nov and at West Charleton Marsh on 8 Dec.

Mortality. The species' run of ill fortune continues, with probable cat victims at Barnstaple and Berry Hd – the fourth year running in which rail remains have been found in the county.

SPOTTED CRAKE *Porzana porzana*

Rare passage and winter visitor. SPEC(4), BoCC (Amber).

One bird – the first since 1999.

A well-watched single was at Bowling Green Marsh 23 – 25 Aug (JBa).

MOORHEN *Gallinula chloropus* — Common Moorhen

Fairly numerous resident breeder and winter visitor.

Table 42: Monthly maxima at main/well-counted sites and WeBS county totals:

	J	F	M	A	M	J	J	A	S	O	N	D
Dartington	7	10	-	-	-	-	-	-	-	11	9	4
Elfordleigh Ponds	11	7	5	4	12	15	9	13	11	12	6	15
Exe Est WeBS	13	27	-	-	-	-	-	25	9	36	26	33
Bowling Green Marsh	*24*	*15*	*10*	*12*	*4*	*6*	*22*	*24*	*15*	*30*	*14*	*20*
Exminster Marshes	*9*	*8*	-	*4*	*2*	*4*	-	-	-	-	-	*4*
Grand Western Canal	55	58	-	-	37	8	4	5	64	49	61	65
Otter Est	13	4	-	2	1	1	2	6	8	3	6	6
Roadford Res	11	2	-	-	-	-	2	38	-	11	7	16
Shobrooke Lake	9	6	-	-	-	-	3	7	4	4	8	4
Slapton Ley	-	5	-	-	11	4	6	-	-	15	12	4
South Huish area	10	11	-	-	-	-	-	-	5	4	8	20
Stover Lake	1	2	3	3	1	4	3	6	5	4	4	6
Taw / Torridge Est	3	8	-	-	-	-	-	-	9	8	4	6
WeBS county total '00	215	204	173	148	88	115	150	176	227	167	168	167
WeBS county total '01	166	175	15	9	14	31	24	71	179	196	189	236

Breeding. Reported from a typical 21 sites: Bowling Green Marsh, Buckfast, Clennon Valley, Dawlish Warren, Grand Western Canal (seven prs), Higher Metcombe, Huntsham Lake, Kenwith NR, Peamore (Alphington), Prawle area, R Carey, Rackerhayes, Radford Lake, Roadford Res, Sidborough NR, Slapton Ley (at least six prs), Sourton Cross (three prs), Stover CP (four prs), Thurlestone, Velator and Yelland.

Maximum counts (untabulated sites). *First winter period* counts > ten: 13 at Peamore on 19 Jan; 12 at Radford Lake on 1 Mar; 14 at Dawlish Warren on 15 Mar. *Post breeding and second winter period* counts > ten: 14 at Clennon Valley on 7 Sep; 18 at Thurlestone on 16 Dec; 18 at Seaton Marshes on 17 Dec; and 20 at Stafford's Bridge, Upton Pyne on 31 Dec.

COOT *Fulica atra* — Common Coot

Not scarce resident breeder and winter visitor.

Table 43: Monthly maxima at main/well-counted sites and WeBS county totals:

	J	F	M	A	M	J	J	A	S	O	N	D
Beesands Ley	55	52	40	27	-	-	-	21	3	3	8	41
Clennon Valley Lakes	9	11	-	-	13	-	-	26	26	20	-	34
Exe Est WeBS	104	50	-	-	-	-	-	60	70	148	64	34
Bowling Green Marsh	*100*	*55*	*56*	*18*	*18*	*26*	*59*	*80*	*37*	*8*	*22*	*24*
Exminster Marshes	90	36	24	23	12	22	16	30	60	160	75	75

	J	F	M	A	M	J	J	A	S	O	N	D
Grand Western Canal	60	72	-	-	15	4	-	-	63	68	67	66
Roadford Res	215	222	-	-	-	-	87	-	-	333	359	455
Shobrooke Lake	35	32	-	-	-	-	19	23	26	30	37	25
Slapton Ley	1	52	-	60	11	-	345	450	870	962	1027	1060
WeBS county total '00	1594	1294	979	181	119	165	231	592	1065	1100	948	641
WeBS county total '01	494	505	6	5	3	4	7	65	239	1508	1517	1413

Breeding. Reported from only 12 sites (*cf.* 19 in 2000 and 14 in 1999): Bowling Green Marsh, Clennon Valley, Dawlish Warren, Exminster Marshes, Grand Western Canal, Huntsham Lake, Roadford Res, Slapton Ley, Stover CP, Thurlestone, Velator and Yelland.

Maximum counts (untabulated sites). The highest counts were: 30 at Goodrington (seafront) on 2 Jan; 50 at Topsham on 29 Jan; 39 on Wrafton Pond on 1 Mar; 26 at Clennon Valley on 4 Aug and 7 Sep; 15 at New Cross Pond on 29 Sep; 30 at Wistlandpound Res on 4 Oct; 67 at Rackerhayes on 1 Dec; and ten at Sherpa Marsh on 12 Dec.

CRANE *Grus grus* — Common Crane

Rare vagrant. BoCC (Amber).

The 2001 record constitutes the 19th for Devon (excluding relocation sightings), in total involving 51 birds; ten of the records have occurred in the within the past 20 years. Past records at Slapton occurred in Nov 1996 (at Plympton next day) and Sep 1966 (earlier at Woodbury and later at Bowcombe Creek).

One at Slapton on 14 Apr was in fields S of Ireland Bay from 08.15h, later spending time at the S end of the Ley before departing N at 11.15h (TM *et al.*). Unlike previous records at Slapton, this bird was not relocated.

OYSTERCATCHER *Haematopus ostralegus* — Eurasian Oystercatcher

Scarce breeder (non-breeders not scarce in summer); fairly numerous winter visitor and autumn passage migrant. BoCC (Amber).

National importance (3,600 birds) was achieved with 3,665 on Exe Est in Aug (WeBS), for the first time since Nov 1996. However, despite this, both winter maxima were rather poor. The Exe supported c.60-70% of Devon's wintering population this year (cf. an average of 73%; see DBR 1999) of which 70-85% tended to roost at Dawlish Warren. Although 835 at Isley Marsh in Aug was better than the low figures of 2000, it was still a relatively poor max count for the Taw/Torridge, which this year supported c.10-20% of Devon's autumn and winter birds.

Table 44:Monthly maxima at main/well-counted sites and WeBS county totals:

	J	F	M	A	M	J	J	A	S	O	N	D
Avon Est	74	83	37	19	12	19	17	37	39	38	44	39
Axe Est	29	27	-	-	-	-	1		8	-	-	23
Dart Est	1	3	-	-	-	-	-	-	-	-	2	-
Erme Est	14	25	-	-	-	-	-	-	32	41	12	29
Exe Est WeBS	2839	1830	-	-	-	-	-	3665	3291	1903	1874	2792
Dawlish Warren	*2300*	*1551*	*1200*	*315*	*340*	*420*	*500*	*1300*	*2240*	*2600*	*1600*	*1250*
Kingsbridge Est	193	135	6	-	-	-	-	200	181	285	229	198
Otter Est	21	15	-	4	11	2	3	9	12	17	17	20
Plym Est	8	6	-	30	12	7	8	1	2	4	4	-
Plymouth Sound	-	-	9	17	-	14	-	-	81	-	2	-
Prawle area	43	-	-	33	40	36	42	58	64	58	36	53
Tamar Complex WeBS	205	129	-	-	-	-	-	-	244	172	133	178
Upper / mid Tamar	*33*	*29*	-	-	-	-	-	-	*47*	*21*	*25*	*136*
Tavy Est	*60*	*23*	-	-	-	-	-	-	*78*	*58*	*68*	*20*
Taw / Torridge Est	271	712	-	42	300	302	420	-	835	350	604	700
Isley Marsh / Instow	*254*	*244*	-	*220*	*6*	*50*	*100*	*50*	*835*	*300*	*25*	*700*
Teign Est	150	-	28	60	120	-	-	108	-	340	30	-
Thurlestone area	40	1	-	-	10	-	37	16	25	-	-	-
Torbay / Hope's Nose	106	53	46	22	23	43	-	-	-	21	-	41
Wembury	34	74	-	-	56	54	63	57	101	104	101	135

continued	J	F	M	A	M	J	J	A	S	O	N	D
Yealm Est	1	-	-	-	-	-	-	-	3	1	3	1
WeBS county total '00	3085	3666	1712	1152	772	796	1098	3356	3531	3034	3833	4376
WeBS county total '01	3527	2886	-	4	11	21	14	3863	4385	2398	2858	3589

Breeding. Very poorly reported, with only a pair at Yelland on 22 May, later seen with yg on 25 June (DC). Compared to an estimated 28-58 widely distributed prs in the *Tetrad Atlas* (1977-85), recent mainland numbers have been low: four (2000), eight (1999); and nine prs (1998).

First winter-period. The Exe Est remains the principal site, producing 2,839 on 21 Jan; in N Devon the max was 712 Taw/Torridge on 11 Feb. Other estuaries/bays with ≥100 included 193 on Kingsbridge Est 21 Jan, 106 at Goodrington on 21 Jan and 150 on the Teign Est on 1 Jan. The max count away from estuaries was 106 at Young Park, nr Clennon Valley on 12 Jan.

Summer. Non-breeding imms probably accounted for most of the *c.*900 birds over-summering in Devon, including 420 Dawlish Warren on 3 Jun, 302 at Yelland on 25 Jun, 54 at Wembury on 26 Jun and 36 at Prawle on 9 Jun. *Inland,* one circled over Tiverton, calling, for *c.*30 mins before flying S on 17 Jul (RJJ).

Autumn/second winter-period. Few during Jul, but a notable increase from early Aug with arrival of passage birds; max counts included 3,665 on the Exe Est on 20 Aug, 835 at Isley on 2 Sep and 81 at Jennycliff on 30 Sep (a high count for there). Numbers typically dropped off during Oct before the influx of wintering birds in (usually late) Nov, but winter-period rather subdued with maxima of only 2,792 on the Exe Est on 9 Dec, 700 at Isley Marsh on 20 Dec and 135 at Wembury. Away from main sites, counts included 22 at Ayrmer Cove, Ringmore, on 1 Dec, ten at Combe Martin on 21 Dec, 15 at Lynmouth on 21 Dec and 31 at Coryton Cove, Dawlish on 26 Dec.

Ringing recoveries. Recoveries of wintering birds ringed at Dawlish Warren indicate breeding in Scotland, Faeroes and Netherlands; see Ringing Report for further details.

AVOCET *Recurvirostra avosetta* — Pied Avocet

Not scarce winter visitor to Exe and Tamar/Tavy Ests; very scarce visitor elsewhere; very scarce outside winter period. SPEC(4/3ʷ), BoCC (Amber).

Nationally important numbers (50 birds) have wintered on the Exe and Tamar Ests since 1980/91 and 1960/61, respectively. Winter max counts on the Exe of c.500 on 22 Feb and 528 on 9 Dec were about as expected. However, the Tamar Complex WeBS max count of 452 in Jan was a record count (previously 380 in Jan 1999) though unless counts from the different sectors are synchronised, high WeBS counts can arise from count duplication; the late year max of 277 on 16 Nov was more typical. As usual, most of the Tamar birds occurred on the upper/mid Tamar, and the Tavy max was only 52 (cf. the record 203 in Jan 2000).

Table 45: Monthly maxima at the two main sites and WeBS county totals:

	J	F	M	A	M	J	J	A	S	O	N	D
Exe Est	285	500	260	1	-	-	1	19	36	40	510	528
Bowling Green Marsh	*20*	-	-	*1*	-	-	*1*	*19*	*30*	*10*	-	-
Tamar Complex WeBS	452	68	-	-	-	-	-	-	-	-	277	252
Upper / mid Tamar	*321*	*68*	-	-	-	-	-	-	-	-	*274*	*246*
Tavy Est	-	*2*	-	-	-	-	-	-	-	-	*10*	*52*
WeBS county total '00	552	472	45	6	-	-	1	-	-	20	422	509
WeBS county total '01	477	434	-	-	-	-	-	-	8	1	589	780

First winter-period/spring. Away from the main Exe Est and Tamar Complex sites: up to four were on the Teign Est 3 Jan - 4 Mar (MKn *et al.*); one on the Kingsbridge Est at Kingsbridge 21 & 26 Jan (PSa); and one *inland* at Roadford Res on 5 Jan, roosting on the dam wall with gulls (GAV). Nearly all had left the Exe by early Mar with the last being five off Topsham on 25th (RJJ). Thereafter, one at Exminster Marshes intermittently 3 Apr - 13 May with two on the last date (MKn, RSPB); two Dawlish Warren 10 Apr (LC); one Bowling Green Marsh 16 Apr (RNK), and the second of the year at Roadford Res, with one on 1 May (FHCK).

Summer. Seven at Dawlish Warren 1 Jun (CC), the first ever Jun record there; also one at Bowling Green Marsh 4-9 Jul (RSPB). Elsewhere, one E past Prawle on 23 Jun (PMM).

Autumn/second winter-period. Aug records, all from the Exe Est were: five at Starcross on 20th

(DJG), 19 at Bowling Green Marsh on 21st (MSW *et al.*) and one on R Clyst on 29th (RJO). The Sep max on the Exe of 36 on 28-30 Sep was preceded by seven on 9th and six on 27th. Birds became regular from 18 Oct, though the first counts ≥100 were not until Nov, and included 100+ at Dawlish Warren on 12 Nov (BG), the largest ever count there. Winter Exe records away from the main Clyst/Topsham/Turf feeding area included six on Exminster Marshes on 2 Dec. *Elsewhere*, one on the Teign Est on 21 & 27 Dec (MKn,TWW,LC).

STONE CURLEW *Burhinus oedicnemus* — Stone-curlew

Rare passage visitor. BoCC (Red).

One record, the first since 1997; details temporarily withheld.

LITTLE RINGED PLOVER *Charadrius dubius* — Little Plover

Very scarce/scarce passage visitor.

With 12 birds in spring and 16 autumn, inc. the two summer birds, the county total is now 410, of which 240 have occurred since 1990. Of the total, 173 (42%) have appeared in spring and 237 (58%) in summer/autumn (see also DBR 1999). At least ten have occurred annually in spring since 1996.

Spring. The first of this year's 12 birds (13 bird-days) was at Passage House Inn, Teign Est on 10 Apr (MKn), then singles Exminster Marshes 13 Apr (RSPB), and another 21 & 22 Apr (RSPB,BBH), three Bowling Green Marsh 14 Apr (RSPB,MSW), one Portworthy Dam 2 May (MDD), one Skern 13 May (RGM), a pr South Huish Marsh 22 May (RBu) and lastly, singles Exminster Marshes (MKn) and the Plym Est (BG), both 1 Jun.

Summer/autumn. A pr at Bowling Green Marsh 4 Jul (RSPB,BBH,MKn) with one remaining to 7th and perhaps the same again on 17th (RSPB,MSW), were followed by a further 14 birds (43 bird-days), the lowest number since 1997, but no records this year from Exeter Riverside CP. The first were singles at Dawlish Warren on 26 Jul (JEF) and Bowling Green Marsh 1 Aug, then a good run of records of singles from Upper Tamar L *(see col plate)* on: 10 & 11 Aug (DC,RD), 7-11 Sep (JEW), 21-24 Sep (DC, RD) and, the *last* of the season, 12 Oct (MAP). *Elsewhere*: two juvs Dawlish Warren 15-27 Aug (LC); singles Portworthy Dam 19 Aug (WeBS), 2 Sep (MDD) & 16 Sep (WeBS); and three Bowling Green Marsh 2 Sep (RSPB) with one 20 Sep.

RINGED PLOVER *Charadrius hiaticula*

Not scarce passage and winter visitor; very scarce breeder. BoCC (Amber).

Low winter counts, but a good autumn passage (inc. inland) and some breeding activity.

Table 46: Monthly maxima at main/well-counted sites and WeBS county totals:

	J	F	M	A	M	J	J	A	S	O	N	D
Avon Est	4	4	-	-	-	-	-	3	-	1	-	-
Axe Est	-	-	-	-	4	-	-	8	-	-	-	1
Exe Est WeBS	59	38	-	-	-	-	-	15	110	112	123	110
Bowling Green Marsh	-	-	-	*4*	-	-	*3*	*9*	*22*	-	-	-
Dawlish Warren	*61*	*62*	*40*	*116*	*18*	*34*	*46*	*227*	*176*	*120*	*75*	*66*
Kingsbridge Est	-	-	-	-	-	-	12	100	4	-	-	-
Otter Est	7	6	-	-	-	-	-	2	-	-	7	5
Plym Est	-	2	-	-	-	-	-	2	6	12	-	-
Prawle area	-	-	-	6	-	-	2	15	6	1	1	-
Tamar Complex WeBS	-	-	-	-	-	-	-	-	6	5	-	-
Taw / Torridge Est WeBS	51	40	-	-	-	-	-	6	72	17	93	44
Braunton / Saunton Sands	-	*49*	*2*	*81*	*91*	*9*	*23*	*61*	*65*	*110*	*21*	-
Instow	-	*85*	-	-	*20*	-	-	*26*	*16*	-	-	-
Northam B / Skern	-	-	-	*2*	*30*	*5*	*32*	*140*	*50*	*100*	-	-
Slapton Sands	9	5	-	5	4	5	-	6	13	7	-	-
Thurlestone area	-	-	-	14	3	-	-	4	3	-	2	-
Torbay: Broadsands	4	-	-	1	-	-	-	4	-	6	7	8
Wembury	-	-	-	-	-	-	1	19	28	-	-	-
WeBS county total '00	95	92	10	6	63	3	17	149	147	116	184	122
WeBS county total '01	117	88	-	-	-	-	-	53	184	145	216	157

Breeding. The first N Devon breeding records since 1996 involved two prs: a pr, nest & four eggs at Northam Burrows 1-20 May (AF,MAP) and a pr holding territory, displaying and feigning injury on Saunton Sands Mar-Jul (RJ). Birds holding territory and nest-scraping at Dawlish Warren have occurred annually since 1992, but egg-laying has not been observed since 1998.

First winter-period. A max count of 85 at Instow on 20 Feb was quite poor, but just 61 at Dawlish Warren, with 62 there on 27 Feb, was the lowest count for this period on the Exe in decades, and none at all were recorded away from these main sites.

Spring passage. After dropping to single figures in early Mar, numbers picked up from 23 Mar rising to a peak period in mid-Apr with max counts of 116 Dawlish Warren 16 Apr and 81 Saunton Sands 15 Apr. *Elsewhere in Apr*: five Slapton 13th and one there 18th; 12 Exminster Marshes 13th; 14 South Huish Marsh 14th with seven there 22nd; four Bowling Green Marsh 16th; one Teign Est 17th; and six Prawle 27th. Numbers fell away for a time, but a second period of passage *mid-May to early-Jun* began with four Axe Est, four Slapton and two South Huish Marsh all on 10 May; max counts were 91 Saunton Sands 28 May and 34 Dawlish Warren 1 June. Only single-figure counts on main sites 15 June - 21 July.

Autumn passage. Influxes from late Jul produced the first autumn records away from main sites with three Bowling Green Marsh 20th, and singles Wembury 26th and Prawle 28 Jul. Peak passage mid-Aug to mid-Sep produced max counts of 227 Dawlish Warren 21 Aug, 140 Skern 24 Aug and 100+ Charleton Bay 31 Aug. *Off headlands*: one S Berry Head 7 Aug, one Start Pt 26 Aug and one Lundy 21 Sep. *Inland*: a tremendous set of records at Upper Tamar L began with one 27 July, two 10 Aug, then regular 16 Aug - 12 Oct with max count of 30 on 14 Sep (FHCK *et al.*), the largest ever inland count for Devon (26 at same site on 24 Aug 1972 being the next highest); five Portworthy Dam 19 Aug; one Fernworthy Res 2 Sep; one Roadford Res 13 Sep, with six there on 15 Sep and one 7 Oct; and three Wistlandpound Res 24 Sep and two 7 Oct.

Second winter-period. Max counts 123 Exe Est 11 Nov and 93 Taw/Torridge Est 4 Nov, with none recorded away from these main sites.

***C. hiaticula tundrae* (Tundra Ringed Plover)**

Breeding in Scandinavia and Russia, members of the *tundrae* race are regular migrants through W Europe to and from wintering grounds in Africa. Up to two were seen amongst influxes of the nominate *hiaticula*-race at Dawlish Warren 3-19 Aug (IL,KRy *et al.*).

KENTISH PLOVER *Charadrius alexandrinus*

Rare passage visitor. SPEC(3).

An annual total of six, all at Dawlish Warren, was the highest total there since 1993 and the best total in Devon since 1996. In total, at least 97 birds have officially appeared in Devon, though at least 85 birds (not all published) are believed to have occurred at Dawlish Warren alone. Dawlish Warren, where the species has occurred annually since 1995, remains one of the most important sites for this species in Britain. The first of the year, on 22 Mar, was the second earliest ever after one 17 Mar 1991, and the late juv was probably the second latest record after one 9-16 Nov 1991.

Spring. Four birds at Dawlish Warren: a ♀ 22 Mar sporting a Belgian colour-ring (D Land,JEF *et al.*) *(see col plate in centre section)*; a pr 26 & 27 Mar (KRy,JEF *et al.*) - the ♀ was unringed; and a ♂ 27 - 29 May (DS *et al.*).

Autumn. Two birds at Dawlish Warren: a very pale juv 22 Sep - 5 Oct (KRy *et al.*) and another 19 - 23 Oct (JEF *et al.*).

DOTTEREL *Charadrius morinellus* — Eurasian Dotterel

Very scarce/rare passage migrant. BoCC (Amber).

A below average year, with just three records bringing the county total since 1990 to 50 records of 94 birds, with the majority on Lundy (see DBR 1999).

Spring. Five on Lundy on 11 May (LFS).

Autumn. One juv at Yes Tor 26 Aug (SN) was followed by a late bird, also on Dartmoor, at Cox Tor (with a flock of Golden Plovers) on 10 Nov (RJL).

GOLDEN PLOVER *Pluvialis apricaria* — European Golden Plover

Fairly numerous winter visitor; very scarce migrant breeder. SPEC(4), Dartmoor BAP.

Table 47: Monthly maxima at main/well-counted sites and WeBS county totals:

	J	F	M	A	M	J	J	A	S	O	N	D
E Dartmoor	150	100	-	80	-	-	-	-	-	500	56	150
S Dartmoor	200	252	-	-	-	-	-	-	32	50	30	300
N Dartmoor	1	115	-	-	-	-	-	-	15	30	1500	332
Exe Est	330	470	30	-	-	-	-	-	-	93	780	650
Bowling Green Marsh	*250*	*280*	-	-	-	-	-	-	-	*24*	*350*	*300*
Prawle area	300	-	-	-	-	-	1	-	-	10	170	130
Taw / Torridge Est	2000	2000	-	-	-	-	-	-	15	350	2000	2000
Tamar Complex WeBS	402	-	-	-	-	-	-	-	-	-	28	700
Upper / mid Tamar	*402*	-	-	-	-	-	-	-	-	-	*28*	*300*
Sum max at all other sites	1307	16	279	245	-	-	-	-	1	221	1476	1270
No of other sites	9	2	1	4	-	-	-	-	1	7	7	6
WeBS county total '00	1439	2145	404	75	-	-	-	-	-	140	2199	1521
WeBS county total '01	1132	2070	-	-	-	-	-	-	1	93	1786	1380

Breeding. The small Dartmoor population (two to eight prs in recent years) was not surveyed this year owing to FMD access restrictions, but two birds alarm-calling at West Dart Hd on 14 Jul were in suitable breeding habitat (MLHS).

First winter-period/spring. On *estuaries*, the Taw/Torridge remains the principal site with a max of 2,000 at Penhill Marshes on 25 Jan (JEW) and 3 Feb (RJ); on other estuaries, high counts include 470 Exe Est 18 Feb and 402 R. Tamar 1 Jan. On *Dartmoor* the max was 600 at Routrundle (RJL). *Elsewhere*, 470 Hendom Cross/Huntsham area 20 Jan and 450 Brendon Common, NW Devon 12 Jan were the highest counts. The few spring records show birds still flocking through to the beginning of May, with *c*.80 at Dunnabridge 20 Apr and probably the same at Lower Cherry Brook 26 Apr followed by the last flock of 170 at Bowling Green Marsh on 1 May (NPi).

Autumn/second winter-period. One in bp at East Prawle 22 July (JCN *et al.*) was an isolated record. The next were singles Willsworthy, W Dartmoor 5 Sep, Skern 8 Sep, two Soar 10 Sep and one Dawlish Warren 17 - 23 Sep. Small flocks gathered from mid-Sep with 15 Skern 16th and 15 Cox Tor 20th being the first, then three-figure counts became regular at usual sites from early Oct with 500 at Rippon Tor on 12 Oct, 350 R Caen 18 Oct and the same number at Skern on 23 Oct the largest in that month, followed by 500 at Roadford Res on 4 Nov. On the *estuaries*, the Taw/Torridge again produced a max of *c*.2,000 at Penhill 16 Nov at Pottington 24 Nov and at Bradiford NR, Taw 20 Dec (JEW,MHC). On other estuaries, high counts included 780 Exe Est 28 Nov and 700 Tamar Est on 1 Dec. On *Dartmoor*, a series of large counts in the W Dartmoor area through the early winter-period produced a max of 1,500 at Cox Tor on 19 Nov (RHi). *Elsewhere*: 700 Winkleigh 20 Nov and 500 nr Kingston on 4 Dec were notable counts.

GREY PLOVER *Pluvialis squatarola*

Not scarce (but very local) winter visitor; very scarce/scarce passage/non-breeding visitor; rare/very scarce inland. BoCC (Amber).

The Exe Est supports about two-thirds of the birds wintering in Devon, with most of the others being on the Taw/Torridge. The winter maxima again fell well short of the qualifying level for national importance (430), last achieved in Feb 1998. Numbers on the Taw/Torridge Est (and the Tamar) also reflected the three-year lull, with 175 the best count there since Dec 1998.

Table 48: Monthly maxima at main/well-counted sites and WeBS county totals:

	J	**F**	**M**	**A**	**M**	**J**	**J**	**A**	**S**	**O**	**N**	**D**
Exe Est WeBS	321	245	-	-	-	-	-	6	7	49	259	338
Dawlish Warren	*347*	*282*	*279*	*3*	*3*	*1*	*1*	*13*	*25*	*87*	*196*	*262*
Kingsbridge Est	30	2	-	-	-	-	-	-	-	-	1	6
continued...	**J**	**F**	**M**	**A**	**M**	**J**	**J**	**A**	**S**	**O**	**N**	**D**
Slapton Sands / Ley	18	-	8	-	-	-	-	1	-	-	-	-
Tamar Complex WeBS	30	-	-	-	-	-	-	-	-	-	10	37
Taw / Torridge Est	144	175	-	28	-	-		1	12	35	41	74

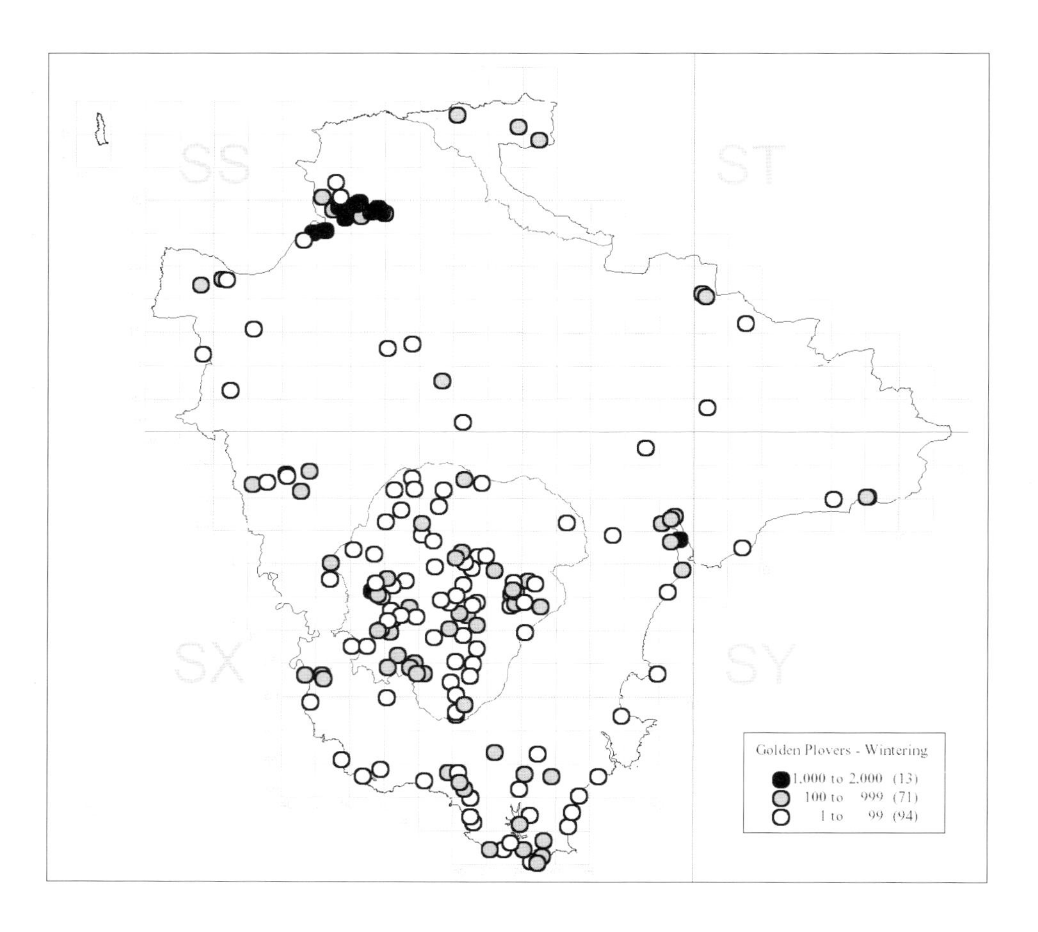

Maps 3: Golden Plover - wintering
Distribution of wintering (Nov-Feb) Map 3 and migrant (Sep-Oct & Mar-May) records Map 4 on following page. Abundance is based on maximum count at each location. Note that late wintering and spring migrant flocks may occur on the moorlands well into spring. On Dartmoor, it is not always possible to distinguish records of spring migrants from small parties of possible breeding birds.
Data: Based upon a ten year period 1992-2001. Source: DBRs) (see Overview on page 25)

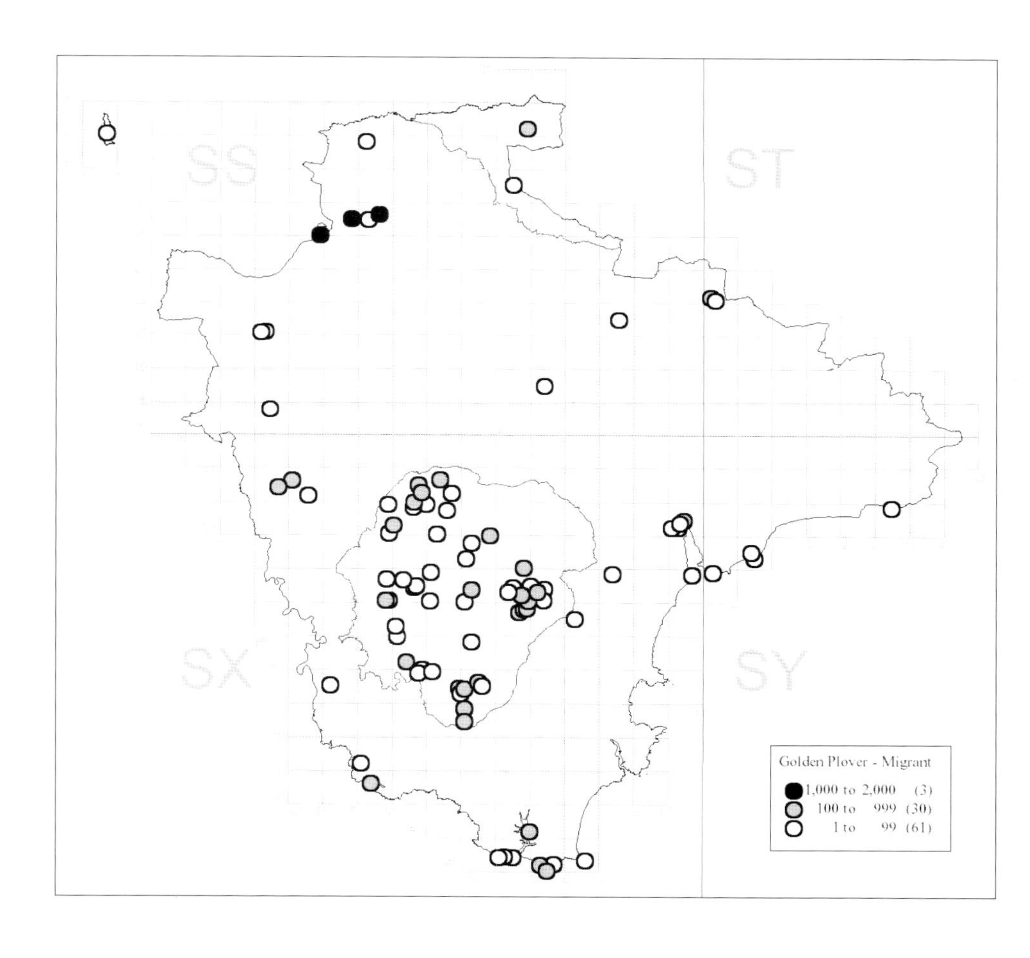

Map 4: Golden Plover - migrant
(see Map 3 on previous page for description)
Data: Based upon a ten year period 1992-2001. Source: DBRs. Source: DBRs) (see Overview on page 25)

continued	J	F	M	A	M	J	J	A	S	O	N	D
Skern	-	-	-	*28*	-	-	-	*1*	*12*	*35*	*24*	*46*
Isley Marsh	*25*	*50*	-	-	-	-	-	*1*	-	*15*	-	*27*
Wembury	7	8	-	-	-	-	-	-	4	-	2	5
WeBS county total '00	357	580	114	29	12	2	3	1	2	31	88	499
WeBS county total '01	495	422	-	-	-	-	-	6	10	60	301	418

First winter-period/spring. On the estuaries, 347 Dawlish Warren 26 Jan and 175 Taw/Torridge 11 Feb were the max counts; elsewhere in winter, 18 at Slapton Sands 3 Jan. Spring passage included 19 Teign Est 15 Mar, 28 Skern 2 Apr where the last departed 18 Apr, and up to three noted regularly at Dawlish Warren until 16 May. Later, two Bowling Green Marsh 1 & 19 Jun, and singles Dawlish Warren 16, 19 & 22 Jun.

Autumn/second winter-period. Singles Dawlish Warren 23 & 26 Jul were the *first* to return; birds became regular here from early Aug. Singles Isley Marsh 1 Aug and Skern 17 Aug were the first on Taw/Torridge and small flocks frequently appeared on main estuaries throughout Sep to mid-Oct. *Non-estuary* records (all singles): Slapton 23 Aug; Prawle 24 Sep; in Plymouth Sound, Jennycliff 30 Sep, Staddon Pt 1 & 9 Oct and Mount Batten 2 Oct; S past Berry Hd 7 Aug and 1 Oct; and Prawle Pt 24 Nov. *Wintering* numbers rose rapidly from early Nov to max counts of 338 Exe Est 9 Dec (with 262 Dawlish Warren 19 Dec), and 74 Taw/Torridge 16 Dec.

LAPWING *Vanellus vanellus* — Northern Lapwing

Fairly numerous winter visitor; scarce breeder. BoCC (Amber).

Table 49: Monthly maxima at main/well counted sites and WeBS county totals (excludes Dartmoor breeding birds):

	J	F	M	A	M	J	J	A	S	O	N	D
Avon Est	250	1	1	-	-	-	3	4	4	4	7	38
Axe Est	1500	900	-	-	-	-	21	10	22	90	97	1500
Exe Est	1795	800	1	1	-	37	64	68	45	113	339	982
Bowling Green Marsh	*375*	*280*	*1*	*1*	-	*37*	*64*	*68*	*45*	*10*	*339*	*407*
Dawlish W - Starcross	*50*	*29*	*15*	-	-	*3*	*4*	*4*	*5*	*11*	*29*	*39*
Exminster Marshes	*1000*	*310*	*89*	*26*	*57*	*37*	-	*27*	*38*	*160*	*240*	*212*
Kingsbridge Est	-	2	-	-	-	-	1	-	-	-	6	300
Otter Est / Valley	93	30	-	-	-	-	2	2	2	1	20	18
Portworthy Dam	52	10	-	-	-	-	-	-	-	-	11	23
Prawle area	30	-	-	-	-	-				1	27	25
Tamar Complex	525	-	-	-	-	-	11	3	7	35	500	350
Upper / mid Tamar	*325*	-	-	-	-	-	-	-	*7*	*13*	*229*	*220*
Tavy Est	*35*	*2*	-	-	-	-	-	-	-	-	-	*45*
Taw / Torridge Est	1656	330	-	-	-	-	-	-	4	40	304	723
Isley Marsh	*1000*	*200*	-	-	-	-	-	-	-	*4*	*26*	*80*
Teign Est	58	-	-	-	-	-	-	-	-	60	31	39
South Huish area	108	2	-	45	-	-	7	1	-	-	-	200
Sum max at all other sites	116	143	13	1	-	4	-	8	9	72	232	584
No of other sites	5	6	1	1	-	1	-	2	2	3	9	14
WeBS county total '00	6092	5759	9	5	5	2	60	43	65	475	2103	4417
WeBS county total '01	5214	1685	-	-	-	-	24	36	48	197	1018	2350

Breeding: *Exminster Marshes* supported 22 prs (RSPB) (*cf.* 19-20 prs1998-2000). *Dartmoor*: an inadequate picture, with at least nine prs located (*cf. DBR* 2000) but some areas (e.g. Peter Tavy) not covered, and coverage of the following areas was limited to roadside observations. Sticklepath Moor (three prs, two birds on nests) (NB); Bonehill area (two prs, an encouraging nine juvs) (SMRY); Rushlade Common (present) (RHi); Thornworthy, nr Fernworthy Res (one pr; significant decline noted here) (DSGp); Wigford Down (pr mobbing corvids) (RJL). Breeding season records, not thought to refer to breeding birds, also came from Blackslade Mire (MD,AR), Cadover Bridge (RJL) and Wallabrook (RHi). *Lundy*: one pr thought to have bred (LFS).

First winter period/spring. Four-figure WeBS counts achieved on three estuaries: 1,795 Exe Est 21 Jan, 1,656 Taw/Torridge Est 14 Jan and 1,500 Axe 21 Jan. *Away from estuaries*, 108 Thurlestone 1 Jan was the highest count. Numbers dropped quickly from mid-Feb and most wintering birds had departed by late Feb/ early Mar, with no records to suggest a spring passage.

Autumn/second winter-period. Dispersal around the Exe began with three Dawlish Warren 18 Jun and 35 Bowling Green Marsh 23 Jun, with the first away from breeding areas being seven Thurlestone 2 Jul. Apart from the Upper Exe Est, only odd ones and small parties were observed until 60 at Passage House Inn 1 Oct and 90 on the Axe Est 16 Oct, followed by a general influx late Oct to early Nov. The main influx occurred from late Nov, numbers rising to max counts of 1,500 Seaton Marshes 1 Dec (IJW), 982 Exe Est 9 Dec and 723 Taw/Torridge 16 Dec.

KNOT *Calidris canutus* — Red Knot

Scarce and very local passage and winter visitor; very scarce in summer and inland. SPEC(3ʷ), BoCC (Amber).

Dawlish Warren is the principal roost site in Devon but numbers always fall well short of achieving national importance (2900 birds). Elsewhere, 41 on Taw / Torridge in Feb was rather poor.

Table 50: Monthly maxima at main sites and WeBS county totals:

	J	F	M	A	M	J	J	A	S	O	N	D
Exe Est WeBS	157	16	-	-	-	-	-	35	3	-	7	94
Dawlish Warren	*194*	*134*	*109*	*21*	*3*	*2*	*3*	*36*	*54*	*25*	*10*	*81*
Bowling Green Marsh	-	-	-	*17*	*11*	*1*	*2*	*50*	*8*	-	*3*	*1*
Taw / Torridge Est	11	41	-	3	-	-	3	50	30	5	3	12
Pottington	-	-	-	-	-	-	-	*18*	*30*	*4*	*3*	*12*
Skern	*46*	-	-	*18*	-	-	*1*	*4*	*12*	-	-	*3*
WeBS county total '00	114	27	47	4	5	1	1	-	35	22	18	69
WeBS county total '01	268	57	-	-	-	-	-	35	33	-	7	94

First winter-period/spring. A max of 194 Dawlish Warren 16 Jan was the highest count there since Jan 1998; 46 Skern 19 Jan was the next best count. Most wintering birds had departed by late Mar, but parties of migrant and lingering birds were regular at key sites until late May, and there were seven on the Plym Est on 16 Apr. Subsequently, two Dawlish Warren 11 Jun, and one Bowling Green Marsh 22–24 Jun were the only records.

Autumn/second winter-period. The first arrivals appeared from 9 Jul, with six Dawlish Warren (and one there 15th and three 19th) and one Skern (also on 11th); also two Bowling Green Marsh 20 Jul and one on 21st. Birds became regular from mid-Aug with autumn peaks of 50 Bowling Green Marsh 24 Aug, 50 Instow 28 Aug and 54 Dawlish Warren 12 Sep. *Elsewhere*: one Axe Est 28 Aug; one Charleton Bay 24, 27 & 28 Aug, with ten there 30th & 31st; three Prawle 17 Aug & two 17 Sep; one *inland* at Upper Tamar L 23 Aug and two Tamar Complex in Sep. Numbers remained modest through Oct & Nov, wintering birds arriving mainly early Dec with 94 on the Exe Est on 9 Dec the season's max count contrasting with just 12 Taw/Torridge Est 20 Dec.

SANDERLING *Calidris alba*

Scarce, but very local, passage and winter visitor; rare/very scarce inland.

Saunton Sands and Dawlish Warren continue to be the two county strongholds. Frequent observations of Saunton Sands produced a very impressive set of records with several monthly maxima being the highest, or amongst the highest, ever recorded in the county. An overall max of 200 at Saunton was the highest count there since 1980, though still below the level required for national importance (230 birds). By contrast, Dawlish Warren counts were rather poor.

Table 51: Monthly maxima at main/well-counted sites and WeBS county totals:

	J	F	M	A	M	J	J	A	S	O	N	D
Exe Est WeBS	30	28	-	-	-	-	-	-	5	14	26	27
Dawlish Warren	*55*	*43*	*30*	*5*	*57*	*78*	*27*	*65*	*23*	*20*	*42*	*39*
Slapton Sands			1	1	84	35	-	-	-	-	-	-
continued	J	F	M	A	M	J	J	A	S	O	N	D
Taw/Torridge WeBS	51	22	40	-	-	-	-	-	109	11	134	45
Saunton Sands	*168*	*190*	*200*	*134*	*104*	*47*	*76*	*93*	*165*	*155*	*144*	-

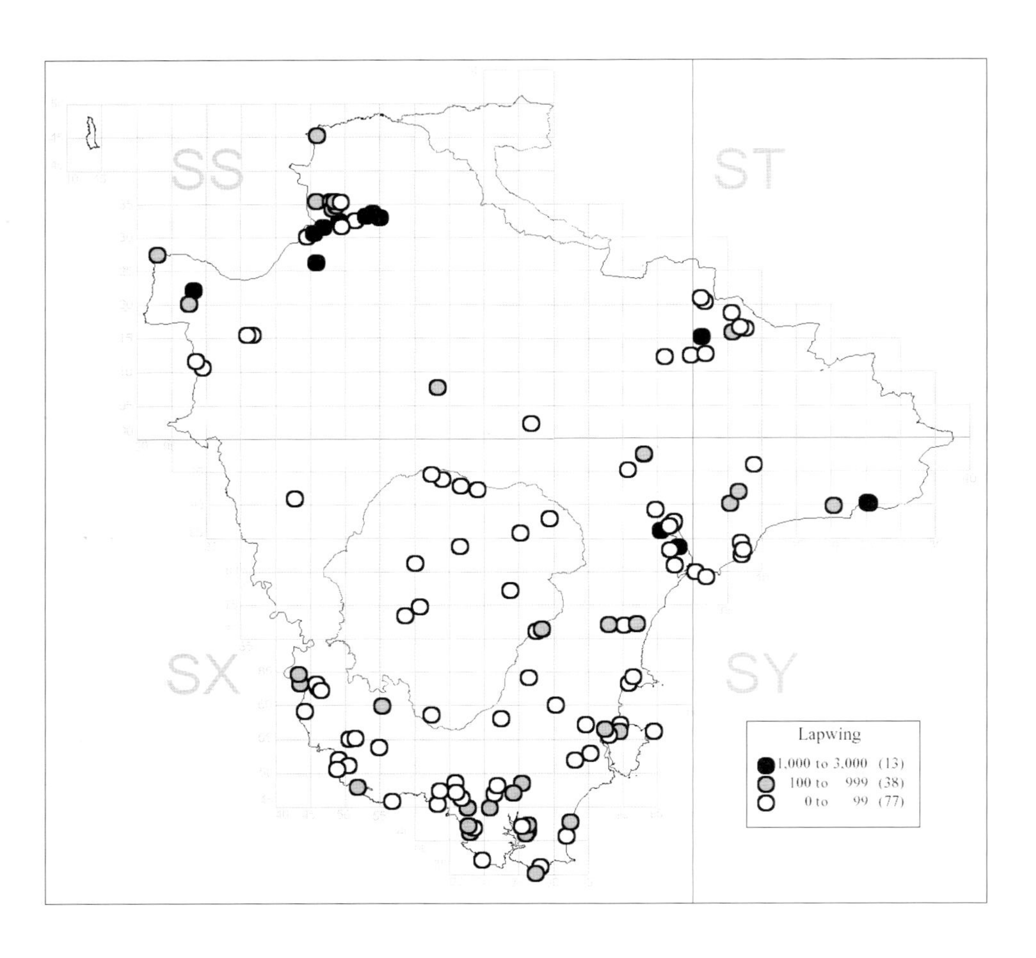

Map 5: Lapwing.
Distribution of winter (Nov-Feb) records. Abundance is based on the maximum count of each location.
Data: Based upon a ten year period 1992-2001. Source: DBRs) (see Overview on page 25)

continued	J	F	M	A	M	J	J	A	S	O	N	D
Wembury	-	-	-	-	15	-	5	1	3	1	-	1
Westward Ho !	-	-	-	-	-	-	42	-	-	2	-	94
WeBS county total '00	25	25	44	2	26	-	12	1	7	1	16	22
WeBS county total '01	81	30	-	-	-	-	-	1	114	25	160	72

First winter-period/spring. Counts >100 were regular at Saunton Sands until 4 May, with a max of 200 on 9 Mar (RJ), but 55 Dawlish Warren 1 Jan was the lowest early year max since 1997. The only *winter* record away from these two areas was one at Broadsands on 20 Mar. In contrast to winter numbers, the *late spring passage* was better represented in S Devon, giving max counts of 84 Slapton Sands 27 May and 78 Dawlish Warren 3 June; there were also two S Huish Marsh 10 May and one flew N from Bowling Green Marsh on 27 May. Late migrants appeared in early Jun with the last being 29 Dawlish Warren on 8th, then just a single on 11th.

Autumn/second winter-period. The first on *autumn* passage were singles Saunton Sands 13 & 20 Jul and 22 Dawlish Warren 17 Jul. Numbers rose quickly from late Jul with 76 on 20th rising to a max count of 165 Saunton Sands 28 Sep, and 65 Dawlish Warren on 1 Aug. *Elsewhere*: one Broadsands 23 Jul; 17 Thurlestone 27 July with two there 19 Aug; two Charleton Bay 1 Aug and singles there 19 Aug & 20 Sep; and one Prawle 19 Aug. *Winter* max counts were 144 Saunton 17 Nov and 42 Dawlish Warren 30 Nov; one at Seaton on 17 Dec was the only record in this period away from key sites.

LITTLE STINT *Calidris minuta*

Scarce autumn passage visitor; rare in winter and spring.

A few wintering birds at the main site, for the fourth year in succession; no spring passage apart from a Jun bird; and a good autumn. Discounting the Pottington birds from 24 Oct on, in total there were 337 bird-days on 53 dates at 15 sites during 24 July - 13 Nov (cf. 82 bird-days in 2000 and 107 bird-days in 1999).

First winter-period/spring. Up to three 14 Jan - 18 Feb at Pottington, Taw Est (RJ,JEW); wintering birds having been annual on the Taw/Torridge since 1998. No spring passage, the first blank spring since 1989, but one at Dawlish Warren on 9 Jun was only the third Jun record there and the first in Devon since 1997.

Autumn/second winter-period. Two at Bowling Green Marsh on 24 Jul (BBH,MSW) was the only *Jul* record. In *Aug* birds only on the Exe Est (one on 12th, two on 26th) and Upper Tamar L (one on 23rd). Only regular from mid-Sep, with three sites supporting the majority: *Dawlish Warren* (117 bird-days on 22 dates, 17 Sep - 22 Oct), *Upper Tamar L* (97 bird-days on 20 dates, 23 Aug - 9 Oct) and *Bowling Green Marsh* (76 bird-days on 32 dates, mostly 5 Sep - 25 Oct). The remaining 13 sites supported 47 bird-days. *Elsewhere*: two at Powderham 10 Oct; max count of 15 on Exe Est on 23 Sep, with eight accounted for at key sites; up to two Axe Est 1 Sep - 1 Oct; one Roadford Res 19 Sep; one Kingsbridge Est 22 Sep with two the next day; one Plym Est 25 Sep; one Wistlandpound Res 25 Sep; one Skern 28 Sep with three 30 Sep - 1 Oct; three Exminster Marshes 1 Oct; and one Pottington 10 Oct. Birds became infrequent after mid-Oct with the *last migrants* being rather late, one (unusually, an ad) S Huish Marsh 3 & 4 Nov, one Bowling Green Marsh 10 Nov and one Roadford Res 13 Nov. In the *second winter-period*, one to three (four on 25 Oct) at Pottington 24 Oct – 31 Dec, with perhaps same elsewhere on the estuary including three Skern 30 Oct, one R. Torridge and one R Caen 27 Nov.

TEMMINCK'S STINT *Calidris temminckii*

Rare passage migrant. BoCC (Amber).

One bird, becoming the 14th record and 15th bird for Devon. Of the 12 records since 1961, seven were in spring (16 May - 3 Jun) and five in autumn (28 Aug - 30 Sep), an increase in the proportion of spring records is in line with the national trend. Site record totals: Exe Est (6), Taw/Torridge Est (4), Lundy (1) and Thurlestone (1) indicate no coastal bias.

One Bowling Green Marsh 31 May, and 2-3 Jun (MRAB,MD,DS,MSW).

BAIRD'S SANDPIPER *Calidris bairdii*

Rare vagrant.

One bird, representing the 8th record for Devon and the 4th for Dawlish Warren; all but one of Devon's records have occurred since 1983 and all have appeared within the period 6-30 Sep, although now two have remained into Oct. Other records were from Northam Burrows, Upper Tamar L and Lundy (two, including the first on 6 - 13 Sep 1974).

A juv at Dawlish Warren 19 Sep - 7 Oct (LC *et al.*) appeared each high tide on The Bight or roosted with other waders on the beach. (One of eight birds in Britain in 2001 accepted by BBRC.)

Baird's Sandpiper - (juvenile) - *Mike Langman*

PECTORAL SANDPIPER *Calidris melanotos*

Rare vagrant.

This rarity has appeared within Devon in 11 of the past 14 years; the county total is now 40 records, involving 45 birds (16 of which were during the l990s – see DBR 1999.

One, perhaps two: a juv Upper Tamar L 3-15 Sep (RD,DC,RGM *et al.*); and one or the same there, 30 Sep - 7 Oct (GPS *et al*).*(see col plate)*

CURLEW SANDPIPER *Calidris ferruginea*

Scarce autumn passage visitor; very scarce in spring; rare in winter.

A good autumn, with 598 bird-days on 71 dates at 15 sites during the period 24 July to 14 Nov (246 bird-days in 2000 and 746 bird-days in 1999); also an above average spring.

Spring. Four records of five birds was above average (see *DBR* 1999): two Dawlish Warren 8 May and one 15th (LC, JEF); one Wembury 15 May (PFG,MDD); and one Bowling Green Marsh 26 May (CC,RSPB).

Autumn. Three sites supported the majority: *Bowling Green Marsh* (201 bird-days on 35 dates: mostly 26 July - 20 Oct); *Dawlish Warren* (161 bird-days on 36 dates: 26 July - 18 Oct), with an additional 55 bird-days on four extra dates distributed around the Exe Est; and *Upper Tamar L* (125 bird-days on 18 dates: 3 Sep - 9 Oct).*(see col plate)*. On *other estuaries*: Kingsbridge Est (24 bird-days, 5 dates); Taw/ Torridge Est (20 bird-days, 13 dates); and Axe Est (13 bird-days, six dates). The only other records were singles at Slapton 5 Sep and Tamar Est 18 Oct. The last migrants of the season were singles Kingsbridge Est 27 Oct, Tamerton Creek 2-3 Nov and Bowling Green Marsh 9-14 Nov. In the *second winter-period*, one Isley Marsh 12 Dec (IT).

PURPLE SANDPIPER *Calidris maritima*

Scarce winter visitor. SPEC(4), BoCC (Amber).

Recorded from 11 regular sites (13 in 2000, 11 in 1999). The annual max of 18 was the highest count since 33 in Dec 1997 (also at Prawle).

Table: Monthly maxima at main sites:

	J	F	M	A	M	J	J	A	S	O	N	D
Brixham	5	6	10	-	-	-	-	-	-	8	8	1
Hope's Nose	6	-	3	6	-	-	-	-	-	1	9	9
Plymouth	2	8	-	6	-	-	-	-	-	-	-	-
Prawle Area	1	6		18	-	-	-	-	-	4	6	2
Sidmouth	9	-	9	-	2	-	-	-	-	-	1	3
Westward Ho!	10	-	7	-	4	-	-	-	-	1	4	3

First winter-period/spring. Regular at six sites with a *max* of ten at Westward Ho! 30 Jan (IM) and ten Brixham 11 Mar (BG); however, a *spring passage* count of 18 Prawle Pt 15 Apr (PMM) was the annual max. Also present in this winter-period were one Otter Est 14 Feb, and five 28 Jan Torquay Harbour which likely came from nearby Hope's Nose. Possible migrants were four at Ilfracombe 23 Mar and two Otter Est 31 Mar; the *last of the season* were two Sidmouth 1 May and four Westward Ho! 3 & 6 May (MPP).

Autumn/second winter-period. The *first arrivals* were four on Lundy and singles at Brixham, Hope's Nose and Westward Ho! - all on 7 Oct. Migrants at other sites included one Berry Hd 5 Nov (GV) and two Ilfracombe 15 Oct (MAP). Five sites became occupied during Nov and the *season's max counts* were nine Hope's Nose 29 Nov & 2 Dec (CJP) and eight Brixham 27 Oct & 20 Nov. Also present in this winter period were: one Wembury 16 Nov, with two there 25 Nov and one 6 Dec; and one Otter Est 26 Nov.

DUNLIN *Calidris alpina*

Fairly numerous winter visitor, not scarce on passage; very scarce migrant breeder. BoCC (Amber), Dartmoor BAP.

The small breeding population of the schinzii race on Dartmoor was not surveyed this year owing to FMD access restrictions; however there was a single record of one at West Dart Hd on 14 Jul (MLHS). Counts on the Exe Est exceeded nationally important numbers (>5,300) in Nov & Dec, and the max of 7,167 was the highest since 1994.

Table 53: Monthly maxima at main/well-counted sites and WeBS county totals:

	J	F	M	A	M	J	J	A	S	O	N	D
Avon Est	-	-	-	-	4	-	3	3	3	3	2	-
Axe Est	48	19	-	3	22	-	1	33	37	20	10	101
Exe Est WeBS	3311	3009	-	-	-	-	-	209	750	849	5780	7167
Dawlish Warren	*4250*	*1800*	*1800*	*150*	*360*	*20*	*320*	*317*	*900*	*1168*	*2700*	*4000*
Bowling Green Marsh	*55*	*77*	-	*8*	*66*	*2*	*41*	*170*	*80*	*8*	*720*	*140*
Kingsbridge Est	1035	520	-	-	-	-	60	120	99	8	185	371
Otter Est	-	-	-	-	-	-	-	10	6	5	9	3
Plym Est	40	10	6	420	-	-	-	3	21	21	250	-
Slapton	-	82	40	-	32	-	-	-	-	-	-	-
Tamar Complex WeBS	2010	1205	-	-	-	-	-	-	63	70	1517	1062
Upper / mid Tamar	*10*	-	-	-	-	-	-	-	*13*	*24*	*17*	*33*
Taw / Torridge Est WeBS	1331	1785	-	-	-	-	-	-	128	22	27	1827
R. Caen	*150*	*80*	-	-	-	-	*8*	*2*	*34*	-	*100*	
Isley Marsh	*400*	*516*	-	-	*50*	-	*42*	*10*	*25*	*10*	*70*	*300*
Northam / Skern	*300*	-	-	-	*17*	*3*	*55*	*60*	*50*	-	*100*	-
Pottington	-	-	-	-	-	-	-	*3*	*30*	-	*400*	*600*
Saunton Sands	-	-	*2*	*23*	*122*	*7*	*4*	*14*	*97*	*1*	-	-
Teign Est	3	1	3	-	-	4	-	17	-	-	-	-
South Huish area	-	-	-	1	18	-	-	3	30	-	1	-
Wembury	1	-	-	-	17	-	10	8	17	-	1	-
WeBS county total '00	5598	7871	458	76	817	3	312	593	380	212	2157	6811
WeBS county total '01	5731	5333	-	-	-	-	2	332	1067	956	6048	9481

First winter-period. Among high counts in the early year, the max of 4,250 at Dawlish Warren on 6

Jan, was the highest count there since Feb 1995. On the smaller estuaries, 48 Axe 21 Jan was the largest count, and 82 Slapton Sands on 27 Feb was the highest non-estuary count (and the largest there since the massive numbers of Jan & Feb 1996).

Spring. With the departure of most wintering birds by the end of Mar, there were few Apr counts until the increase in migrants from late Apr. *Max counts* were 150 Exminster Marshes 30 Apr, 420 Plym Est 1 May and 360 Dawlish Warren 4 May; elsewhere only from Prawle (max five 13 May) and Portworthy Dam (two on 2 May). Migrants dwindled to very small numbers in late May and Jun.

Autumn. Modest numbers arrived through Jul with a sudden influx of 320 Dawlish Warren 27 July and 60 Kingsbridge 31 July, then a lull followed by another influx in late Aug with 317 Dawlish Warren 21 Aug and 93 Kingsbridge 27 Aug. Less intense passage in early Sep was followed by influxes of 521 Dawlish Warren 19 Sep and a short-lived max of 1,168 Dawlish Warren 2 Oct, most of which had departed by the next day. *Elsewhere*, small numbers recorded from Torbay, Sidmouth, Prawle, Dart Est and Beesands (with a max of six at Prawle on 3 Aug). *From headlands*: nine past Hope's Nose 10 July, five 17 July and one 18 Aug; five Berry Hd 17 July and six 18 Aug; four Staddon Pt, 17 Sep, six 23rd and seven 29th. Two *Lundy* 21 Sep. *Inland*: two over Lipson, Plymouth 20 July; one Portworthy Dam 22 July; bird(s) heard migrating overhead after dark at Lee Moor on 26 Jul; Roadford Res one 10 Aug, four 13 Sep and on 4 Nov; one Fernworthy Res 25 Aug (EM); one Wistlandpound 10 Sep. Upper Tamar L had a fantastic set of records on 27 dates 27 July - 12 Oct with max counts of 50 on 3 Sep (DC) and 46 on 17 Sep - the largest ever inland counts for Devon, beating 35 in Sep 1993.

Second winter-period. The Exe Est max of 7,167 on 9 Dec (WeBS) was the largest count since Dec 1994. The largest count from the smaller ests was 250 Plym 28 Nov, and <10 were recorded from the Dart, Erme and Yealm Ests. Singles at South Huish Marsh and Wembury in Nov were the only non-estuary counts.

Dunlin and Baird's Sandpiper - (Dawlish Warren) - *Steve Young*

RUFF *Philomachus pugnax*

Scarce passage and winter visitor. SPEC(4), BoCC (Amber).

An average year overall with 183 bird-days (excluding unrecorded dates of long-stayers); relatively poor on the Exe Est with 77 bird-days (64 of them at Bowling Green Marsh), the Taw/Torridge was as expected with 33 (30 in the Velator area) but an excellent additional 73 bird-days were produced from seven other sites.

Table 54: Monthly bird-days at main sites and number of sites

	J	F	M	A	M	J	J	A	S	O	N	D
Exe Est	3	-	2	2	-	1	11	14	23	20	1	-
Taw / Torridge Est	1	-	14	-	1	-	-	6	7	3	1	-
Sum bird-days, all other sites	-	-	-	1	-	-	-	4	31	22	-	15
No. of other sites	-	-	-	1	-	-	-	3	4	2	-	2

Map 6: Jack Snipe.
Maximum count of migrant (Sep-Nov & Mar-Apr) and wintering birds (Dec-Feb) at each location.
Data: Based upon a ten year period 1992-2001. Source: DBRs) (see Overview on page 25)

First winter-period. One at Velator on 5 Jan (JEW) remained until 5 May; one Bowling Green Marsh 26 Jan (RSPB) may have moved to Exminster Marshes (MKn) the next day and again onto Starcross golf course on 28 Jan (DJG). (This species has wintered annually around the Exe since 1949, rising to a peak of 100 in Feb 1971; with a total absence from the whole county during the same period last year, and just two birds this year, the decline of this fantastic bird has gone relatively unnoticed.)

Spring. Fifteen birds from five records roughly fits the pattern of alternating good and bad years. The first was one Exminster Marshes 8 Mar (RSPB); then a flock of 11 circling Velator 23 Mar (RJ), the largest count in the county since 14 at Bowling Green Marsh in May 1994, one or two remaining until 27th; one Dawlish Warren 30 Mar (LC); ♀ South Huish Marsh 17 Apr (MDD); and a ♀ Bowling Green Marsh 26 Apr (RSPB,BBH).

Autumn. A total of 150 bird-days on 65 dates at 12 sites within the period 22 June - 4 Nov. The principal site, *Bowling Green Marsh*, produced 60 bird-days on 41 dates: 22 June (the first of the autumn) to 20 Oct, mostly ones and twos, up to three 28 Sep - 7 Oct with a max of four 30 Sep. *Elsewhere around the Exe*: singles Dawlish Warren 1 & 18 Sep; two Exminster Marshes 4 Jul, one 4 Oct and two 26 & 28 Oct; and singles Exe Est 26 Aug and 21 Oct. *Upper Tamar L* produced a good set of records with 38 bird-days on 17 dates: 10 Sep - 4 Nov (the last of the season), usually one to three with a max of four 30 Sep. The *Kingsbridge Est* produced 20 bird-days on 17 dates: 22 Aug - 31 Oct with a max of three West Charleton Marsh 17 Sep. The *Taw/Torridge Est* had 18 bird-days on seven dates, 19 Aug - 21 Oct with a max of six Velator 23 Sep. *Elsewhere*: ♀ Axe Est 26 Aug, one Teign Est 28 Aug; and a juv Wistlandpound Res 15 Sep.

Second winter-period. Four singles: Bowling Green Marsh 21 Nov (RSPB); Braunton Marsh 27 Nov (RJ); West Charleton Marsh 4 & 8 Dec (PSa, BW); and Seaton Marsh 17 - 29 Dec (mo).

JACK SNIPE *Lymnocryptes minimus*

Scarce winter visitor, but greatly overlooked. SPEC(3[w]).

A total of 37 records at 18 sites.

Table 55: Monthly sums of site maxima and number of sites:

	J	F	M	A	M	J	J	A	S	O	N	D
Sum max at all sites	16	7	1	-	-	-	-	-	3	4	7	20
Number of sites	9	4	1	-	-	-	-	-	2	2	4	6

First winter-period. Records from nine sites, involving ones and twos except a max of seven near Okehampton on 31 Jan, and four there on 13 Feb (RHi). After 19 Feb, just a single at Dawlish Warren on 2 Mar (JEF).

Second winter-period. Records, again from nine sites began with one at Staddon Pt, Bovisand 25 Sep (SGe) and one Upper Tamar L 3 Oct (GPS). Again one or two at most locations, with max counts of six West Charleton Marsh 22 Dec (PSa) from three dates, and at least that number near Okehampton on 31 Dec from seven dates (*per* RHi).

SNIPE *Gallinago gallinago* — Common Snipe

Not scarce winter visitor; scarce breeder. BoCC (Amber).

Table 56: Monthly maxima at main/well-counted sites and WeBS county totals (excluding breeding birds):

	J	F	M	A	M	J	J	A	S	O	N	D
Avon Est	56	32	-	-	-	-	-	-	-	-	7	28
Axe Est	104	49	-	-	-	-	3	5	30	19	51	41
Clennon Valley	8	4	-	-	-	-	-	-	-	-	9	1
Exe Est WeBS	46	30	-	-	-	-	-	2	-	16	23	71
Dawlish to Starcross	*46*	*22*	*15*	-	-	-	*1*	*1*	-	*10*	*17*	*71*
Exminster Marshes	*37*	*78*	-	*1*	*3*	-	-	-	*1*	*15*	*11*	*6*
Bowling Green Marsh	*2*	-	*4*	-	-	*1*	*3*	*2*	*2*	*7*	*3*	*3*
Otter Est	20	8	-	-	-	-	-	-	-	-	6	10
Roadford Res	43	30	-	-	-	-	-	-	-	1	16	6
Stoke Canon	60	62	-	-	-	-	-	-	-	4	-	17

continued	J	F	M	A	M	J	J	A	S	O	N	D
Stover Lake	-	3	1	1	-	-	-	-	-	1	2	
U & L Tamar Lakes	1	16	-	-	-	-	-	-	1	-	4	5
Tamar Complex WeBS	6	10	-	-	-	-	-	-	-		4	5
Taw / Torridge Est	37	30	-	-	-	-	-	1	5	1	10	23
Bradiford NR	-	*36*	*13*	*6*	-	-	-	-	-	-	*6*	*8*
Isley Marsh	-	*30*	-	-	-	-	-	-	-	-	*12*	*14*
Teign Est	20	2	3	8	-	-	-	-	-	-	-	2
South Huish area	30	12	-	-	-	-	3	-	-	-	-	20
Sum max at all other sites	61	16	1	1	-	-	4	11	6	21	55	105
No. of other sites	13	7	1	1	-	-	2	3	5	5	7	14
WeBS county total '00	202	162	143	8	3	-	-	2	29	85	221	239
WeBS county total '01	306	195	1	1	-	-	3	8	6	38	123	187

Breeding. A handful of records from Dartmoor, just nine birds from six sites but access limited by FMD outbreak, and the DMBBS estimated 100-200 prs in 2000; see also *DBR* 1997.

First winter-period/spring. A count of 104 Axe 12 Jan (WeBS) was possibly the highest ever there; other early year max counts over 50 include 78 Exminster Marshes 15 Feb, 60 & 62 Stoke Canon in Jan & Feb and 56 Avon Estl Jan. Away from main sites, 16 at Stover CP on 30 Jan was the highest count. The last double-figure count on 3 Mar was followed by a trickle of records through to mid-Apr with the last being one Efford Marsh, Plymouth on 13th , three Bradiford NR, on 14th, and a late isolated record of three Exminster Marshes 1 May (RSPB).

Autumn/second winter-period. The first records away from breeding areas were one Bowling Green Marsh 24 Jun, three South Huish Marsh 2 July and one Dawlish Warren 6 July, followed by a scattering of records through Jul and Aug, and only becoming more frequent from late Sep. In winter-period, only three counts > 50: 71 Dawlish Warren to Starcross on 9 Dec; 61 Seaton Marsh & Axe 17 Dec; and 50 Sourton Cross 1 Dec.

WOODCOCK *Scolopax rusticola* — Eurasian Woodcock

Probably not scarce winter visitor, but much overlooked; rare breeder (last bred 1988). SPEC(3ᵂ), BoCC (Amber).

Above average numbers from the usual sources, but an incredible extra 116 birds were noted on just three dates (17 Jan, 28 Nov & 1 Dec) counted as 'rises' from shoots on Dartmoor - a useful reminder of the extent to which numbers can be under-estimated for species that are widespread but at low densities.

Table 57: Monthly sums of site maxima and number of sites (excluding data from shoots):

	J	F	M	A	M	J	J	A	S	O	N	D
Sum max at all sites	25	11	3	-	-	-	-	-	-	-	25	19
Number of sites	15	7	2	-	-	-	-	-	-	-	15	14

First winter-period. Recorded on 22 dates from 25 sites, mostly as ones and twos, rarely threes, but seven Clennon Valley 8 Jan (ML) and four Telegraph Hill 21 Jan (DS). The last were one Huccaby 2 Mar with two 7th (RHi), and one at Man Sands 19 Mar (MI).

Second winter-period. One Lundy 15 Oct was the first arrival (LFS), followed by four at Sourton Cross on 3 Nov and one at Venford Res the next day (RHi). Recorded on 24 dates from 25 sites, mostly as ones and twos, rarely threes, but six on Lundy on 3 Nov (LFS), at least eight at Mel Tor-Luckey Tor on 28 Dec (JMW) and nine at Sourton Cross 17 Nov (*per* RHi).

BLACK-TAILED GODWIT *Limosa limosa*

Not scarce, but very local, winter visitor; very scarce inland. SPEC(2), BoCC (Red).

Several monthly maxima on the Exe Est exceeded the qualifying level for national importance (70) and the overall max of 709 on 18 Feb exceeded the qualifying level for international importance (700). The Exe supported around 90% of the Devon total, and Bowling Green Marsh, with seasonal max counts of 655 on 11 Jan and 550 on 19 Oct & 26 Nov, is a major site; 460 on Exminster Marshes on 11 Dec was also an exceptional count there, perhaps reflecting recent habitat enhancement. On the Upper/mid-Tamar sector of the Tamar Complex, numbers reached national importance Oct - Dec with 105 on 16 Nov the highest count since Dec 1997. The Plym also continues to attract this species, with a max of 50 in Feb.

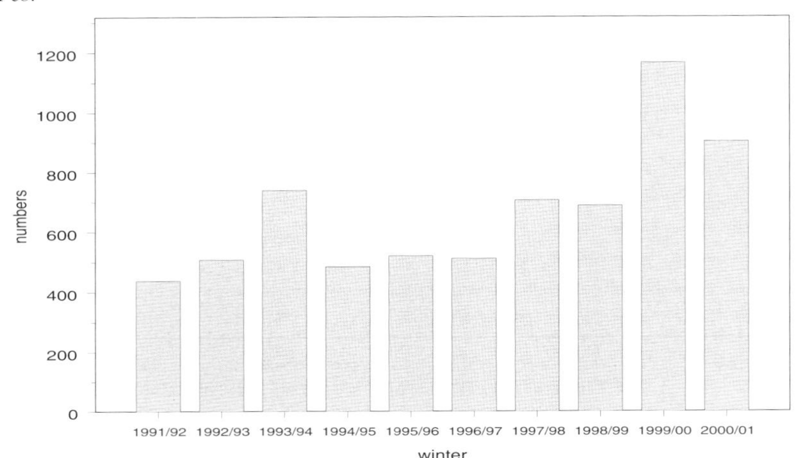

Fig. 21: Black-tailed Godwit. Maximum WeBS County Totals for the 1991/92 - 2000/01 winters (Source: WeBS data in DBRs)

Table 58: Monthly maxima at all regular sites and WeBS county totals:

	J	F	M	A	M	J	J	A	S	O	N	D
Axe Est	-	-	-	-	2	-	-	14	5	-	-	6
Exe Est WeBS	**654**	**709**	-	-	-	-	-	**403**	**524**	699	680	676
Dawlish to Starcross	*240*	*10*	*100*	*6*	*9*	-	*20*	*15*	*180*	*10*	*47*	*36*
Exminster Marshes	*115*	-	*8*	*8*	-	-	-	-	*1*	*75*	*58*	*460*
Bowling Green Marsh	*655*	*490*	*150*	*30*	*30*	*27*	*240*	*520*	*450*	*550*	*550*	*520*
Plym Est	35	50	-	-	-	-	-	3	2	2	-	-
Tamar Complex WeBS	27	64	-	-	-	-	-	-	12	**101**	**106**	**84**
Upper / mid Tamar	*2*	-	-	-	-	-	-	-	*10*	*98*	*105*	*72*
Taw / Torridge Est	2	2	-	2	5	1	4	8	15	4	4	4
WeBS county total '00	948	681	213	157	51	12	242	384	685	513	632	902
WeBS county total '01	689	825	-	-	-	-	-	404	537	799	785	755

First winter-period/spring. *Winter records* away from the principal sites were: singles Isley Marsh and Northam Burrows on seven dates 1 Jan - 21 Feb with two 19 Jan; one Kingsbridge Est 7 & 8 Jan; and one Yealm Est 2 Feb. Numbers dropped notably through Mar with 100 Dawlish Warren 29 Mar the last three-figure count. *Spring passage*: up to 30 Bowling Green Marsh 7 Apr & 8 May; *one –two* Taw/Torridge Est 14 Apr – 11 May; and one Axe Est, 10 May. Up to three at Bowling Green Marsh 31 May - 22 Jun blurred the distinction between spring and autumn passage.

Autumn/second winter-period. Autumn passage began with 27 Bowling Green Marsh 23 Jun rising to 91 there 7 Jul. One Taw Est 27 Jun and one Skern 6 Jul were the first autumn records away from the Exe, followed by: two Thurlestone 11 Jul; one Avon Est 13 Jul; at Upper Tamar L on nine dates 15 Aug - 14 Sep with max five 6 Sep; and a high count of 13 South Huish Marsh 26 Aug (PAS), perhaps same as 13 W past Prawle on the same day (AKS). *Winter* records away from the principal sites were: up to eight, Axe Est 16-22 Dec; up to four, Taw/Torridge Est on nine dates 9 Nov - 22 Dec; and one Stoke Gabriel 28 Nov.

BAR-TAILED GODWIT *Limosa lapponica*

Not scarce winter and passage visitor; rare or very scarce inland. SPEC(3^w), BoCC (Amber).

The principal site, the Exe Est, produced seasonal maxima of 436 on 21 Jan and 529 on 9 Dec, the latter just below the qualifying level for national importance (530) which was last achieved in 1999.

Table 59: Monthly maxima at all regular sites and WeBS county totals:

	J	F	M	A	M	J	J	A	S	O	N	D
Avon Est	-	-	-	-	2	-	-	-	1	1	-	-
Exe Est WeBS	436	329	-	-	-	-	-	22	95	249	402	529
Dawlish Warren	*350*	*208*	*200*	*64*	*100*	*7*	*18*	*42*	*90*	*130*	*395*	*275*
Bowling Green Marsh	*2*	*17*	-	-	*20*	*19*	*3*	*30*	*2*	*1*	*45*	*30*
Kingsbridge Est	4	1	-	-	-	-	-	5	4	-	-	1
Plym Est	-	-	-	-	-	-	-	3	7	3	1	-
Tamar Complex WeBS	15	11	-	-	-	-	-	-	23	4	24	25
Taw / Torridge Est	8	15	-	4	14	-	-	1	68	11	9	15
Pottington	-	*1*	-	-	-	-	-	*6*	*35*	*11*	*9*	*15*
Skern	*1*	-	-	*1*	-	-	*1*	*8*	*13*	*7*	-	-
Wembury	-	-	-	-	6	-	-	-	1	1	1	-
WeBS county total '00	31	382	162	9	20	-	32	9	77	246	76	287
WeBS county total '01	443	345	-	-	-	-	-	25	181	260	411	536

First winter-period/spring. On the Exe Est, 350 Dawlish Warren 5 Jan was the highest count from one site. Elsewhere, the only *winter* records were: up to 61 Upper/mid-Tamar 1 Feb; up to 15 Taw/Torridge Est 11 Feb; four Kingsbridge Est 24 Jan and one 18 Feb; and one Teign Est 1 Jan. Up to 200 Dawlish Warren 2-16 Mar and 100 Dawlish Warren 10 May were large counts indicating *spring passage*, and on the upper Exe, up to 37 Exminster Marshes 8-13 Apr, was the best count. Away from Exe & Taw/Torridge Ests: South Huish area, singles on 14 Apr & 6 May, followed by 18 on 7th, 26 on 10th and 15 on 12th; two Avon Est 3 May with one there 16 May; one Watermouth, Ilfracombe 14 May; one over Plymouth Sound 21 Apr, 20 on 27th and one 28th; two Slapton 22 Apr with one 29 Apr and five there 10 May; 20 Berry Hd 9 May flew low over the bungalow in fog; four Axe Est 10 May; nine Leasfoot Beach 13 May. From 4 Jun – 5 Jul, the only records were single-figure counts from Dawlish Warren and Bowling Green Marsh.

Autumn/second winter-period. First indication of *autumn passage* was 21 Dawlish Warren 6 July, but early records elsewhere included: one Roadford Res 15 July; and singles Skern 26 & 30 July, then three there 2 Aug. Although 18 at Dawlish Warren 24 Jul, not regular there until mid-Aug, and elsewhere: one S past Berry Hd 18 Aug; seven Prawle 24 Aug, then several records of singles there 1 - 17 Sep; two Preston seafront 10 Sep; one Erme Est 12 Sep; four Stoke Gabriel 16 Sep; and singles Avon Est 16 Sep, and 7 & 19 Oct. *Winter* numbers rose from early Oct on the Exe. In N Devon the max was 15 Pottington 16 Dec and the only other records were singles at Wembury 14 Nov, Plym Est 17 Nov and Kingsbridge Est 14 Dec.

WHIMBREL *Numenius phaeopus*

Not scarce passage visitor; rare in winter. SPEC(4), BoCC (Amber).

The annual bird-days total of 5369 was better than last year (3659 in 2000). The max count of 287 at Bowling Green Marsh was the largest count in the county since the massive passage off Thurlestone in spring 1996.

Table 60: Monthly maxima at main/well-counted sites and WeBS county totals:

	J	F	M	A	M	J	J	A	S	O	N	D
Avon Est	1	1	1	3	22	1	4	3	2	2	-	-
Axe Est	-	-	-	70	5	-	-	1	-	1	-	-
Exe Est	-	-	-	287	100	11	48	48	3	1	1	-
Dawlish Warren	-	-	-	*116*	*18*	*3*	*16*	*20*	*2*	-	*1*	-
Exminster Marshes	-	-	-	*75*	*6*	-	-	-	-	-	-	-
Bowling Green Marsh	-	-	-	*287*	*100*	*11*	*48*	*48*	*3*	*1*	-	-
Prawle area	-	-	-	10	12	2	7	6	-	-	-	-
Tamar Complex WeBS	-	-	-	-	-	-	-	-	-	-	-	-

continued	J	F	M	A	M	J	J	A	S	O	N	D
Taw / Torridge Est	-	-	-	65	70	3	20	22	9	-	-	1
Instow	-	-	-	*65*	*14*	*3*	-	*6*	*9*	-	-	-
Slapton	-	-	-	65	8	-	1	1	-	-	-	-
Teign Est	-	-	-	1	3	-	-	-	-	-	-	1
Wembury	-	-	-	-	11	-	2	5	2	-	-	-
WeBS county total '00	1	1	-	13	73	3	98	14	17	1	1	1
WeBS county total '01	1	1	-	-	-	-	4	36	16	1	-	1

First winter-period. One Avon Est 14 Jan, 1 & 11 Feb and in Mar (RWBo), and two Stoke Pt 7 Jan, were the only records.

Spring. A total of 4,033 bird-days, 1 Apr - 1 June. The *first* was one over Slapton 1 Apr (RBe), followed by singles Exminster Marshes 8-12 Apr, with 35 Dawlish Warren on 16th the first sizable flock. Numbers quickly increased from mid-Apr to a *peak passage* through late Apr. On the *Exe Est*, Bowling Green Marsh had 1,830 bird-days on 26 dates 13 Apr - 1 June (max 287 on 26 Apr, BBH) and Dawlish Warren had 538 bird-days on 28 dates 13 Apr - 24 May (max116 on 29 Apr). *Away from the Exe*, Slapton had 290 bird-days on 13 dates 18 Apr - 6 May (max 65 on 23 Apr), and Plymouth Sound 164 bird-days on six dates 20 - 28 Apr (max 58 in off sea on 27th). Other site maxima were: 70 Axe Est 28 Apr; 65 Instow 26 Apr; and 70 Frogmore Creek 7 May. Thirty passed Hope's Nose 27 Apr. Numbers tailed off from early May in S Devon and from mid-May in N Devon with 54 still present Yelland 11 May. Small flocks at main sites and scattered records elsewhere after mid-May with one Barnstaple 30 May being the last mainland bird away from the Exe. Lundy held up to three 24 Apr-4 Jun. In *Jun*, up to five, Bowling Green Marsh and up to three Dawlish Warren; one inland over Headland Warren 11 June, two Isley Marsh 12 June, and perhaps same amongst three at Instow 16 June, were the only other 'summer' records.

Autumn. Passage comprised 1,283 bird-days, 4 Jul - 8 Oct. Visible *passage began* with four Prawle 5 July, then one 7th and three 16 Jul; also 20 Taw/Torridge Est 16 Jul; off Berry Hd six 17 July and singles 27 Jul, 4 Aug, then six 18 Aug. On the *Exe Est*, Bowling Green Marsh had 521 bird-days on 43 dates 6 Jul - 8 Oct, and Dawlish Warren had 521 bird-days on 37 dates from 4 Jul - 3 Oct. *Peak passage* through late Jul to late Aug culminated in *max counts* of 48 Bowling Green Marsh 5 Aug, 20 Dawlish Warren 20 Aug and 20 Skern 28 Aug. Lundy had up to eight 24-27 Aug. Numbers dropped suddenly in Sep with only one-three at most sites, except for nine Instow 5 Sep and six Kingsbridge and Taw/Torridge Ests, both on 16 Sep. The last stragglers were: two Wembury 25 Sep and one there 27th; one Dawlish Warren 29 & 30 Sep and 3 Oct, with perhaps same Bowling Green Marsh 6 & 8 Oct; one Axe Est 7 Oct; and two Avon Est 19 Oct.

Second winter-period. Four birds: one on Lundy on 1 Nov (LFS), one Dawlish Warren 11 Nov (JBa), one Taw/Torridge Est 16 Dec (WeBS) and one Teign Est 21 Dec (MKn).

CURLEW *Numenius arquata* — Eurasian Curlew

Fairly numerous winter and passage visitor; scarce breeder. SPEC(3^w^), BoCC (Amber), Devon BAP.

Table 61: Monthly maxima at main/well-counted sites and WeBS county totals (excluding breeding birds):

	J	F	M	A	M	J	J	A	S	O	N	D
Avon Est	60	43	4	-	1	15	38	41	41	51	51	42
Axe Est	175	93	-	-	-	-	37	70	40	62	30	47
Dart Est	41	1	-	-	-	3	-	-	99	-	19	47
Erme Est	-	55	-	-	-	-	-	50	46	65	61	29
Exe Est WeBS	685	937	-	-	-	-	-	931	1656	1667	975	587
Dawlish Warren	*146*	*274*	*238*	*60*	*51*	*390*	*610*	*690*	*750*	*600*	*236*	*320*
Exminster Marshes	*410*	*250*	*180*	*308*	*4*	*1*	-	-	-	*25*	*120*	*280*
Bowling Green Marsh	*280*	*213*	*80*	*166*	*39*	*52*	*327*	*390*	*360*	*420*	*417*	*400*
Kingsbridge Est	324	192	-	-	-	-	-	302	395	270	270	200
Otter Est	26	17	2	-	-	-	-	-	1	4	21	31
Plym Est	106	4	45	-	176	34	64	77	77	100	151	74
Prawle area	-	-	-	-	-	-	1	1	3	2	4	3
Tamar Complex WeBS	314	302	-	-	-	-	-	-	457	394	320	404

Upper / mid Tamar	*99*	*133*	-	-	-	-	-	-	*131*	*177*	*124*	*88*
Tavy Est	-	*54*	-	-	-	-	-	-	*20*	*28*	*55*	*88*
Taw / Torridge Est WeBS	992	877	-	-	-	-	-	-	605	437	604	674
Bradiford/Pottington	-	*50*	*60*	-	-	-	-	*60*	*140*	*250*	*100*	*400*
Isley Marsh	*150*	*135*	*60*	*40*	*80*	*200*	*300*	*176*	*230*	*80*	*15*	*35*
Skern	-	-	-	-	-	*26*	*183*	*220*	-	-	-	-
Teign Est	30	-	22	3	31	-	2	70	-	40	-	30
Wembury	-	-	-	-	-	17	3	24	15	94	72	30
Yealm Est	33	50	-	-	-	-	-	175	84	24	67	5
WeBS county total '00	2126	2312	1566	789	137	490	1902	2673	3187	3303	1591	2121
WeBS county total '01	2412	2379	-	-	-	-	75	1374	3093	2823	2370	1772

Breeding. Reports of only nine prs/territories reflected FMD access restrictions. *NW Devon*: two prs Bursdon Moor (NWCo). *Culm Measures*: one present Knowstone Moor 16 June (BRE), a pr Witheridge Moor (RJJ) and a pr on farm nr West Anstey, where reputedly has bred for the last six yrs (*per* NCW). *E Dartmoor*: probable total of three prs in the Bagtor Down/Blackslade Mire/Rippon Tor area (same as 1999 and 2000), with two small chicks seen at Bagtor on 23 Jun (NB) and a single half-grown chick at Blackslade Mire on the same day (MD,AR). *W Devon*: two in flight at Ivyhouse Cross on 29 May (HM) could possibly have been nearby breeders. *Blackdown Hills* (all info from RMHa): prs returned to Blackdown Common and Whitechapel, and there was again evidence of breeding from Stockland. *E Devon Commons*: a pr attempted breeding in the Hawkerland/Woodbury Castle area. No reports from *Exmoor*.

First winter-period/spring. Max WeBS counts were 992 Taw/Torridge 14 Jan with 877 there on 11 Feb and 937 Exe Est 18 Feb. Numbers dropped quickly from early Mar. During spring passage, four Hope's Nose 11 Apr and one past there 14 Apr, 12 Slapton 21 Apr and three Roadford Res 27 Apr were sightings away from estuaries.

Autumn/second winter-period. Post-breeding parties appeared from mid-Jun, with notable influxes from early Jul, and a protracted peak passage period lasting through to late Oct/early Nov. Max counts of 1,656 Exe Est 23 Sep with 1,667 there 21 Oct were the highest on the Exe since Aug 1997, and exceed the *qualifying level for national importance* (1,200 birds). In N Devon max counts were 605 Taw/Torridge 16 Sep with 604 there 4 Nov.

Coastal. Off *headlands*, one past Berry Hd and Hope's Nose 17 July, then singles Hope's Nose 12 Aug, 26 Sep, three 1 Oct and two 15 Oct. On *open coast*, most frequent at Wembury with max counts of 17 on 16 June and 25 on 4 Oct (*cf.* Prawle with up to three 14 July - 4 Oct).

SPOTTED REDSHANK *Tringa erythropus*

Scarce or very scarce passage and winter visitor. BoCC (Amber).

A below average year with 149 bird-days (177 in 2000, and a 1995-99 average of 180). This is partly a result of poor figures from Bowling Green Marsh which produced only 38% of records this year (c.60% in 2000 & 1999), a poor early spring and poor late-year recording.

Table 62: Monthly maxima at main sites and WeBS county totals:

	J	F	M	A	M	J	J	A	S	O	N	D
Bowling Green Marsh	1	1	4	-	1	-	-	-	3	2	-	-
Tamar complex WeBS	6	4	-	-	-	-	-	-	-	1	-	1
Upper / mid Tamar	*4*	*3*	-	-	-	-	-	-	-	*1*	-	*1*
Taw / Torridge Est	3	2	-	1	-	-	1	2	1	2	1	1
Number of sites	6	4	2	1	2	1	2	7	8	8	1	3
WeBS county total '00	2	-	-	-	-	-	-	1	2	-	6	8
WeBS county total '01	9	6	-	-	-	-	-	-	-	1	-	2

First winter-period/spring. Eleven *winter* birds, from Tamar, Kingsbridge, Taw/Torridge, Exe and Erme Ests: four Upper/mid-Tamar 1 Jan with three there 1 Feb; two Bowcombe Creek 1 Jan with one until 29 Jan; one or two Isley Marsh 3 Jan - 21 Feb, with three whole Taw/Torridge Est 14 Jan; one lower Exe Est 5-16 Jan, perhaps same as one Bowling Green Marsh 1-2 Jan & 1 Feb; and one Erme Est 17

Feb. *Spring passage* possibly involved 12 birds, though some records could relate to wintering birds: singles Bowling Green Marsh 1 and 11 Mar with four 13 Mar, and later, singles 2 and 8 May; also singles Exmouth 17 Mar and Exminster Marshes 12 and 13 May and another 29 May. One at Isley Marsh on 1 Apr was the only spring bird away from the Exe Est.

Autumn/second winter-period. *Autumn* passage involved an estimated 88 bird-days on 48 dates, 28 June - 28 Oct. At the main site, *Bowling Green Marsh*, there were 46 bird-days on 27 dates 2 Sep - 17 Oct, mostly ones and twos, but three 11 and 16-19 Sep. *Elsewhere on the Exe Est* were: two Dawlish Warren 28 June (first June record there since 1983), with singles 26 July and 20 Aug; one Exton 1 Aug; and one Exminster Marshes 17 Sep. Elsewhere: on the *Taw/Torridge Est* (one Pottington 15 Aug; one R. Caen area 26 Aug with two 31st, one 26 Sep and two 18 Oct; singles Penhill Marsh 19 Aug and Northam 19 Sep; and Isley 7 & 21 Oct); on the *Kingsbridge Est*, singles 20 Aug and 23 Sep with one Bowcombe Creek 14 Oct; and singles at South Milton Ley 17 Sep, Bigbury 17 Sep - 3 Oct, Upper Tamar L 18 Sep, Upper/mid-Tamar Est 18 Oct, and Avon Est 28 Oct. *Winter* records comprised four singles on four estuaries: Isley Marsh 10 Nov - 12 Dec, Dawlish Warren 4 Dec, Upper/mid-Tamar Est 16 Dec and Bowcombe Creek 27 Dec.

REDSHANK *Tringa totanus* — Common Redshank

Not scarce winter visitor (passage status uncertain); scarce breeder. SPEC(2), BoCC (Amber).

Table 63: Monthly maxima at main/well-counted sites and WeBS county totals:

	J	F	M	A	M	J	J	A	S	O	N	D
Avon Est	7	4	4	-	1	-	2	2	2	4	6	12
Axe Est	55	36	-	-	-	-	17	12	21	55	57	38
Dart Est	11	8	-	-	-	-	-	-	-	-	30	2
Erme Est	3	-	-	-	-	-	-	-	-	1	-	1
Exe Est	370	397	100	-	-	-	-	305	435	325	341	382
Dawlish Warren	*112*	*131*	*70*	*1*	-	*16*	*157*	*229*	*302*	*220*	*315*	*230*
Bowling Green Marsh	*236*	*236*	*50*	*123*	-	*126*	*277*	*220*	*170*	*106*	*360*	*64*
Kingsbridge Est	250	164	7	-	-	-	32	89	288	110	129	115
Otter Est	4	5	1	-	-	-	-	4	-	1	5	5
Plym Est	76	4	69	57	-	-	62	97	70	200	150	-
Prawle area	-	-	-	-	-	-	-	5	2	-	1	2
Tamar Complex WeBS	588	506	-	-	-	-	-	-	176	532	430	261
Upper / mid Tamar	*464*	*377*	-	-	-	-	-	-	*48*	*318*	*270*	*141*
Tavy Est	*6*	*10*	-	-	-	-	-	-	-	-	-	*38*
Taw / Torridge Est	196	254	126	-	-	-	22	16	112	168	329	213
Isley Marsh	*70*	*100*	-	*16*	-	*15*	*30*	*6*	*30*	*55*	*25*	*24*
Teign Est	35	1	15	-	28	-	-	-	-	12	44	8
Yealm Est	12	20	-	-	-	-	-	-	1	1	8	12
WeBS county total '00	1063	1358	790	67	25	11	556	539	1045	1179	1472	1049
WeBS county total '01	1024	1265	-	-	-	-	19	509	909	981	1256	936

Breeding. At the main site, Exminster Marshes, nine prs displaying, but no details of success (RSPB). No records from elsewhere.

First winter period/spring. Max counts of 588 Tamar Complex in Jan (inc 464 on Upper/mid-Tamar) and 397 Exe Est 18 Feb. One Wembury 28 Feb was the only non-estuary record. Numbers dropped on ests from late Feb although parties appearing through Mar and Apr blurred the distinction between winter and passage birds; a more obvious passage bird was one at South Huish Marsh on 22 May.

Autumn/second winter-period. One Dawlish Warren 18 Jun was the first of *autumn*, with small parties appearing from 25 Jun at various sites and numbers soon increasing rapidly with 215 Bowling Green Marsh and 62 Plym Est on 3 July. Steady increases thereon led to max counts of 532 Tamar complex in Oct, 435 Exe Est 23 Sep and Kingsbridge Est 288 16 Sep. *Away from estuaries*: one Wembury 25 June; one Burrator Res 13 July; one Roadford Res 15 July; singles Upper Tamar L 23 & 27 July; five Prawle 27 Aug and two 1 Sep; and one Broadsands 22 Oct. After the main influx in early Nov, max counts achieved were 430 Tamar Complex in Nov and 382 Exe Est in Dec: away from estuaries, in Dec up to six at Wembury and two at Prawle.

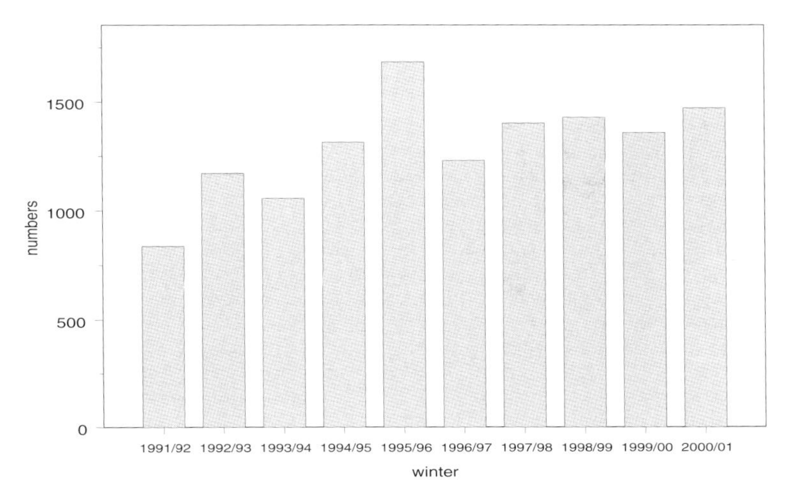

Fig. 22: Redshank. Maximum WeBS County Totals for the 1991/92 - 2000/01 winters (Source: WeBS data in DBRs)

GREENSHANK *Tringa nebularia* — Common Greenshank

Scarce passage and winter visitor.

A total of 3622 bird-days, with the Exe Est as the principal site with 2049 bird-days (mostly at Bowling Green Marsh). The Taw/Torridge Est had 810 bird-days (mostly R Caen and Isley Marsh), and other estuary totals were: Kingsbridge (323); Upper/mid-Tamar (122); Yealm (114); Avon (65), Plym (41); Axe (38); and Teign (16). Excluding Upper Tamar L (with 16 bird-days), the remaining 12 areas/sites produced a total of 25 bird-days.

Table 64: Monthly maxima at main/well-counted sites and WeBS county totals:

	J	F	M	A	M	J	J	A	S	O	N	D
Avon Est	4	4	5	5	-	2	1	4	4	4	6	6
Axe Est	-	-	-	-	-	-	2	20	3	3	-	-
Erme Est	-	1	-	-	-	-	-	-	2	-	-	1
Exe Est WeBS	4	9	-	-	-	-	-	37	27	35	2	14
Dawlish Warren	*4*	*5*	*2*	*1*	*1*	*1*	*1*	*2*	*6*	*7*	*7*	*6*
Bowling Green Marsh	*8*	*8*	*11*	*14*	*8*	*3*	*37*	*55*	*60*	*40*	*27*	*2*
Kingsbridge Est	14	11	1	-	-	-	3	30	36	14	11	8
Plym Est	2	1	1	1	-	-	-	3	3	2	3	1
Tamar Complex WeBS	26	26	-	-	-	-	-	-	37	31	28	20
Upper / mid Tamar	*19*	*20*	-	-	-	-	-	-	*25*	*18*	*22*	*15*
Taw / Torridge Est WeBS	9	19	-	-	-	-	-	-	23	5	14	5
Braunton M / R Caen	*7*	*14*	*9*	-	-	-	*5*	*8*	*23*	*24*	*12*	-
Isley Marsh	*20*	*20*	-	-	-	-	*20*	*11*	*8*	*19*	*11*	*15*
Teign Est	3	-	-	4	-	-	-	2	-	-	-	4
Yealm Est	7	6	-	-	-	-	-	27	17	9	5	7
WeBS county total '00	94	53	42	48	21	2	71	96	169	89	61	55
WeBS county total '01	47	70	-	1	-	1	-	70	138	85	60	56

First winter-period-spring. *Winter counts ≥ten* included: 19-20 Upper/mid-Tamar in Jan and Feb; 20 Isley Marsh 6 Jan, 5 & 20 Feb, perhaps some the same as 14 Braunton 7 Feb; 14 Kingsbridge Est 21 Jan with 11 there on 18 Feb; and 13 Powderham 4 Jan. There were no records away from tabulated sites. Numbers generally dropped quickly from late Feb with few present in Devon during most of Mar. A small *spring influx* late Mar/early Apr included 12 Powderdam 21 Mar, but passage most notable at Bowling

Green Marsh 30 Mar - 15 Apr with a max of 14 on 2 & 5 Apr. Elsewhere: four Coombe Cellars 9 Apr; and singles Plym and Teign Ests17 Apr. Few records after mid-Apr: one Dawlish Warren 8 and 10 May; and up to six Bowling Green Marsh, the last on 10 May; one Exminster Marshes 1 Jun; and two Avon Est 2 Jun.

Autumn/second winter-period. First *autumn arrivals* assumed to be singles at Dawlish Warren 18-22 Jun, Avon Est 20 Jun and Exminster Marshes 21-22 Jun. Birds appeared regularly through late Jun with influxes from early Jul, including an early high count of 24 *Bowling Green Marsh* 9 July. The prolonged peak passage from early Aug to early Oct (see also Coastal and Inland below) produced max counts at Bowling Green Marsh of 55 on 19 Aug, 60 on 3 Sep and 52 on 30 Sep. On *estuaries*, max counts were 27 Yealm Est 20 Aug, 20 Axe Est 26 Aug, 36 Kingsbridge Est 16 Sep, 25 Upper/mid-Tamar 16 Sep and 23 Braunton 19 Sep. Apart from 19 Isley Marsh 29 Oct and 14 there 4 Nov, there were few in the county late Oct to late Nov when counts rose at key sites. Double-figure *winter* counts included: 27 Bowling Green Marsh 26 Nov; 22 Upper/mid-Tamar 16 Nov and 15 there 13 Dec; 11 Isley Marsh 19 Nov, with 15 there 8 Dec, perhaps involving 12 Braunton 29 Nov; and 11 Kingsbridge Est 18 Nov.

Coastal and inland. Off *headlands*: singles at Hope's Nose 18 Aug, Start Pt 19 Aug and Prawle 25 Aug. From other *coastal sites*: one Dawlish town 25 July; three Plymouth Sound 23 Aug and one Thurlestone 27 Aug. *Inland*: on seven dates at Upper Tamar L 27 July - 29 Aug, with five on 23 Aug; singles Burrator 17 & 28 Aug, 17 Sep, and 1 & 3 Oct; singles Portworthy Dam 19 Aug and 16 Sep; one Meldon Res 24 Aug; and late singles at Roadford Res on 12 Oct and 4 Nov. One on a small lake at Throwleigh on 5 Dec is thought to be the first Dartmoor winter record (DSGp).

LESSER YELLOWLEGS *Tringa flavipes*

Rare vagrant

The 15th record for Devon (cf. DBR 1989) of this N American vagrant was the first since 1989 and the first ever inland. Past records involved six on Taw/Torridge Est, three on Exe Est, one Teign, (later on Exe) and singles on Avon Est, Tamar Est (Plymouth), Kingsbridge Est and one on Lundy. Most records have been in the autumn.

One Roadford Res 16-17 Sep (JTi *et al.*) frequented the bay in the SW corner of the reservoir, not far from the dam where it was watched by many observers. On 17th it was seen to fly high to the NE out of sight – a particularly interesting observation, since photographic evidence reveals that this bird, with a distinctive pattern of moult on the upperparts, was in Norfolk 17-21 Sep (Rogers *et al.* 2001). (One of six birds in Britain in 2001 accepted by BBRC.)

GREEN SANDPIPER *Tringa ochropus*

Scarce late summer passage and winter visitor. BoCC (Amber).

Quite a good year, with 443 bird-days mostly attributable to heavy autumn passage and good observations at well-known sites (cf. 660 in 2000, 776 in 1999, 275 in 1998 & 310 in 1997, though the very high 1999 figure resulted from estimating all days present by wintering birds Jan – Mar).

Table 65: Monthly maxima at main/well-counted sites and WeBS county totals:

	J	F	M	A	M	J	J	A	S	O	N	D
Avon Est	4	5	3	4		1	6	6	4	6	6	3
Exe Est	3	1	-	-	-	-	-	2	2	4	3	1
Exminster Marshes	-	-	-	-	*1*	*2*	*4*	*1*	-	*1*	-	-
Bowling Green Marsh	-	-	-	-		*1*	*5*	*12*	*3*	*1*	-	-
Riverside CP area	*4*	*5*	*5*	*4*	*1*	-	-	-	*4*	-	-	*1*
Roadford Res	-	-	-	-	-	-	1	1	3	1	3	-
Tamar Complex WeBS	2	1	-	-	-	-	-	-	1	-	1	1
Taw / Torridge Est	-	-	-	-	-	-	3	5	2	1	-	3
Total no. of sites	9	4	2	4	3	4	12	18	13	7	3	6
WeBS county total '00	7	6	6	3	5	2	15	10	7	9	2	11
WeBS county total '01	9	6	-	-	-	-	7	14	8	11	13	8

First winter-period/spring. At least 18 birds: four Exeter Riverside CP to 8 Apr, five 7 Mar (JRD), the last departing 15 Apr; five Aveton Gifford into Apr (RWBo); two Tavy Est Jan (one Feb); one

Landcross, with two 3 Apr; and singles during Jan at Exmouth, Powderham, Upton Pyne and Bishop's Tawton. Likely spring migrants were singles at Higher Metcombe 22 Apr (GHG), Exeter Riverside CP 1 May (NPi) and Exminster Marshes 9 May (RSPB).

Autumn. There were 314 bird-days from 29 sites within the period 20 Jun - 10 Oct (excluding late arriving DSGp records). Records of singles at Higher Metcombe and Aveton Gifford in early *Jun* were unseasonal, but one Bowling Green Marsh 20 Jun probably the first to return, followed by two Exminster Marshes 28 Jun & 2 Jul, one 4 July, and at Bowling Green Marsh one 5-7 Jul. Frequent from *mid-Jul*, with *peak passage generally late July to mid-Aug*, max counts being: eight Axe Est 26 Aug; six Avon Est 18 Jul, 19 Aug, in Sep and on 7 Oct; and six Bowling Green Marsh 19 Aug. Away from tabulated sites: one 19 Jul Axe Est with five 20th and eight 26 Aug; one Brixton 22 Jul; four Plym Est 26 Jul; a total of seven Upper Tamar L 27 Jul - 9 Oct with two on 3 Aug; singles Dawlish Warren 30 Jul and 4 Aug. In *Aug*, max count was five at South Huish Marsh on 18 Aug; and singles were recorded from Batworthy Mire, Cadover Bridge, Exmouth, Goosemoor NR, Hatch Bridge, Lee Moor, Otter Est, Phear Park, Prawle (rare there), Redhill Quarry, Stoke Canon and Uffculme. In *Sep*, singles were at Sourton Cross early in the month, Kennick Res on 11th, Tamar Complex on 16th and Mill Leat, Sandford on 25th. In *Oct*, records were frequent until 10 Oct with stragglers thereafter comprising four Exe Est 21 Oct, perhaps including the singles seen at Bowling Green Marsh on 20 Oct and Exminster Marshes on 26th.

Second winter-period. At least 23 birds: four Avon Est (from 4 Nov); four Landcross (from 13 Nov); three Roadford Res (from 2 Nov); three Exe Est on 1 Nov with one thereafter; three Taw/Torridge Est on 16 Dec; two Stoke Canon (from 11 Nov); two Grand Western Canal on 16 Dec; one Tamar Est (from 16 Nov); and one Bishops Tawton on 15 Dec.

WOOD SANDPIPER *Tringa glareola*

Very scarce passage migrant. SPEC(3), BoCC (Amber).

Apart from 1998, this species has been annual in spring since 1987, and numbers have increased steadily; the c.38 birds in 2001 is the highest ever recorded in Devon.

Spring. Three birds, a typical turnout: one Taw Est 4-5 May (RJ); one Matford Pools 12-13 May (JGa,MKn *et al.*); and one Bowling Green Marsh 6 Jun (BBH,MKn,K&JWM).

Autumn. Autumn passage 23 Jun – 10 Sep, comprised 49 bird-days on 26 dates at 14 sites, involving an estimated *c.*35 birds. The first, at Drakelands 23 Jun (PJS,JGS), was very early, and the first in Jun since 1997. *Around the Exe Est*, all singles: Bowling Green Marsh, 6 Jul (RSPB), 21-22 Jul (MKn,MSW *et al.*), 27 Jul (MKn), 6 Aug (BBH), 22 & 24 Aug (RSPB,CC *et al.*) and perhaps another 27 Aug (BBH,KGB); Dawlish Warren, 23 Jul (IL,KRy); and Exminster Marshes, 31 Aug (RSPB). *Elsewhere*: singles on R Culm at Burlescombe 23 Jul and Uffculme 21-27 Aug (PC); a good set of records from Upper Tamar L with one 27 Jul (RD,DC), three juvs there 20 (GPS) & 29 (RGM) Aug and one 3-4 Sep (RD,JEW); one Wistlandpound Res 27-29 Jul (MAP), one West Charleton Marsh 30 Jul (PSa); one Velator 8 Aug (JEW); one South Huish Marsh 18 Aug (RBu), with another 26th (PAS); two Axe Est 20 Aug (WeBS), with four there 26th (DWH); one Taw Est 30 Aug (RJ); and lastly, three Roadford Res 10 Sep (FHCK).

2000 correction. Delete record of one on 3 May in Exminster village.

COMMON SANDPIPER *Actitis hypoleucos*

Scarce passage visitor; very scarce in winter; rare/very scarce breeder.

An average year with 1077 bird-days (995 in 2000, 1329 in 1999, 995 in 1998.

Table 66: Monthly maxima at main/well-counted sites and WeBS county totals:

	J	F	M	A	M	J	J	A	S	O	N	D
Avon Est	-	-	-	2	7	-	9	4	4	2	2	1
Axe Est	-	-	-	2		-	6	12	-	8	2	1
Burrator Res	-	-	-	-	-	-	2	2	1	1	-	-
Dawlish Warren	-	-	-	1	1	-	4	3	-	1	-	-
Bowling Green Marsh	-	-	-	1	1	1	4	4	3	-	3	1
Countess Wear area	1	-	1	2	1	-	-	-	-	-	-	-
Dart Est/ Totnes	1	1	1	-	-	-	1	1	-	1	1	1
Erme Est	-	2	-	-	-	-	-	-	2	2	1	-

continued	J	F	M	A	M	J	J	A	S	O	N	D
Kingsbridge Est	1	1	1	-	-	-	1	7	1	1	1	3
Otter Est	-	-	-	3	2	-	2	4	3	-	-	-
Plym Est	-	-	-	3	-	2	7	10	5	2	2	2
Portworthy Dam	-	-	-	-	-	3	3	1	1	-	-	-
Prawle area	-	-	-	1	1	-	1	6	2	-	-	-
Roadford Res	-	-	-	-	-	-	5	6	2	2	1	-
Slapton Ley	-	-	-	3	4	-	1	1	-	1	1	-
Tamar Complex WeBS	3	3	-	-	-	-	-	-	9	5	4	5
Upper / mid Tamar	*1*	-	-	-	-	-	-	-	*4*	-	*1*	*2*
Tavy Est	*2*	*1*	-	-	-	-	-	-	*5*	*3*	*2*	*2*
Taw / Torridge Est	-	-	-	12		2	-	8	4	-	3	2
Braunton/Sherpa/Velator	*1*	*1*	*1*	*5*	*1*	-	*11*	*8*	*1*	*2*	*2*	*1*
Pottington	-	-	-	*4*	-	-	*5*	*3*	*2*	*1*	-	-
Teign Est	2	1	2	3	-	-	13	1	-	-	1	2
Wembury	1	-	-	-	1	-	4	-	1	1	1	2
Total no. of sites	9	9	6	22	13	4	29	35	24	22	17	13
WeBS county total '00	7	9	3	19	22	6	64	64	37	15	8	7
WeBS county total '01	7	6	-	3	-	3	20	30	25	19	13	13

Breeding. Pr with at least one juv, Torycombe Valley, nr Lee Moor on 21 Jun (RJL) and other birds present in the breeding season included two Cadover Bridge 7 Jun, two Portworthy Dam 11 Jun and one Newpark Waste 19 Jun.

First winter-period/spring. At least 15 birds *wintered* at 12 of the usual sites in ones, rarely twos; some remaining into Mar, and rarely perhaps to early Apr; at non-tabulated sites, singles at Bishop's Tawton and Topsham playing field. *Spring passage* comprised 171 bird-days at 28 sites 1 Apr - 6 June. One Velator 29 Mar was perhaps the *first*, followed by one Stover CP 1 Apr, two Lopwell Dam 5 Apr, then regular from 11th onwards with a max of 12 Bideford Quay 14 Apr (RGM), seven Penhill Marshes 25 Apr and seven Avon Est in May. *Additional sites/records in Apr*: singles Beesands 12 & 19 Apr and two 27th; one Portworthy Dam 15 Apr; singles Baggy Pt and Bicton on 17 Apr; one Jennycliff 27 Apr; and one Huccaby 30 Apr. Into *May*, all singles: Littlecombe Shoot 5th; South Milton Ley 13th; Ilfracombe 14th; and Dawlish town 27th. The last migrants were one Barnstaple 30 May, two Bideford 1 Jun and one Bowling Green Marsh 6 Jun.

Autumn/second winter-period. *Autumn passage* 21 Jun – 31 Oct comprised 751 bird-days at 45 sites. *Bowling Green Marsh* was the key site with 123 bird-days on 60 dates 5 Jul - 22 Sep; other main site bird-day totals include *R Caen area* (90; 4 Jul - 9 Oct), Plym (71; 21 Jun - 19 Oct), *Upper Tamar L* (39; 16 July - 14 Sep); *Dawlish Warren* (35; 9 July - 21 Oct) and *Wistlandpound Res* (33; 22 July - 22 Sep). One Plym Est 21 Jun with two there 28th, and one Broadsands 29 June were the *first* to return, with birds frequent from 3 Jul onwards and peak passage from late Jul to early Aug reaching a max of 13 Teign Est 14 Jul (MKn), 11 R Caen 26 July and ten Plym Est 1 Aug. Some *untabulated records* include: one Dartmeet 3 Jul; three Littlecombe Shoot 14 Jul and one there 29th; a total of six at Fernworthy Res 25 July - 21 Sep; five Broadsands 1 Aug; five singles Stover CP 3 - 24 Aug; one Grand W Canal 13 Aug; one Shobrooke L 24 Aug; one Galmpton 25 Aug; two Stoke Canon 27 Aug; one Kennick Res 20 Sep; and three Bishop's Tawton 5 Oct. Still frequent until early *Oct*, thereafter scarcer, with the *last migrants* being singles Pottington 22 & 29 Oct, two Erme Est 30 Oct, one Roadford Res 30 Oct and one Slapton 2 & 3 Nov. Most *wintering* birds arrived from late Oct to mid-Nov, with at least 26 birds, at 16 of the usual sites; mostly ones and twos, but three at Bowling Green Marsh 5 Nov and Landcross 13 Nov.

TURNSTONE *Arenaria interpres* — Ruddy Turnstone

Not scarce passage and winter visitor; very scarce in summer; rare inland. BoCC (Amber).

Table 67: Monthly maxima at main/well-counted sites and WeBS county totals:

	J	F	M	A	M	J	J	A	S	O	N	D
Exe Est WeBS	136	79	-	-	-	-	-	-	-	76	88	111
Dawlish Warren	*62*	*43*	*50*	*35*	*15*	-	*8*	*5*	*4*	*25*	*20*	*25*

Starcross	*66*	*161*	*157*	*21*	*8*	*-*	*3*	*46*	*123*	*50*	*103*	*2*
Plym / Plym. Hoe	20	10	13	25	-	-	-	-	8	10	-	-
Prawle area	18	-	-	7	2	-	-	6	1	1	6	6
Tamar Complex WeBS	14	10	-	-	-	-	-	-	-	6	3	3
Upper / mid Tamar	*10*	*9*	-	-	-	-	-	-	-	-	-	-
Taw / Torridge Est	21	31	-	-	2	-	6	8	2	-	21	21
Torbay	7	18	13	-	3	-	-	1	8	17	7	20
Wembury	90	79	-	-	24	-	2	19	41	76	77	145
Westward Ho !	4	-	-	-	7	4	-	-	49	51	15	-
WeBS county total '00	84	173	144	37	8	-	19	1	5	187	134	163
WeBS county total '01	153	322	-	-	-	-	-	-	2	77	106	121

First winter-period/spring. *Maximum counts*: Starcross again confirmed as the single most important roost site in the county with a max of 161 on 21 Feb (DJG); 90 Wembury 20 Jan (PFG) was the largest non-estuary count; and 31 Taw/Torridge Est 11 Feb was the best count from the N Devon coast. *Elsewhere,* two Otter Est 1 Jan and three there 14 Feb. A count of 59 at Shaldon 1 Jan perhaps involved Starcross birds. *Spring passage* peaked in Mar, resulting in 157 Starcross 22 Mar (DJG) and lighter passage through Apr and May producing 35 Dawlish Warren 13 Apr with 24 there 24 Apr and 11 May. Elsewhere, one Bowling Green Marsh 30 Apr, two South Huish Marsh 10 May; the last of the season were two Prawle 27 May and two Dawlish Warren 28 May. The only *Jun* records were four at Westward Ho! on 20 Jun.

Autumn/second winter-period. One Dawlish Warren 7 Jul was the first of *autumn passage*, followed by two past Berry Hd 17 Jul with sightings frequent thereafter. Small parties arrived from early/mid-Aug, with peak passage mid-Sep to mid-Oct producing a max of 123 Starcross 16 Sep, with 70 Wembury 9 Oct the coastal max and 51 Westward Ho! 7 Oct the best N Devon count. Small numbers also recorded from Otter and Teign Ests, Charleton Bay and Thurlestone. The only *inland* record was one Portworthy Dam 7 Oct. *Winter* maxima included: 145 Wembury 1 Dec; 111 Exe Est 9 Dec; and only 21 from the N coast, at Isley Marsh, 19 Nov and 23 Dec. The only untabulated records were one Kingsbridge 3 Nov with two there 16 Dec.

GREY PHALAROPE *Phalaropus fulicarius*

Very scarce autumn passage visitor; rare in winter and inland.

A total of 17 birds, all in autumn, was the second best tally since 1988 (after 30 in 1995); all between 30 Sep and 19 Oct, and most 7-9 Oct.

The first were: singles Dawlish Warren 30 Sep (IL,KRy) and 1 Oct (KRy); one Plymouth Sound 1-2 Oct (VRT,SRT); and one on the sea at Hope's Nose on 4 Oct (GV). One off Teignmouth seafront in the evening 6 Oct was the first of a small influx on the 7th: one Axe Est on a small pool remaining to 9th (IJW,GV); one Hope's Nose (LC,AK); one Dawlish Warren lingered offshore all day; probably two Plymouth Sound, with one next day (VRT,SRT); and one inland at Portworthy Dam (WeBS). Other 'wrecked' birds after this storm found the next day, 8 Oct, included one Kennerleigh, Crediton on a small pool (JMr) and an ad Upper Tamar L (DC,RGM) which remained to 9 Oct *(see col plate)*. Next were singles off Prawle 11 Oct (BMc,ML), Lannacombe 15 Oct (PBo,BW) and lastly off Slapton 19 Oct (PSa).

POMARINE SKUA *Stercorarius pomarinus*

Very scarce/scarce passage visitor; rare in summer and very scarce in winter.

Another poor year with 88 birds counted (cf. 62 in 2000), this total relating to probably 80 individuals allowing for duplication at neighbouring sites on the same date (although, as with all skuas on successive days, it is often impossible to estimate how many of the birds are the same individuals seen the day, or even days, before). Oct and Nov are normally the best months of the year for this species so the complete lack of counts during Nov was notable - a result of poor 'seawatching' weather. N Devon faired only slightly better than the previous two years with just two birds, both during Oct.

Table 68: Monthly totals (birds) from individual sites:

	J	F	M	A	M	J	J	A	S	O	N	D
Berry Head	1	-	-	-	-	-	4	1	8	38	-	1

continued	J	F	M	A	M	J	J	A	S	O	N	D
Dawlish / Exmouth	-	-	-	-	1	2	3	-	6	-	-	-
Hope's Nose	-	-	-	-	-	-	1	1	2	5	-	-
Prawle Point	-	-	-	-	-	1	1	2	1	-	-	1
other sites	-	-	-	-	-	-	-	3	-	4	-	1
Total 2000	2	4	-	4	6	-	8	-	21	11	6	3
Total 2001	1	-	-	-	1	3	9	7	17	47	-	3

First winter-period. Just one, at Berry Hd on 2 Jan (MD).

Spring. A very poor period, with just five recorded on four dates 21 May - 29 June; *May*, on 21st a single at Dawlish Warren (IL); in *Jun* one at Prawle Pt on 2nd (JCN), one off Teignmouth seafront on 10th (LC) and two at Dawlish Warren on 29th(JEF).

Autumn. Nine non-breeding or imm birds were seen during *Jul* on six dates: singles on 8th and 9th at Orcombe Pt (MSW); an ad W at Prawle Pt on 16th; on 17th three at Berry Hd, an ad, sub-ad and 3y (MD,ML,BMc), also one at Hope's Nose (CJP); then singles at Berry Hd on 21st (MD) and at Dawlish Warren a dark phase bird on 22nd (LC). In *Aug*, seven recorded on two dates: two at Prawle Pt on 9th (JLb), then on 18th three at Slapton (NCW) and a full sp ad south past both Hope's Nose (CJP) and Berry Hd (MD,AR,WJD). In *Sep*, 17 recorded on ten dates, with four being the monthly maximum count on 30th at Berry Hd (mo); counts of more than one were three at Berry Hd on 13th (DH), two off Exmouth on 17th (ARB) and two at Hope's Nose 30th (CJP,LC). *Oct* was an excellent month with 47 recorded on 14 dates, the max count being 15 S at Berry Hd on 5th (MD,ML,BMc); counts of more than one were all at Berry Hd, with four on 1st (MD,AR,ML), two on 7th (MD,AR,ML), three on 8th (BBH,ML) and 15th (G&BJ) and six on 26th (MD,AR). No records in *Nov*.

Second winter-period. Singles noted on just three dates in *Dec*, at Berry Hd on 5th (MD,AR), at Prawle on 9th (PMM) and at Wembury on 28th, where a Herring Gull was forced to disgorge its food (VRT,SRT).

ARCTIC SKUA *Stercorarius parasiticus*

Scarce passage visitor; very scarce in winter and summer.

A good year involving 1,080 birds counted (cf.689 in 2000 and 750 in 1999), this total relating to around 993 individuals allowing for duplication at neighbouring sites on the same date.

Table 69: Monthly totals (birds) from individual sites:

	J	F	M	A	M	J	J	A	S	O	N	D
Berry Head	-	-	-	-	-	2	7	69	210	246	-	-
Dawlish / Exmouth	-	-	-	3	6	11	81	36	75	68	1	-
Hartland Point / Lundy	-	-	-	-	-	-	-	2	6	-	-	-
Hope's Nose	-	-	-	-	-	-	6	18	54	61	-	-
Prawle Point	-	-	-	-	-	-	12	10	9	15	-	-
other sites	-	-	-	2	10	4	2	10	14	30	-	-
Total 2000	1	-	-	8	11	9	25	53	456	136	19	8
Total 2001	-	-	-	5	16	17	108	145	368	420	1	-

Spring. In *Apr,* recorded on four dates at three sites - all singles. In *May,* on seven dates at five sites, the max count being two or more off Teignmouth on 27th; one noted chasing Herring Gull at Saunton Sands on 16th (RJ). In *Jun,* recorded on eight dates, nearly all from Dawlish Bay with a max of three on 18th & 19th, and two at Berry Hd on 29th.

Autumn. During *Jul*, recorded on 17 dates from seven sites, the only double figure counts being from Dawlish Warren, with 12 on 17th and ten on 22nd. In *Aug*, recorded on 18 dates from nine sites; double-figure counts only on 18th, when 56, mostly juvs, went S at Berry Hd 0615 - 1400 hrs (MD,AR) and 17 passed Hope's Nose. In *Sep*, records came from 16 dates and 11 sites, with double-figure counts as follows: 36 at Berry Hd on 13th; 15 at Dawlish Warren on 15th, this count coincided with several other moderate counts during freshening NW winds; 20 on 16th at Berry Hd; 11 off Exmouth on 17th; 11 at Berry Hd on 18th; and the year's best count of 143 at Berry Hd 0700 - 1630hrs (mo), and 46 at Hope's Nose, both on 30th. *Oct* was similar, with records on 16 dates from eight sites, but this was the best month for double-figure counts: at Berry Hd, 54 on 1st, when one was noted chasing an Alpine Swift low over the sea (ML,MD *et al.*); 23 on 4th; 71 on 5th; 37 on 6th; 29 on 7th; 18 on 8th; 21 on 15th; and 12 on 27th. At Hope's Nose there were 25 on 6th, and at Dawlish Warren, ten on 1st, 15 on 6th and 21 on 7th. Two birds were noted well up the Exe Estuary, off Starcross, on 1 Oct during stormy conditions (DJG). *Nov* was extremely poor with just a single sighting off Dawlish Warren on 2nd.

2000 Addition: A dark phase ad flew in off the sea and continued high northwards over Plymouth, during the evening of 22 Apr (SGe). This bird appeared to be on active migration and provides evidence of overland passage of some seabirds.

Arctic Skua with Alpine Swift - *Mike Langman*

LONG-TAILED SKUA *Stercorarius longicaudus*

Rare/very scarce, mainly autumn passage visitor.

A poor to average year with only four records (cf. six in 2000 and 23 in 1999). The lack of S and SE winds during the autumn, very similar conditions to 2000, dictated the numbers of this rare skua being blown close enough to shore to be seen well; note the number of unidentified Long-tailed/Arctic Skuas. Spring/Summer sightings are very rare and especially noteworthy. All observers should be reminded that a full description is required for this highly variable and tricky species.

Spring/summer. The first of the year was an ad in full bp which sheltered with Kittiwakes in appalling weather and strong SSE winds off Shoalstone, Brixham on 14 Jun (ML), similar weather conditions on 17 Jul brought in another ad with a broken tail off Hope's Nose (CJP).*(see col plate)*

Autumn. Two records, both from Oct: a 2y bird at Berry Hd on 5th (MD,ML); and a dark juv at Dawlish Warren on 18th (JEF).

GREAT SKUA *Catharacta skua*

Scarce passage and wintervisitor; very scarce in summer. SPEC(4), BoCC (Amber).

A good autumn (but poor spring and winter periods) resulted in an annual total of 484 birds (cf. 286 in 2000 and 266 in 1999), relating to around 439 individuals, allowing for duplication at neighbouring sites on the same date.

Table 70: Monthly totals (birds) from individual sites:

	J	F	M	A	M	J	J	A	S	O	N	D
Berry Head	-	-	-	3	-	-	13	81	74	94	3	-
Dawlish Warren	-	-	-	3	-	-	2	5	35	6	-	1
Hartland Point / Lundy	-	-	-	-	-	-	-	-	13	-	-	-
Hope's Nose	-	-	-	4	1	-	6	25	13	17	-	-
Prawle Point	-	1	-	-	-	-	4	18	29	3	1	-
other sites	1	-	-	-	-	1	2	4	15	6	-	-
Total 2000	1	2	-	8	3	-	6	29	147	72	18	6
Total 2001	1	1	-	10	1	1	27	133	179	126	4	1

First winter-period. One bird noted close inshore attacking Herring Gull at Baggy Pt on 4 Jan, the only other record being one W at Prawle Pt on 10 Feb.

Spring. A poor passage, with mostly single birds recorded on four dates from three sites in *Apr* but three noted at both Berry Hd and Dawlish Warren on 24th. The only one seen in *May* was at Hope's Nose on 13th, and in *Jun* just a single at Dawlish Warren on 29th.

Late summer/autumn. In *Jul,* singles logged on seven dates at five sites, the best counts coming on 17th from Berry Hd with 12 S and Hope's Nose with six. During *Aug*, recorded on 14 dates from five sites, the best counts at Berry Hd with eight on 13th and 62 on 18th, when Hope's Nose also recorded 22. A good *Sep*, with records on 15 dates from eight sites, the best counts being at: Prawle Pt, with six on 2nd then nine on 7th; Bolt Hd, with nine on 4th; Berry Hd, with 18 on 13th then 11 on 15th, 12 on 16th and 31 on 30th; Hartland Pt, with six on 8th; and Exmouth, with seven on 16th. *Oct* records were on 15 dates from eight sites, the best counts being at Berry Hd with 29 on 1st, 32 on 5th and eight on 6th and 7th; the next best count was five at Dawlish Warren also on 7th. A poor *Nov* produced singles on just four dates from two sites. The only Lundy records were singles on 4 & 7 Oct.

Second winter-period. One sighting, from Dawlish Warren on 13th Dec.

SKUA species *Stercorarius* spp.

Although skua identification can be very difficult, still just ten observers, all regular 'seawatchers', submitted records of unidentified Skuas.

Unidentified skuas, either Pomarine or Arctic: 16 recorded, all during the autumn. Most of the records refer to singles except five at Prawle on 16th *Aug*; monthly breakdown: *Aug* ten, *Sep* two and *Oct* four.

Unidentified skuas, either Long-tailed or Arctic: this category also includes records submitted but not accepted by the Records Committee as Long-tailed Skua. Eleven records, mostly singles, except at Berry Hd, two on 30th *Sep* and three on 5th *Oct*; monthly breakdown: *Jun* one; *Aug* one; *Sep* four; *Oct* five.

Unidentified skuas, not specified: 23 records distributed by months as follows: *Feb* two; *Mar* one; *Apr* two; *Jun* one; *Aug* six; *Sep* five; *Oct* five; *Dec* one.

MEDITERRANEAN GULL *Larus melanocephalus*

Scarce post-breeding and winter visitor; very scarce/rare inland. BoCC (Amber), SPEC(4).

This year saw over 500 records submitted of this species - the most ever - although the estimated total of 117 birds was lower than the 133 in 2000. There were a max of 62 ads, 23 2y, and 44 1y (inc. 12 juvs, as in 2000, and the joint highest annual total). County abundance continues to reflect the increases in both UK (90-109 prs at 28 sites in 2000) and near Continent breeding populations, in passage birds, and in observer interest in ringed birds.

Table 71: Monthly distribution of birds at main sites and county totals (figures do not allow for long-staying individuals):

	J	F	M	A	M	J	J	A	S	O	N	D
Exe Est / Dawlish Bay	9	4	5	1	3	10	8	6	5	2	3	2
Dawlish Warren	*9*	*4*	*3*	-	-	*1*	*7*	*5*	*5*	*2*	*3*	*2*
Bowling Green Marsh	-	-	-	*1*	*3*	*9*	*1*	-	-	*1*	-	-
Plymouth area	1	2	3	1	-	1	2	-	1	3	2	4
Slapton	3	-	1	-	-	-	1	-	-	1	2	2
Taw / Torridge / West. Ho!	1	2	-	-	-	-	12	15	12	7	2	1
Teign Est	3	3	5	5	-	-	-	-	-	-	-	-
Torbay area	4	5	6	1	-	-	7	7	6	5	1	2
All other sites	2	4	2	-	-	-	1	2	1	3	6	7
County total 2000	14	9	13	4	3	6	19	16	16	12	9	12
County total 2001	23	20	22	8	3	11	31	30	25	21	16	18

Maximum counts. The maximum day count was 11 (7 ads, 4 juvs) at *Instow* (a Taw/Torridge Est record count) on 23 Aug (DC,RGM); counts of eight or more were made on this estuary on five dates during 13 Jul-11 Sep, including several ringed birds (see below). On the *Exe*, *c*.58 birds (*cf.* 41 in 2000; 46

in 1999) and the *Torbay* total of *c.*44 (*cf.* 32 in 2000 & 1999). In *Plymouth*, *c.*22 birds reversed the recent decline in sightings (*cf.* 11 in 2000). Max counts included: eight at Bowling Green Marsh (five 1s, one 2s) on 16 Jun (BG); seven (four ads, three 2s) at Three Beaches area, Torbay, on 4 Jul (ML,BMc); five (unaged) at Passage House Inn, Teign Est, on 15 Mar (GJ); and at Dawlish Warren on four dates in Jan (10-23rd) - a probable total of seven birds in various permutations (IL,KRy). Seawatches produced maxima of six at Hope's Nose on 2 Aug (3 ads, 2 juv/1w flew S, and one 1w at outfall) (CJP); and two-five ads flew S past Berry Hd on 7 Oct (ML *et al.*). *Elsewhere,* there were multiple counts of two-four at another 17 widespread coastal sites, including four ads at Axe Est on 18 Nov (BG). *Inland*: an ad bp was at Burrator Res on 11 Feb (MRAB,KEM), and one (2y) at Stover CP on 22 Oct (JDA).

Feeding behaviour and abnormalities. A 1s hawked insects with Hobbies at Exminster on 23 May (MD,AR); and four fed on small fish displaced by Bass, at Westward Ho! on 2 Aug (BG). A 2w at Skern on 31 Oct had no feet (RGM).

Ringed birds. A total of 11 darvic-ringed birds were noted this year (*cf.* 10 in 2000), of which eight were of known origin, and at least two of these returned to the same wintering area. The Colour Ringing Programme is now starting to give interesting results, thanks mainly to greater observer awareness, greater ringing effort and the increasing numbers of birds visiting Devon. The following details only refer to birds whose ring numbers were read.

Returning birds:

- ad **39W** (white-ringed as pullus Solvay, Antwerp, Belgium, 27 Jun 1998) returned to Plymouth for the third consecutive winter and was faithful to the same area, often feeding on handouts from the public; it also remained into 2002 (SRT,VRT);
- ad **11Z** (white-ringed as pullus Solvay, Antwerp, Belgium, 1 Jun 1998) returned to Torbay for the third consecutive winter; it also remained into 2002 (ML,BMc).

Others (history given where known):

- ad **64W** (white-ringed as 3cy Antwerp, Belgium, 25 May 1999; previously white-ringed 37L as pullus Netherlands, 8 Jun 1995) was on Taw Est 17 Jul - 23 Aug (RGM)
- juv **87J** (green-ringed as pullus Brussels, Belgium, 23 May 2001) was on Taw Est 8 Aug - 3 Sep (DC,RM)
- ad **49S** (white-ringed as pullus Antwerp, Belgium, 17 Jun 1998) was on Taw Est 7 Aug - 20 Sep (DC,RM)
- ad **61Y** (white-ringed as 2cy Oost-vlaanderen, Belgium, 8 Jun 1996) was on Taw Est 10-28 Aug (DC,RM)
- juv **2C31** (yellow-ringed as pullus at Langstone Harbour, Sussex, 26 Jun 2001) was at Instow, Taw Est, on 9 Aug (DC,RGM); the first positively identified British bred individual in Devon.
- ad **3A6** (red-ringed as pullus at Szeged-Fehértó, Csongrád, Hungary, 23 Jun 1998) was on Taw Est 16-23 Aug (DC,RGM); only the second positively identified individual from Hungary in Devon.

Record request*. Multiple sightings of long-staying and mobile individuals complicate analysis; this task would be easier, and the estimate more reliable, if all observers provide information on age, plumage features, rings, and time of observation.*

Presumed MEDITERRANEAN GULL / BLACK-HEADED GULL hybrids

Once again, the regular winter ad remained at Preston, Torbay; last seen on 11 Feb. It returned for its sixth winter season on 26 Jul, remained until at least 25 Nov, and was seen roosting off Broadsands during Nov (ML,GJ,MD,AR).

FRANKLIN'S GULL *Larus pipixcan*

Rare vagrant

The 1w found on 16 Dec 2000 was rediscovered at Goodrington on 19 Jan (ML) and regularly observed in Torbay until 17 Mar, then later at the Hope's Nose outfall (WJD) and Teign Est (MKn) from 31 Mar until its departure on 21 Apr. During its stay it utilised the gull roost off Broadsands or Goodrington. Most of its diurnal feeding grounds during its period of stay remained a mystery, but it faithfully returned to the gull roost during late afternoon. Initially, in Torbay, it roosted with Black-headed Gulls, but from mid-Feb it roosted every night with Common Gulls. This was the third record of this Nearctic gull in Devon and the second long-stayer. (The only one in Britain in 2001 accepted by BBRC, so no new arrivals).*(see colour plates)*

Franklin's Gull - (with Black-headed gulls, Broadsands) - *Steve Young*

LITTLE GULL *Larus minutus*

Scarce passage and winter visitor; very scarce/rare inland. SPEC(3).

Perhaps near normal numbers, after last year's autumn influx, with only 49 birds recorded (75% immature). Of this total, 11 appeared in Jan, ten in the Mar-May period, and 26 in Sep-Oct. All but one were recorded from coastal locations, mainly along the S coast from Slapton to Dawlish.

Table 72: Monthly distribution of birds at main sites and county totals (long-staying individuals counted once in each month present):

	J	F	M	A	M	J	J	A	S	O	N	D
Exe Est / Dawlish Bay	2	1	3	2	1	1	1	-	1	5	3	1
Slapton / Beesands	2	-	2	-	-	-	-	-	3	1	-	-
Torbay area	2	-	1	-	-	-	-	-	1	5	1	-
Plymouth area	3	-	1	-	-	-	-	-	-	1	-	-
All other sites	-	-	-	1	-	-	-	-	2	2	-	-
County total 2000	14	9	13	4	3	6	19	16	16	12	9	12
County total 2001	9	1	7	3	1	1	1	-	7	14	4	1

First winter-period. A small influx in Jan with three in Plymouth (inc 1 ad), and singles at Goodrington (1w) and Dawlish Warren on 1 Jan, an ad at Broadsands on 5 Jan, and further single imms at Start Bay, Dawlish Warren, and Hope's Nose, with two at Slapton (ad w and 1w) on 23 Jan. During Feb, several records of a 1w around Exe Est probably referred to the same bird.

Spring/summer. In Mar, an ad wp was at Starcross on 8th; 1w birds were at Dawlish Warren, Hope's Nose, Topsham (26 Mar-9 Apr); and an ad in poor condition on Plym Est on 30th. A 1s was on the Teign Est on 11 Apr, and two-three individuals, including a 2s, were at Dawlish Warren 22-27 Apr. A 1s *summered* at Exe Est, mainly reported from Bowling Green Marsh 28 Apr - 30 Jul.

Autumn. Unusually there were no Aug records. Up to three juvs were at Slapton Ley 5-9 Sep and a 1w was at Dawlish on 11th. Other singles were at Skern (1w) on 14 Sep, Otter Est (ad) on 23rd, Slapton (1w) on 26th, and Hope's Nose (juv) on 30th. Four flew S past Berry Hd on 1 Oct with another 1w at Hope's Nose later the same day. During an unsettled start to Oct at least two 1w were in Plymouth 1-8th, another was at Exminster Marshes on 3rd, and one at Lower Tamar L on 13th was the *only inland record.* During late Oct: two ads at Dawlish Warren on 18th, with three (ad, 1w, 2w) there on 26th; and one at Bideford on 21st. During Nov, three ads were in Dawlish Bay 14-20th, and one utilised the gull roost in Torbay on 24th.

Second winter-period. Just a single record of one at Dawlish Warren on 5-10 Dec.

SABINE'S GULL *Larus sabini*

Very scarce autumn passage visitor; rare at other times.

Another year with lower numbers; only six to eight birds recorded. The records this year were split between three areas involving a single long-staying bird at Hope's Nose. Another juv at Westward Ho! and a storm-driven influx of up to five individuals at Plymouth. The running total since 1994 now stands at 62-64, Torbay continuing to be the favoured location with 40-41 birds, or 66% of the total records.

The first sighting was a 1s S past Berry Hd on 17 Jul (ML,MD); then a juv flew past Westward Ho! on 30 Aug (RGM), and another at Hope's Nose on 30 Sep (LC,CJP) *(see front cover)* lingered in the area, feeding at the outfall, until 8 Oct. On 7-8 Oct there were three-five birds in the Plymouth area: during severe SW gales, a juv sheltered in Cattewater, Plymouth (SRT,VRT); it with another, or two different juvs, were in Plymouth Sound on 8 Oct (SGe); and one, possibly two, ads, were also seen in Plymouth Sound on 8 Oct (VRT). These birds more than doubled the tally for this locality since 1950 (Tucker, 1995).

Sabine's Gull - *Ren Hathway*

BONAPARTE'S GULL *Larus philadelphia*

Rare vagrant

A 1s bird ranged widely around the Exe Est 19-23 May (DS *et al.*). The bird frequented Bowling Green Marsh but was believed to roost out at sea with a very large flock of Black-headed Gulls during its stay. This was the sixteenth Devon record of this Nearctic species. (One of six in Britain in 2001 accepted by BBRC.)

BLACK-HEADED GULL *Larus ridibundus*

Numerous/abundant winter visitor, not scarce in spring and summer; former breeder (last bred 1961). BoCC (Amber).

Similar WeBS county totals and site counts to 2000 for the winter months.

Table 73: Monthly maxima at main sites (recording >500) and WeBS county totals:

	J	F	M	A	M	J	J	A	S	O	N	D
Axe Est	260	585	-	-	-	-	176	169	246	115	237	180
Dart Est	148	350	-	-	-	-	-	-	708	-	508	88
Exe Est	1000	600	-	-	-	-	-	-	-	-	-	-
Bowling Green Marsh	-	-	-	-	*244*	*1000*	*2000*	*1200*	*300*	*8*	*7*	*3*
Kingsbridge Est	1391	263	-	-	-	-	-	1827	1825	3006	2420	890
Otter Est	760	568	-	-	-	-	107	175	375	467	138	685
Plym Est	200	700	-	510	52	56	715	-	400	-	906	-
Tamar Complex WeBS	4832	2246	-	-	-	-	-	-	4729	2763	5453	2838
Upper / mid Tamar	*1282*	*967*	-	-	-	-	-	-	*1153*	*1040*	*751*	*1589*
Tavy Est	*1200*	*600*	-	-	-	-	-	-	*500*	*500*	*350*	*1300*
Taw / Torridge Est	2109	1053	-	-	-	-	-	5000	1582	1478	1567	2260
Torbay	-	5000	-	-	12	-	-	-	-	115	1500	-
WeBS county total '00	8849	8272	1209	422	232	947	6345	10238	7157	7804	7216	8065
WeBS county total '01	8175	5051	51	5	-	11	400	2359	7331	7443	8445	6945

Maxima at principal sites (≥ 500). As usual, the highest counts came from the large estuaries and Torbay (but no counts of the customary winter roost on the upper Exe Est since 1999 when it held over 11,000 birds, and not included in Exe Est WeBS counts), the max being 5,000 at Broadsands on 5 Feb (DIJ). In addition to the tabulated sites, high counts were at: Slapton Ley, 3,000 in Jan; Braunton Burrows, 1,500 anting in flight on 14 Aug; Burgh Island, 1,000 on sea in Dec; Exminster Marshes, 970 in Aug; Saltram Tidal Pool, 893 in Aug; Stover CP, 800 in Dec; Riverside CP, 749 in Mar; Thurlestone Bay, 750 in Dec; Weare Giffard, 648 in Dec; Trenchford Res, 570 in Aug; and Roadford Res, 500 in Dec.

Elsewhere. *Sites and subsites with ≥ 200*: Dawlish Warren, 400 in Aug; Teign Est, 381 in Aug; Avon Est, 350 in Nov & Dec; Lower Tamar L, 350 in Jan; R Caen, 278 in Jul; Erme Est, 260 in Jan; Thurlestone Valley, 250 in Apr; Upper Tamar L, 220 in Jan; and West Charleton Marsh, 200 in Feb. *Sites with 100-199*: Beesands Ley; Clennon Valley; Shobrooke L; South Huish Marsh; Wembury; and the Yealm Est. *Less than 100* recorded at several other sites, inc. Fernworthy Res with 23 in Dec (*cf.* a max of nine in 2000).

Juveniles. The first juv was at Bowling Green Marsh on 20 Jun, followed by other Jun records at Berry Hd, Dawlish Warren, Saltram Tidal Pool and the Taw/Torridge Est. A useful count on the R Caen on 31 Jul gave 71 juvs in a total flock of 278.

Aberrant and ringed individuals. A leucistic bird returning to the Torbay area for its eighth year was present until 22 Mar and from 25 Nov (mo); recorded from Clennon Valley and Broadsands, and also from Stoke Gabriel on 29 Jan. Three colour-ringed birds on the Plym Est on 1 Nov were from Sjaelland, Copenhagen (TF), and one ringed in Plymouth in 1996 was seen nr Copenhagen in Mar 2001 (see Ringing Report for further details). An ad at Instow on 23 Aug had been ringed as a pullus in Germany (Nordrhein-Westfalen) in 1999 (DC,RGM).

RING-BILLED GULL *Larus delawarensis*

Very scarce winter and spring visitor/vagrant, rare at other times.

A good year with 15 individuals, again including the returning adult at Barnstaple.

First winter-period/spring. The annually wintering ad at *Barnstaple* entered its twelfth calendar year of residence on Taw/Torridge Est, and was last recorded on 18 Mar (JEW,MSS,DC). Two birds, a long-staying 1w (KGB,RD) and a 2w (DC,MAS *et al.*) were at *Northam Burrows* on 1 Jan, but only the 2w was subsequently reported, until 24 Feb (IM,RGM). An ad at the same locality on 30 Jan (RGr) may have been the Barnstaple bird. In the south, an ad roosted at *Dawlish Warren* on 10 Jan (JEF,KRy), and was reported subsequently until 5 Mar (BG). A 1w was at *Riverside CP, Exeter*, on 8 Mar (JRD). At *Torbay*, another 1w roosted with Common Gulls on 23 Feb (ML,GJ). On the *Teign Est*, two ads were at Coombe Cellars on 25 Mar (MKn,JEF); one remained until 2 April, the other until 24th (JEF) *(see col plate)*.

Autumn/second winter-period. Records from three localities included the returning ad at *Barnstaple* beginning its twelfth season from 19 Aug to the year end (JEW,MSS,DC); and in the Plymouth area, an ad and a 1w were on the *Plym Est* on 12 Dec (TF).

COMMON GULL *Larus canus* — Mew Gull

Fairly numerous winter visitor and passage migrant; scarce in summer. SPEC(2), BoCC(Amber

Much lower second winter-period max for the Exe Est, and some higher first winter counts compared to 2000, but otherwise a similar pattern in terms of numbers and sites.

Table 74: Monthly maxima at main/well-counted sites and WeBS county totals:

	J	F	M	A	M	J	J	A	S	O	N	D
Avon Est	9	-	17	-	-	-	2	-	-	4	8	12
Axe Est	2	3	-	-	-	-	-	-	-	-	26	12
Exe Est (mostly Dawlish Warren)	100	39	92	30	4	1	3	30	39	63	42	65
Kingsbridge Est	12	10	-	-	-	-	-	17	-	14	31	28
Portworthy Dam	330	121	-	-	-	-	-	-	-	-	98	12
Saunton Sands	142	207	80	10	-	-	16	18	8	29	-	-
Tamar Complex WeBS	175	136	-	-	-	-	-	-	4	17	42	25
Tavy Est	*2*	*7*	-	-	-	-	-	-	-	*2*	*24*	*8*
Taw / Torridge Est	12	19	-	20	-	1	4	-	24	38	44	24
Teign Est	21	20	280	40	-	-	-	-	-	-	-	16
Torbay	350	500	9	-	-	-	-	-	-	23	80	-
WeBS county total '00	361	161	43	4	-	-	1	1	5	33	100	170
WeBS county total '01	384	167	-	-	-	-	-	17	24	58	187	206

First winter-period. The highest count of the period (and year) was 1,000 at Slapton Ley on 12 Jan (PSa), but as usual, Torbay roosts also produced high counts, with a max of 673 at Broadsands on 6 Feb (GJ); substantially higher than the max of 450 in 2000, but WeBS county totals only slightly higher. Apart from the above sites, high counts also occurred at Portworthy Dam and Saunton Sands (newly tabulated) and there was the usual spring build-up on the Teign Est with 280 on 24 Mar and 250 on 26th. In addition to the tabulated counts, 210 flew S past Dawlish Warren on 8 Mar, and a max of 40+ were at Hope's Nose on 29 Mar. All other counts were <10 with site maxima of: eight at Prawle & Erme Est; seven Stoke Gabriel & Stover CP; six at Upper Tamar L; and five at South Huish and Yealm Est. Only singles occurred at Roadford Res, Shobrooke L & (surprisingly) the Plym Est.

Autumn/second winter-period. One or two imm birds persisted on the Exe (Bowling Green Marsh & Dawlish Warren) and Taw/Torridge (Skern) Ests, May-Jul, but probably the first returns were single ads at Saunton Sands on 7 Jul and at Dawlish Warren on 10th. WeBS county totals for Aug-Dec were higher than in 2000, but maxima for the main sites were lower, only slightly for Portworthy Dam and Torbay, but substantially so for the Exe Est where 414 were counted in 2000, and 706 in 1999. Elsewhere the only counts received were: *c.*100 at Man Sands on 29 Dec; 22 at Huntsham on 14 Dec; a max of eight at Fernworthy Res; three in a field at Holne on 16 Dec and two at Burrator Res on 12 Dec. Again, the Plym only produced a single.

LESSER BLACK-BACKED GULL *Larus fuscus*

L. f. graellsii* (western race): *not scarce migrant breeder, passage and winter visitor. SPEC(4), BoCC (Amber).

*(The main part of text refers to **L. f. graellsii**, or those not identified to subspecies.) The Seabird 2000 census confirms Lundy as the main breeding site, with most of the 20 mainland prs on Exeter roofs. Portworthy Dam and Roadford Res continue to hold nationally important numbers.*

Table 75: Monthly maxima at main/well-counted sites and WeBS county totals:

	J	F	M	A	M	J	J	A	S	O	N	D
Avon Est	36	60	8	19	-	2	7	5	42	36	86	200
Erme Est	11	25	-	-	-	-	-	-	10	13	75	21
Exe Est WeBS	-	5	-	-	-	-	-	8	1	1	-	-
Dawlish Warren	*6*	*11*	*20*	*6*	*2*	*3*	*5*	*6*	*11*	*10*	*3*	-
Kingsbridge Est	10	19	9	-	-	-	-	-	11	3	7	-
Plym Est	-	-	98	13	6	7	-	-	3	1	4	1

continued	J	F	M	A	M	J	J	A	S	O	N	D
Portworthy Dam	105	3	-	-	-	18	150	**900**	**1000**	**855**	**2000**	100
Roadford Res	100	9	-	-	-	-	-	-	-	31	50	**2650**
Tamar Complex WeBS	34	7	-	-	-	-	-	-	21	56	2	1
Tavy Est	-	-	-	-	-	-	-	-	-	*2*	*1*	*2*
Taw / Torridge Est	7	40	-	-	-	-	2	5	3	50	3	1
Teign Est	2	-	62	-	1	-	-	-	-	1	-	-
Thurlestone area	10	12	-	-	-	-	-	-	-	-	2	-
WeBS county total '00	103	115	87	119	81	118	314	791	397	238	266	584
WeBS county total '01	195	197	98	7	-	20	160	919	1075	908	2168	184

Breeding. The *Seabird 2000* census recorded 426 prs for 1999-2001. Of these, 406 occurred on Lundy (*cf.* 443 prs counted in 2000) where the main increase from the all time low of 40 prs in the 1950s appears to have occurred in the last two decades, since the cessation of control measures (DJP). Twenty mainland prs comprised 15 from Exeter (roof sites) and single prs at Burgh Island, Dartmouth, Hope's Nose, Start Pt and Torquay. In 2001, at least 20 prs nested on roofs in an Exeter industrial estate (RAda), one at Prawle (nest destroyed) and two on Burgh Island.

Maxima at principal sites (>100). Compared to 2000, the WeBS county totals were higher for most months, and in the second winter-period were mostly higher than in 1999, mainly due to some high autumn counts from Portworthy Dam with a max of 2,000 (*cf.* 750 in 2000). Once again, however, the peak count was at Roadford Res with 2,650 on 8 Dec (*cf.* 4-5,000 in 2000) (GAV,JWV). This site continues to qualify as one of *national importance* (given a generally accepted threshold of 500), yet the latest WeBS report (Musgrove *et al.*, 2001) lists *Portworthy Dam*, with a mean of WeBS maxima for the last five years of 1,043, as the only nationally important Devon site. The only other sites with >100 were: Slapton Ley, with 500 on 12 Jan; the Avon Est, with 200 in Dec; Slapton Sands, with 100+ feeding on Sprat on 23 Aug with other gulls; and Nether Exe, with 125 on 9 Aug.

Elsewhere. On 11 Jan, 22 flew over Okehampton. In Mar, 28, the Clennon Valley max for the year occurred on 6th, and there were 35 in Thurlestone Valley on the 27th. It would be interesting to know whether the 61 (mostly ads, bathing in river) at Stafford's Bridge on 4 May and 82 at Stoke Canon on 14th (DJJ) involved any of the Exeter breeding population. High Dartmoor counts included 79 in a silage field nr Cox Tor on 25 Jul, and 75 W over Mary Tavy on 7 Sep, a max for Fernworthy Res of 20 on 21 Sep and 30 were in a field at Moorshop, Tavistock on 14 Nov. Elsewhere: 60 at Bampton Down on 22 Sep; 50 S over Dawlish Warren on 28 Oct; a Prawle max of 35 in Nov; and 32 at Broadsands on 3 Dec, apparently the max for Torbay.

L. f. intermedius* (southern Scandinavian race) *(scarce passage and very scarce winter visitor).

*(Not as black above as those of the nominate race **fuscus,** these birds are nevertheless of a darker grey above than those of the **graellsii** race normally found in most of Western Europe.) Only four birds recorded, compared with over forty in 2000.*

Four singles: one on the Teign Est on 31 Jan (BNR), perhaps the same as an *intermedius/fuscus* there on 10 Feb (MKn); one in the Clennon Valley on 6 & 11 Mar (ML,MRAB); and singles at Skern on 31 Jul and 21 Nov (ArSt).

HERRING GULL *Larus argentatus*

L. a. argenteus* (British race): *fairly numerous resident breeder, migrant status uncertain. BoCC (Amber).

*(The main part of text refers to **L. a. argenteus** or those not identified to subspecies) The Seabird 2000 census has confirmed that unlike most seabirds in Devon, most breed on the mainland, 50% of them on roofs. Not regularly counted at all sites, but WeBS county totals generally higher than in 2000, and some large numbers associated with feeding bonanzas.*

Table 76: Monthly maxima at main/well-counted sites and WeBS county totals:

	J	F	M	A	M	J	J	A	S	O	N	D
Avon Est	450	400	350	120	150	78	150	600	350	800	550	400
Axe Est	200	400	-	-	-	-	116	122	720	263	325	320

Erme Est	250	37	-	-	-	-	-	-	100	204	28	30
Dart Est	16	-	-	-	-	-	-	-	41	-	50	14
Exe Est (Dawl. Warren only)	1000	-	600	236	1450	930	1500	3000	3000	-	-	-
Kingsbridge Est	175	239	-	-	-	-	-	5000	2700	1357	339	305
Otter Est	540	469	-	176	200	316	386	528	652	690	153	765
Plym Est	1000	300	-	2450	850	2400	630	-	600	-	55	-
Portworthy Dam	92	9	-	-	-	15	1000	900	2000	950	1000	110
Slapton / Beesands Leys	1000	41	-	-	81	-	-	-	-	-	-	-
Tamar Complex WeBS	409	100	-	-	-	-	-	-	1052	394	353	307
Upper / mid Tamar	*23*	*18*	-	-	-	-	-	-	*168*	*20*	*31*	*102*
Tavy Est	*17*	*18*	-	-	-	-	-	-	*101*	*46*	*38*	*47*
Taw / Torridge Est	168	91	-	-	-	-	-	-	171	59	196	681
Teign Est	-	-	1000	350	-	-	650	174	15	750	-	-
Wembury	-	135	-	300	36	50	90	55	45	200	470	-
Yealm Est	180	204	-	-	-	-	-	-	138	150	325	130
WeBS county total '00	1593	3270	2006	2076	1484	2344	1718	340	2412	3006	2404	2625
WeBS county total '01	3094	2273	17	176	202	409	1652	2556	7953	4560	3662	2951

Breeding. The *Seabird 2000* census revealed a total of 3,839 prs for 1999-2001, of which about 20% (777) were on Lundy. On the mainland, the 3,062 prs were recorded from all coastal sectors (see separate article), with Torbay the main area (with 1,257) and Thatcher Rock (190) and Ore Stone (120) the largest colonies. Just over 50% of the mainland birds (1,695 prs) nested on the roofs of 32 towns and villages, with the largest concentration in Torbay (747 prs) including c.200 each in Brixham, Paignton and Torbay, and only ten in N Devon (all at Ilfracombe). In *2001*, however, two prs were noted on roofs at both Bideford and Northam; there were also 20 at Starcross (*cf.* eight in *Seabird 2000*). At Prawle, 30 prs produced only two chicks as a result of nest destruction; Lesser & Great Black-backed Gulls were also affected, and the Police Wildlife Liaison Officer informed (PMM). An unusual nest site in the Clennon Valley was at the top of an ivy covered pine stump about 12m high and in the middle of a wooded area (ML).

Maximum counts. The WeBS county totals were generally higher than in 2000 in Jan and Aug-Dec, but as usual, this species is not regularly counted at some important WeBS sites (eg. Exe Est) and in coastal areas such as Torbay. On the Exe Est, the tabulated counts are almost entirely derived from the monthly maxima at Dawlish Warren. The highest count of the year was *c.*5,000 feeding on Sprat in the Kingsbridge Est on 29 Aug, and at least 1,000 were also recorded from Dawlish Warren, Plym Est, Portworthy Dam, Tamar Complex and Teign Est among the tabulated sites. *Elsewhere* the highest counts were: 1,510 roosting at Trenchford Res on 5 Aug; 1,000+ flushed off stock fields at Staddon Heights just before dawn on 25 Oct; *c.*1,000 on Slapton Beach on 23 Aug and *c.*1,000 on Slapton Ley in Jan; and 700 bathing on Wistlandpound Res on 3 Sep.

Aberrant and ringed individuals. A leucistic individual on the Plym Est throughout most of the year (TM *et al.).* Colour ringed birds: an ad on Dawlish seafront on 1 Jul had been ringed at Heathfield landfill site on 15 Nov 2000 as an ad (DJG); nine birds on the Plym Est in autumn and winter were traced to Suffolk (one, probably), Guernsey (three 1w and one 2w), Jersey (3w) and Bristol (two 1w and one 2w) (TF).

L. a. michahellis* '(Western)Yellow-legged Gull'** ***(scarce visitor mainly in autumn):
Now regarded by many authorities (e.g. British Birds) as a separate species, these birds are still not yet split by the BOU. This year there were no submitted records from Portworthy Dam, the main location in recent years. Fewer records than in 2000, after several years of increase.

Jan-Mar. Four birds: two ads at Langston, Kingston on 9 Jan (ASCB); another at Bowling Green Marsh on 6 Feb (MSW); and a 1y at Dawlish Warren on 24 Mar (TM).

Apr-Jun. Four birds: a 3y on the Teign Est on 20 Apr (MKn); a 3s at Bowling Green Marsh on 30 Apr (MSW); an ad at Topsham on 2 Jun (MKn); and one at Dawlish Warren on 6th (DHa).

Jul-Sep. A slight influx in Jul with six birds, ads unless otherwise stated: one at Hope's Nose on 17 Jul (CJP); one at Aveton Gifford 21 Jul - 9 Sep (JCN,RWBo); a 2s and a juv at Upper Tamar L on 27 Jul, the former staying until 3 Sep and moulting into ad plumage (DC,RD,RGM); one at Topsham on 30 Jul (MKn); and another at Skern on 31st (ArSt). Apart from the two lingerers, the only other was an ad on the Plym Est on 10 Sep (BG).

Oct-Dec. At least another eight birds: on 18 Oct, one ad on the Plym Est (BG) and three at Slapton Sands (NCW), with one there on 27th (PSa); in Nov, one at Prawle on 11th (TM), and an ad at Horsely Cove on 17th (JCN); and in Dec, the same or another at Prawle on 11th & 18th (JLt), an ad at Hope Cove on 19th (BHS) and a 1w at Sutton Hbr, Plymouth on 20th (BG).

L.a. smithsonianus* ('American' Herring Gull). *Rare vagrant.
1998 addition. A 1s on the Plym Est on 26 April 1998 (M K Ahmed) was a first record for Devon. (Only the seventh in Britain accepted by BBRC, but several others are under consideration.)

ICELAND GULL *Larus glaucoides*

L. g. glaucoides: very scarce winter visitor; rare in summer.

Another average year with eight to eleven imm birds reported during the first winter period, all except one were first years, and another four imms in the second winter period.

First winter-period/spring. First-year individuals were recorded at Coombe Cellars, Teign Est, on 14 Jan (BG), Broadsands 25-29 Jan (ML), Instow on 8 Feb (RGM), Plym Est on 10 Mar (MKn), and Dawlish Warren 24 Mar - 10 Apr (JEF,KRy). On 9 Apr, two 1s birds (MRAB,ML) were at Hope's Nose (one may have been the same as Dawlish Warren), with one present again on 10 Apr. There was a second multiple occurrence at Slapton Ley; a 1s present on 14 Apr was joined by a 2s bird 15-18 Apr (PBo,JCN *et al.*). Further individuals were at Hope Cove on 16 Apr (JHB) and Clennon Valley, Paignton, on 18 Apr (ML). An accurate assessment of the total number of birds present was complicated by the probable movement of individuals along the coast between Torbay and Dawlish Warren, and a leucistic Herring Gull, which was probably responsible for several spring records of 'Iceland Gull' on the Plym Est.

Second winter-period. A worn 1w was in the Torbay area from 29 Nov - 5 Dec (ML,MD,AR), and this or another 1w was at Man Sands on 2 Dec (RBe). Finally, a 1w was at Landcross on 15 Dec (DC).

2000 correction. Delete the 2w Plym Est bird, 8 Nov – 30 Dec; this was the leucistic Herring Gull.

GLAUCOUS GULL *Larus hyperboreus*

Very scarce winter visitor, rare in summer.

A below average showing with eight to nine individuals, including up to two ads, all Jan-Apr.

First winter-period/spring. The first was a 1w at Slapton Ley on 12 Jan (PSa), which was also seen at Beesands on 14 Jan. An ad then frequented Cattewater, Plymouth, 16-20 Jan (PHA,SRT,VRT), and it, or another ad, was on the Chelson landfill on 24 Mar (MKn). Other 1w birds were at: Torbay 11-18 Feb (ML,GJ); Dawlish Warren on 11 Feb (IL); South Huish Marsh on 25 Feb (PAS); and Passage House Inn, Teign Est, on 9 Mar (JWn). A 1s was on the Plym Est (TF,SRT,VRT) on 21 Mar. Presumably, both Mar individuals stayed into Apr; the former was seen feeding at sea off Dawlish Warren on 29 Mar (DC,RD) and later frequented the Teign Est until 16 Apr, and the latter was last reported over Mt Gould, Plymouth, on 11 Apr.

GREAT BLACK-BACKED GULL *Larus marinus*

Not scarce resident breeder and numerous winter visitor. SPEC(4).

Some higher counts than in 2000 with an indication of at least local movements, and nationally important numbers on the Exe Est, Plymouth area and Slapton.

Table 77: Monthly maxima at main/well-counted sites and WeBS county totals:

	J	F	M	A	M	J	J	A	S	O	N	D
Avon Est	13	12	1	1	3	1	3	3	12	13	12	20
Axe Est	11	5	-	-	-	-	16	1	26	11	17	14
Erme Est	55	150	-	-	-	-	-	-	21	218	46	29
Exe Est	230	-	-	-	-	-	-	311	**510**	240	64	19
Dawlish Warren	*150*	-	*30*	*20*	*104*	*25*	*72*	*319*	*634*	*657*	*145*	*150*
Kingsbridge Est	20	23	-	-	-	-	-	19	26	22	42	95
Otter Est	4	4	-	230	5	11	-	3	9	78	-	-
Plym Est	19	56	9	60	38	135	67	-	133	70	17	**1000**
Tamar Complex WeBS	31	12	-	-	-	-	-	-	31	29	29	18

Upper / mid Tamar	*7*	*7*	-	-	-	-	-	-	*11*	*3*	*8*	*10*
Tavy Est	*8*	*9*	-	-	-	-	-	-	*10*	*8*	*21*	*11*
Taw / Torridge Est	100	37	-	-	-	25	25	-	80	45	47	58
Teign Est	-	6	18	31	-	-	-	-	-	32	-	-
Wembury	-	44	-	-	37	36	50	49	230	4	-	220
Yealm Est	11	57	-	-	-	-	-	-	20	14	13	30
WeBS county total '00	109	156	87	130	91	156	136	227	147	577	287	410
WeBS county total '01	183	285	-	5	5	12	22	339	636	523	278	291

Breeding. According to the *Seabird 2000* census (see separate article) there were 163 prs distributed between Lundy (34) and the mainland (129) in 1999-2001; thus, as with Herring Gull (but in contrast to Lesser Black-backed Gull), Lundy holds about 20% of the Devon breeding population). On the mainland, there were only eight N coast prs, and on the S coast only seven were outside the Plymouth-Prawle and Torbay sectors. The largest colonies were on Thatcher Rock (40) and Wembury Mewstone (25). In 2001, single prs reported from Burgh Island, Kingswear (very exposed flat roof), Plymouth Mt Gould Hospital (copulation on roof, 4 Apr) and Prawle (pr nested but chicks destroyed).

Maximum counts. The WeBS county totals for Jan & Feb, Aug & Sep were higher than in 2000, especially in Sep as a result of a high count on the Exe Est. The criterion for *national importance* (400) was achieved at Dawlish Warren for the tenth year in succession, and also on the Plym Est in Dec with the largest count of the year (1,000) described as 'a sudden and unusual arrival' (SRT,VRT); there were also 500 at Slapton Ley on 2 Jan and 400 on the Plymouth Breakwater at dawn on 17 Sep. The only other untabulated counts >100 were: 300 at Slapton Sands on 9 Jan (the max there), and 120 on 17 Oct; 270 at Start Pt on 12 Oct, and 260 there on 11 Sep; 151 on the Ore Stone on 30 Dec; 150 at Bolt Tail on 4 Dec; and 110 at Roundham Hd (the max for Torbay) on 7 Oct. *Inland,* max counts of two at Roadford Res on 27 Apr and one at Fernworthy Res on 16 Nov, are a reminder of the scarcity of this species away from the coasts.

KITTIWAKE *Rissa tridactyla* — Black-legged Kittiwake

Fairly numerous migrant breeder and passage/winter visitor; very scarce/rare inland. BoCC (Amber).

Scant breeding data broadly support the results of the Seabird 2000 census. Breeding again unsuccessful at Berry Hd, but not at Hallsands. Lower numbers recorded at sea than in 2000, particularly on the N coast.

Table 78: Monthly maxima for coastal areas:

	J	F	M	A	M	J	J	A	S	O	N	D
N coast	24	15	-	-	11	40	-	40	45	11	100	-
S coast (W of Dart)	-	-	-	-	76	483	350	52	-	9	77	500
S coast (E of Dart)	50	1	150	170	100	200	160	50	330	728	212	47

Breeding. The *Seabird 2000* census gave a county total of 964 prs, 237 (25%) on Lundy and 727 (75%) on the mainland. Only 29 prs on the N coast: 22 at Wringapeak, five at Woody Bay and two at Wringcliff Bay. The 698 S coast prs were at: Hallsands (495), Straight Pt (165), Berry Hd (27) and Ore Stone (11). *Counts in 2001* broadly supported the census results: on Lundy, 200+ aon in colony on 13 May (RJJ); 164 aon at Straight Pt in May (and 50 on cliffs there in Feb)(MKD); a max count of 61 birds and 26 aons at Berry Hd, but again no successful breeding (BHNNR); and 3-400 birds at Hallsands, with perhaps 200 juvs, in Aug. Evidence of Carrion Crow predation at Hallsands: seen removing object from nesting area, and broken egg shells found under a tree about 500m away.

Maximum counts (non-breeding). Counts, some presumably feeding parties of breeding birds, from Dawlish Warren (max 200+ in Jun), Prawle (max 483 in Jun) and Hartland Pt (max 100 in Nov), mainly used to compile the tabulated data. In addition, there were counts from Berry Hd (hourly rates of 330 on 30 Sep, and 728 on 1 Oct), Hope's Nose (212 in three hours on 17 Oct), Slapton (100+ at sea on 18 Aug, and a max of 40 on the Ley on 8 Jun) and Westward Ho! (24 in Jan, 15 in Aug).

SANDWICH TERN *Sterna sandvicensis*

Not scarce non-breeding/passage visitor; rare in winter. SPEC(2), BoCC (Amber).

Exe Est/Dawlish Warren counts again exceeded the criterion for national importance (200); the spring peak was lower than in 2000 and the autumn peak higher.

Table 79: Monthly maxima at main/well-counted sites and WeBS county totals:

	J	F	M	A	M	J	J	A	S	O	N	D
Exe Est (mostly Dawl.Warren)	-	-	13	40	47	34	374	300	335	35	-	-
Bowling Green Marsh	-	-	-	*1*	*20*	*30*	*30*	*3*	*6*	-	-	-
Kingsbridge Est	-	-	-	-	6	-	7	18	-	1	-	-
Plymouth Sound	-	-	2	5	2	-	-	23	-	3	-	-
Prawle area	-	-	-	3	1	5	9	14	3	5	-	-
Start Bay	-	-	3	8	7	-	9	50	6	8	-	-
Taw / Torridge Est	-	-	-	-	3	1	11	16	5	12	-	-
Teign Est	-	-	-	-	15	6	-	8	-	1	-	-
Torbay	-	2	8	59	10	1	19	140	42	76	-	-
WeBS county total '00	-	-	-	30	22	15	210	96	74	2	-	-
WeBS county total '01	-	-	-	5	6	-	-	162	85	2	-	-

Feb-Mar. The *first arrivals* were very early with an ad w and fw at Broadsands on 12 Feb (ML). In *Mar* there were almost daily records at Dawlish Warren starting with one on 1 Mar, but 13 on 22nd was the only double-figure count.

Apr-Jun. In *Apr* there were ten double-figure counts, including: 18 off Hope's Nose and 59 at Three Beaches on 1st; 10 at Hope's Nose on 12th; 16 there on 22nd; *c*.30 at Goodrington on 23rd; and 40 at Dawlish Warren on 27th. Ten double-figure *May* counts included 47 at Dawlish Warren on 19th, 15 at Teignmouth on 10th and 10 at Broadsands on 11th. *Jun* double-figure counts numbered 12, all from the Exe Est.

Jul-Sep. Double-figure counts on the Exe Est alone increased to 22 in *Jul* including the annual max of 374 on 30th (IL); elsewhere, 13 were at Hope's Nose on 17th, 10 at Berry Head on 21st and 11 at Skern on 30th. In *Aug*, the largest counts away from the Exe Est (which peaked at 300 on 7th) were: 13 at Skern on 1st; 25 at Hope's Nose on 2nd; 50 at Berry Head, 30 at Preston & 12 at Prawle Pt on 7th; 10 at Prawle on 8th; 61 S at Hope's Nose on 12th and 57 S there on 14th; 72 S at Berry Head & 140 S at Hope's Nose on 18th; 25 at Torcross on 21st; 23 in Plymouth Sound on 23rd; and 10+ on the Kingsbridge Est on 28th, with 18 there the next day. Among the 18 double-figure counts in *Sep*, 42 were at Preston on 2nd and 22 flew S at Hope's Nose on 29th.

Oct-Nov. *Oct* most of the 11 double-figure counts came from headlands rather than the Exe Est: 15 at Hollicombe Head on 1st; 12 W at Westward Ho! on 2nd; 19 at Preston on 3rd; 60 at Berry Hd on 5th; 76 S at Hope's Nose on 6th; 29 S at Berry Hd & 21 S at Hope's Nose on 7th; and 26 at Berry Hd on 8th. The last in Oct was one at Dawlish Warren on 28 Oct, and the last of the year were two in Thurlestone Bay on 20 Nov (VRT,SRT), the only Nov record.

ROSEATE TERN *Sterna dougallii*

Very scarce passage visitor. SPEC(3), BoCC (Red), UK BAP.

The mouth of the Exe Est is Devon's principal site for this species, with occasional coastal records from other parts of the county. There was an excellent series of summer records from Dawlish Warren.

Apr-Jun. The *first arrival* was a single at Dawlish Warren on 1 May (KRy), followed by nine further records from there until 22 May, with peak counts of three on 15th and 18th and four on 19th (IL,KRy,LC). Elsewhere an ad was off Slapton Sands on 13 May (PSa).

Jul-Oct. There was a remarkable run of almost daily records from Dawlish Warren from 4 Jul to 15 Aug (40 records); careful observation of ages, moult patterns and rings by JEF suggested that ten birds were involved. (see Box on final page and drawings on inside rear cover). There were peak counts of six on 10 Jul, and five (three ad s, one ad w and a 1s) on 22 Jul (JEF,LC). Elsewhere, there were: three at Bowling Green Marsh on 4 Jul (MSW) and an ad there on 7th (BBH); an ad at Berry Head on 17th (ML,MD); and on 8 Aug, an ad at Prawle (PMM), the only Aug record. Then none until the *last*, an ad at Hope's Nose on 19 Oct (CJP).

COMMON TERN *Sterna hirundo*

Not scarce passage visitor; casual breeder (last bred 1978).

Table 80: Maximum monthly counts from main/well-counted sites:

	J	F	M	A	M	J	J	A	S	O	N	D
Dawlish Warren / Bay	-	-	-	30	40	16	20	160	15	15	-	-
Start Bay	-	-	-	-	30	-	1	-	2	1	-	-
Torbay	-	-	-	-	2	25	35	155	8	15	-	-
Plymouth Sound	-	-	-	-	-	-	-	-	3	4	-	-
Taw / Torridge Est	-	-	-	-	13	-	1	40	1	-	-	-

Apr-Jun. The *first arrival* was one at Dawlish Warren on 13 Apr (JEF). There were six *Apr* records (peak counts: six at Dawlish Warren on 17th, 30 there on 24th and six at Starcross on 29th). Of twenty nine *May* records, 18 were from Dawlish Warren, including eight double-figure counts, with a max of 40 on 17-18th. Peak May counts away from Dawlish Warren were 18 at Prawle and 23 at Thurlestone on 12th, and 22 at Thurlestone, and 13 from the Lundy Crossing, on 13th. Among 13 *Jun* records, peak counts were 14 on 5th & 16 on 17th at Dawlish Warren, and 25 at Berry Head on 29th.

Jul-Sep. In *Jul* there were 37 records (24 from Dawlish Warren) with eight counts between 10 and 20 throughout the month at Dawlish Warren, and 35 passed S at Berry Hd on 17 Jul. There were 48 records in *Aug* (14 from Dawlish Warren, nine from Skern and nine from Instow) with 13 counts of ten or more. Peak counts included 155 past Berry Hd and 29 past Hope's Nose on 18th, 160 at Dawlish Warren on 25th and 35 at Hartland Pt on 30th. Only three of the 37 *Sep* records involved double-figure counts: 22 at Hartland Pt on 8th, 15 at Dawlish Warren on 10th and 12 there the next day. *Inland* one was at Upper Tamar L on 30 Sep.

Oct. There were 18 *Oct* records with the only double figure counts on 7 Oct during SW gales (15 S at Dawlish Warren, 15 at Broadsands & 10 S at Berry Head) and one *inland* at Underdown, Haldon on 8 Oct (JGa). The *last record* of the year was of one at Slapton Ley on 19 Oct (PHA).

ARCTIC TERN *Sterna paradisaea*

Scarce passage visitor. BoCC (Amber).

Table 81: Maximum monthly counts from main/well-counted sites:

	J	**F**	**M**	**A**	**M**	**J**	**J**	**A**	**S**	**O**	**N**	**D**
Dawlish Warren / Bay	-	-	-	-	2	-	2	2	2	-	-	-
Start Bay	-	-	-	1	-	-	-	-	-	-	-	-
Torbay	-	-	-	-	-	-	-	1	1	2	-	-
Plymouth Sound	-	-	-	-	-	-	-	-	-	2	-	-
Taw / Torridge Est	-	-	-	-	-	-	2	11	2	1	-	-

Apr-Jun. The *first arrival* was one at Slapton on 22 Apr (MRAB). The year's highest count came from Thurlestone with 29 on 13 May (PAS). At Dawlish Warren/Exmouth, only one on 1-2 May (KRy,THS), two on 18th (IL) and one on 22nd (KRy).

Jul-Sep. The majority of *Jul* records came from Dawlish Warren, with seven records of singles and one of three birds, on 21st (IL). Elsewhere, two ads were at Skern on 24 Jul (RGM). *Aug* was the peak month, with 16 records, the majority coming from the Taw/Torridge area, mostly at Skern: juv on 5th (RD); one 10-11th (ArSt,DC,RGM); two on 13th (RGM); 11 on 19th (ArSt); three on 20th (RGM); five on 24th (ArSt,RGM); two on 26th (ArSt); and two ads were at Saunton Sands on 2 Aug (RJ). In S Devon: one S at Hope's Nose on 14th (CJP); a juv at Dawlish Warren on 16th (JEF); a juv at Berry Head (RWBu), a juv at Dawlish Warren (IL) and one at Hope's Nose (CJP) all on 18th; a juv at Dawlish Warren on 20th (KRy), and two ads there on 21st (GV); and a juv E at Prawle Pt on 29th (NPR). There were six *Sep* records: an ad at Beer on 5th (RJO); two at Skern on 10th (ArSt); two at Dawlish Warren on 15-16th (LC,IL) and one there on 23rd (LC); and a juv at Hope's Nose on 30th (TWW,CJP).

Oct. On 1st, two went S at Berry Head (MD,ML,AR), a juv was at Skern & Westward Ho! (IM,RGM) *(see col plate)* and one was on the Exeter Canal (RSPB); on the 7th, one was at Dawlish Warren (IL) and two juvs were at Cattewater, Plymouth (SGe *et al.*); three were at Hartland Pt on 9th (MSS); and the *last of the year* was on the Torridge Est on 15 Oct (IM).

LITTLE TERN *Sterna albifrons*

Scarce passage visitor. SPEC(3), BoCC (Amber).

The Exe Estuary was the principal site this year.

Apr-Jun. In *Apr*, 15 at Dawlish Warren on 24th (IL) were the *first arrivals* and the largest flock of the year, followed by one there on 28th. In *May* there were eight records from the mouth of the Exe Est, with a max of seven on 11th, all other counts 1-27 May being of two or less. The only record elsewhere was one off Lundy on 13th May. All *Jun* records were from the Exe Est, with one on 3rd and 23rd at Bowling Green Marsh and one at Exmouth on 23rd.

Jul-Aug. In *Jul* there were five records of singles at Dawlish Warren between 11-24 Jul, while in N Devon four were at Skern 20-21st. In *Aug*, three were at Westward Ho! on 4th, and at Dawlish Warren there were three on 18th, a juv on 21st and three, the last of the year, on 24th (PBo).

2000 addition. A group of 24 off mouth of R Otter on 23 Aug (RJJ).

Unidentified TERNS (*Sterna* spp.)

Unidentified terns, usually Common/Arctic included: 30 on 13 May at Thurlestone; 72 on 15 May at Bideford; 15 on 26 Aug off Exminster seafront; and seven on both 26 Sep and 7 Oct at Hope's Nose. There was also one inland at Lee Moor on 7 Aug.

BLACK TERN *Chlidonias niger*

Scarce passage visitor. SPEC(3).

An average year, with fewer spring records than 2000. The peak autumn passage was in Oct, especially along the S coast on 7 Oct during SW gales.

Apr-Jun. *Two* at Dawlish Warren on 24 Apr (LC,JEF,KRy) were the *first arrivals*. In *May* there were three at Thurlestone Bay on 12th (RWBo,PSa), two there the next day (PAS) when six were also seen from the Lundy Crossing (ML,TWW) and one off Exmouth on 14th (GV).

Jul-Sep. In *Jul,* two, an ad and juv, off Orcombe Pt on 17th (MSW,BBH) and one off Dawlish Warren on the same day (KRy) with two wp ads off Starcross on 30th (DJG). In *Aug,* two passed Hope's Nose on 17th (TWW), one at Dawlish Warren on 26th (LC), a 1y at Torcross on 27th (AKS) and two off Hartland Pt on 30th (MSS,DC). In *Sep*, there were: two past Hartland Pt on 3rd (MD,AR); two on Lower Tamar L on 13th (GPS); one at Dawlish Warren on 17th (JEF,LC); one past Hope's Nose on 26th (CJP); two at Dawlish Warren on 27th (IL); a juv on the Plym Est on 29th (PHA); and one at Berry Hd on 30th (MD,AR).

Oct. *Berry Hd* produced the majority of the records in this peak month: one on 1st (MD,AR) and 6th (MD), six S on 7th (mo), one on 8th (ML) and two on 15th (G&BJ). *Elsewhere*: one at Dawlish Warren on 4th (GV); on 7th there were singles at the Axe Est (IJW), Broadsands (RBe) and Prawle Pt (TM) and two passed Hope's Nose (CJP,BBH); on 8th one was in Torbay (BBH); on 9th a moulting ad at Slapton Ley (BW) with another there on 12-13th (DC,BW); two at Lower Tamar L on 13th (GPS); one at Dawlish Warren on 14th (IL,ARo), and two there the next day (D Wood); one at Seaton on 19th (IJW); one at Beesands on 20th (JHB); and a juv at Slapton Ley on 20-22nd (mo). The *last record* was of one at Dawlish Warren on 23 Oct (JEF,KRy).

GUILLEMOT *Uria aalge* — Common Guillemot

Not scarce breeding, passage and winter visitor. BoCC (Amber).

Table 82: Monthly maxima of passage birds at main sites

	J	F	M	A	M	J	J	A	S	O	N	D
Berry Head	-	-	-	-	-	-	-	-	9	-	180	-
Hope's Nose	-	-	-	165	65	-	-	-	-	59	-	-
Prawle Point	-	-	-	-	40	53	3	-	-	30	66	305
Hartland Point	-	-	-	-	-	-	-	-	-	426	-	-

Breeding. The estimated county breeding population 1999-2001 from the *Seabird 2000* census (see separate article) was 2,630 prs, of which 1573 (about 60%) were on Lundy. The 1057 mainland prs were mainly in two sectors: Braunton-Foreland Pt, with 579 prs (291 at Wringapeak, 151 at Cow & Calf and

137 at Woody Bay); and Start Pt-Berry Hd with 476 prs (all at Berry Hd). Torbay was credited with three prs on the Ore Stone, but this would seem to be a gross underestimate considering the counts of birds there in 2000 (max of 47) and 2001 (max of 106 on 7 Apr (WJD)). At Berry Hd in 2001, the max count of ads was 1,125 (BHNNR), clearly compatible with the *Seabird 2000* estimate and very close to the count of 1,086 in 2000; no productivity data this year, but the first egg was laid on 1 May and the last bird on cliffs was on 5 Jul.

Non-breeding. Records came from all months, but were scarce in Feb and Mar. Max counts were from Hartland Pt, including: *c.*50 on 18 Jan (IM); 271 (in 2h) on 3 Oct and 426 (in 3h) on 9 Oct (MSS); and 200 (in 3h) on 7 Nov (DC). The only other counts >10 were: 18 in Plymouth Sound on 1 Jan; 13 W past Start Pt on 20 Oct; and 18 past Prawle Pt on 11 Nov. Otherwise, mostly one-three, from: Abbotsham, Avon Est, Beer, Bideford, Bolt Hd, Brixham Harbour, Dawlish Town, Dawlish Warren (max 5), Exe Est, Hallsands, Orcombe Pt, Plymouth Cattewater (max 4), Slapton, Soar Mill Cove, Teign Est and Thurlestone Bay (max 6). An unspecified number were involved with other seabirds in a feeding frenzy off Saunton Down on 15 Nov. A bridled individual was noted among the Berry Hd population on 17 Apr (MKn). Counts were generally lower than those for Razorbill.

Corpses, oiled birds and recoveries. Corpses included four freshly dead on the saltmarsh at Passage House Inn on 6 Jan and four at Elberry Cove on 13th. The SDST at Teignmouth (Jack's Patch) took in 93 oiled Guillemots (out of a total of 110 oiled seabirds) from Devon beaches in 2001, and successfully released 69 of them; the casualties were mostly moribund birds in Jan. The monthly totals Jan-Dec were 34,5,9,8,1,1,2,6,0,2,2,23, and the birds came from the Plymouth area (6), Wembury – Dartmouth (38), Brixham – Oddicombe (39) and Teignmouth – Dawlish (10). One bird was recovered as a corpse in Scotland in 2001, over four years after rehabilitation and release from the SDST; see Ringing Report for further details of this and other recoveries.

RAZORBILL *Alca torda*

Not scarce breeding, passage and winter visitor. SPEC(4), BoCC (Amber).

Breeding. The estimated county breeding population for 1999-2001 from the *Seabird 2000* census was 762 prs, of which 637 (*c.*84%) were on Lundy. The 125 mainland prs were all on N coast between Braunton and Foreland Pt (95 at Woody Bay, 12 on Cow & Calf, ten at Wringapeak and eight in Elwill Bay). In view of the absence of breeding on the S coast, it was surprising that one of the few breeding season records in 2001 was of seven on the sea below the seabird colony at Berry Hd on 11 Jun (BHNNR).

Non-breeding. The only counts >10 were at: Berry Hd, with 18 past (in 4h) on 30 Sep, 48 past (in 2h) on 4 Oct and 32 (in 4.5h) on 7 Oct (JBa); at Hartland Pt, with 30 on 18 Jan (IM) and 12 SW (in 8h) on 8 Sep; and 11 past Hope's Nose on 10 Jul. Otherwise, mostly one-three, at: Baggy Pt, Beesands, Bideford Bay, Torbay, Budleigh Salterton, Dawlish Warren (max seven in Jul & Nov), Exmouth seafront, Langstone Rock, Man Sands, Plymouth Cattewater, Plymouth Sound, Salcombe, Start Bay and Torbay (max six in Jan).

Unusual. Two flying over the A38 at Marsh Mills from Plymouth to the Plym Est on 20 May (PMM), was appropriately described by the observer as a very extraordinary sighting!

GUILLEMOT/RAZORBILL *Uria aalge/Alca torda*

Almost all records concern large numbers of auks passing headlands in less than ideal weather conditions. The prime interest is in the scale and timing of the movements, but those in 2001 were much smaller than in 2000, with a max of 426 at Hartland Pt in Oct, compared with an overall max of 9000 in 2000 and four monthly maxima >500. The highest passage rate was 165 past Hope's Nose (in 3h) on 22 Apr.

BLACK GUILLEMOT *Cepphus grylle*

Rare vagrant/ winter visitor. SPEC(2), BoCC (Amber).

Records for the second year running, after a two year absence, and a good year with four birds.

'Residents'. A 1w in *Plymouth Sound* 14 Jan – 8 Oct (SGe, PHA, VRT,SRT *et al.*), initially in wp but gained partial bp, and could possibly have been one of the Torbay birds from 2000; recorded from Cattewater, Hooe, Oreston, Sutton Hbr and West Hoe as well as on the Cornish side, but was often elusive, with only single Devon records in Jul and Aug and thereafter only on 6-8 Oct. Another appeared off *Dawlish Warren* as a 1s on 14-21 & 29-30 Jun (IL *et al.*) and returned, moulting into wp, on 2 Sep and seen again on 20th & 24th (LC *et al.*).

Passage. One in wp past Hope's Nose on 9 Sep (JBa), and another W past Prawle on 6 Oct (PMM).

2000 correction. Delete one off Broadsands on 12 Mar.

LITTLE AUK *Alle alle*

Very scarce passage and winter visitor.

With sixteen live records, and perhaps fifteen different birds, a better year than both 2000 (nine) and 1999 (one) and the best since 1998. All but one occurred in the period 11-30 Nov, and all but one were from the S coast.

Second winter-period. Fourteen records, all in Nov as in 2000: on 11th, one SW past Hartland Pt (MD,AR), one off Dawlish Warren (LC,DJSu), and another or same later up Exe Est at Turf (BG); on 14-16 Nov off Dawlish Warren, one on sea on 14th (JEF),one past on 15th (KRy) and one offshore on 16th (KRy); on 17th, one alive on beach at Broadsands *(see col plate)* after second killed by gull (PLe); on 21st, one past Berry Hd (ML); on 24, 25 and 29 Nov, singles at Hope's Nose (CJP), with one past Berry Hd on 25th (MD); and on 30th, one W at Prawle Pt (MD).

PUFFIN *Fratercula arctica* — Atlantic Puffin

Scarce passage/summer visitor; scarce/very scarce breeder on Lundy. SPEC(2), BoCC (Amber).

A slight increase in ads on Lundy, but still no unequivocal evidence of breeding there. A shorter period of passage compared to 2000, with no Apr records, but several in Aug, and a peak in Jul.

Breeding. The official breeding population for Lundy (and therefore Devon) is now seven prs, according to the *Seabird 2000* census. This is perhaps a *slight* underestimate, with max counts of 19 ads in 2000 and 26 on 11 Jun 2001 (LFS), but there is no doubting that this species is at a very low ebb. There is also still no current evidence (eg. juvs, or ads carrying food) of actual breeding, even though three prs were seen entering burrows in May and Jun 2001 at St Philip's Stone, and birds were seen mating and carrying nesting material (LFS).

May-Jun. The first of four birds was a 1s on 11 May at Three Beaches (ML), the next passed Start Pt on 27 May (ML,BMc) and there were only two Jun records: one on 16 Jun past Prawle (TM) and three on 29th past Berry Hd (ML).

Jul-Aug. At least 17 birds, with records on six dates in *Jul*: 1 Jul, two W past Prawle (JCN,TM); 8 Jul, three W past Prawle (JCN); 10 Jul, two S past Berry Hd (ML); 17 Jul, four S past Berry Hd in morning (ML,MD) and one W past Prawle in evening (PMM); 21 Jul, three S past Berry Hd (MD), two W past Prawle (JCN) and four W past Start (ML), possibly involving the same birds: and 28 Jul, one W past Prawle (JCN). At least five birds on four dates in *Aug*: 2 Aug, two S past Hope's Nose (CJP) and one S past Berry Hd (MD,AR) could have been one of same; 7 & 11 Aug, singles W past Prawle (TM; JLb,JCN); and on 28 Aug, one at Orcombe Pt (MSW).

Sep-Oct. Three birds on three dates: 8 Sep, one SW past Hartland Pt (MD); 13 Sep, one imm at Hartland Pt (MSS); and 5 Oct, one S past Berry Hd (ML *et al.*).

ROCK DOVE *Columbia livia* — Rock Pigeon

Columbia livia* var. 'Feral Pigeon': *fairly numerous resident (naturalised feral) breeder.

Again under-recorded, though twice as many records received (14) as last year.

Maximum counts. Two records of counts exceeding 100, both in Bideford, with 110 on 27 Feb and 104 on the quay on 5 Dec. The latter site also produced counts of 80 on 10 Jul and 3 Nov.

Predation. Studies of the diet of Peregrines in Exeter have revealed the importance of this species as a prey item (see separate report). Also of interest was a record of one brought down by a Peregrine at Bowling Green Marsh on 6 Oct, but then escaping after Crows harassed the falcon (RJJ).

STOCK DOVE *Columba oenas* — Stock Pigeon

Not scarce breeding, passage and winter visitor. SPEC(4), BoCC (Amber).

A strong autumn passage well recorded on the south coast.

Breeding. Confirmed records came from the Prawle area, Roadford Res and Velator, with prs displaying or seen near probable nest sites at Bellever, Clennon Valley, Huccaby, Lopwell Dam and the Tiverton area. Records also from a further 12 widely distributed sites May-Jun. *Post-breeding* flocks: 62 feeding in a cornfield at GW Canal on 14 Aug.

First winter-period/spring. Flocks ≥10: 250 at Bridford on 6 Jan; 16 in the Bigbury area on 14 Jan; 15 at Sourton Cross on 13 Feb; 51 at Bulland, Landscove on 30 Mar; 18 in the Prawle area on 16 Apr, ten there on 27 Apr and 60 at Netton on 8 May.

Autumn/second winter-period. *Passage* very well recorded this year, with observations at *Staddon Point* (SGe) producing a half-monthly total of 1,270 passing mainly W/NW 1-15 Nov, with the peak passage in the first nine days of Nov, and a max day count of 416 on 1st (see separate report). Other S coast sites were also well represented by passage records, with counts of >10 as follows. *Prawle area*: 12 on 17 Oct, 20 next day, 26 on 21st, 46 next day, 42 on 27th, 60 on 1 Nov, *c.*100 on 4th, 110 next day, 700 on 10th (JCN), *c.*500 next day (PMM), 60 on 18th, *c.*100 on 25th, 20 on 1 Dec. *Soar area*: 30 on 20 Oct, 38 next day, 46 on 22nd, 250+ off sea on 3 Nov (SRT,VRT), 44 on 4th, 60+ on 20th and 15 on 22 Dec. *Start Pt*: 200 N off sea on 21 Oct. The only inland second winter-period count of >10 was *c.*30 nr Bridford on 13 Nov.

WOODPIGEON *Columba palumbus* — Common Wood Pigeon

Abundant resident breeder, passage and winter visitor. SPEC(4).

A bumper yield of migrating birds on the south coast in autumn.

Breeding. Confirmed only at Berrynarbor, Higher Metcombe, Huccaby, Northam and Prawle area. The only count of breeding prs came from Dawlish Warren, where five prs were estimated.

First winter-period. Counts of >50 (all in Jan): 80 in woodland at Grenofen on 18th, 150 in Roughtor Plantation, Burrator on 21st, 103 at Arlington Court Estate on 25th and a max of 360 to roost in Brimpts Plantation on 29th.

Autumn/second winter-period. Observations at *Staddon Point* (SGe) produced a remarkable estimated total of 80,835 birds passing W/NW during the autumn, including 23,885 in just two hours on 10 Nov. This survey produced the largest day and autumn totals of migrating Woodpigeon ever in the county (see separate report). Other coastal counts of >50 involving migrants as follows. *Exmouth*: 740 between 08.00-09.00h on 31 Oct, with 600 over the same period the next day. *Dawlish Warren*: 392 on 31 Oct. *Start Point*: 75 N off sea on 21 Oct. *Prawle area*: 150+ on 26 Oct and 4 Nov, with 100 on 25th. *Soar area*: 300+ on 3 Nov. *Wembury*: 153 on 6 Nov.

Elsewhere (flocks ≥50). Post-breeding, migrating and wintering flocks: 80 at Nether Exe on 12 Aug; 100 at Stoke Canon on 26 Aug; 100+ at Huccaby on 22 & 26 Oct and 3 Nov with 65 on 23 Nov; 190 at Bowling Green Marsh on 3 Nov; 60+ on Lundy on 4 & 10 Nov; 54 at Exminster Marshes on 21 Nov; 327 at Roborough Down on 26 Nov.

COLLARED DOVE *Streptopelia decaocto* — Eurasian Collared Dove

Fairly numerous resident breeder and rare passage visitor (first bred 1962).

Breeding. Confirmed records: Barton Pines (pr and yg); Braunton; Dartmeet; Exmouth (pr mating); Kenwith NR (pr); Lundy (pr with two clutches); Netton (pr and 2 yg); Sticklepath, Okehampton (3 prs, yg seen); Plymouth City Centre (pr nesting on floodlight bracket on shop in Cornwall St and 3 yg at Armada shopping centre); Stoke Canon (2 prs produced 2 yg each). There were also records of prs from a further 13 sites Apr-Jun. Evidence of continuing colonisation of higher ground on Dartmoor, with the first record for Huccaby on 27 Apr, the first known breeding record for Dartmeet, and an increase in records at Princetown (450m above sea level).

Non-breeding. Outside breeding season counts of >10: 26 at Bigbury on 9 Jan, 73 feeding on grain at Wembury on 22 Sep (SGe), 17 at Grand Western Canal on 7 Oct and 23, presumably migrants, at Pig's Nose valley on 3 Nov. Also worth noting were five migrants at Staddon Pt on 25 Sep.

TURTLE DOVE *Streptopelia turtur* — European Turtle Dove

Scarce migrant breeder and passage visitor. SPEC(3), BoCC (Red), UK BAP.

49 records involving a minimum of 34 birds in possible breeding areas, with a minimum of 13 migrant individuals at coastal sites in spring and five in autumn.

Table 83: Monthly distribution of all passage records:

	J	F	M	A	M	J	J	A	S	O	N	D
Sites	-	-	-	1	5	1	-	2	2	-	-	
Birds	-	-	-	1	11	1	-	3	2	-	-	

Breeding. Most records came from usual areas around Exeter, with a minimum of eight (nest sites located) and perhaps 10-12 prs on the *Haldon Ridge* (RK); at least three breeding prs at a farm at Kennford with a max of 14 birds on 26 Jun (likely to include some Haldon birds) (RAda). In this area records also came from Dunchideock (four birds) and Starcross (single). On the *E Devon Commons* there were records from Bicton Common and Hawkerland Valley (both singles), Higher Metcombe (two) and Venn Ottery Common (up to two present, no breeding). *Elsewhere* a trio of records from the Kingswear area May-Sep, including an imm on 17 Sep, might possibly indicate breeding nearby.

Spring. The *first* was one at Prawle on 28 Apr (PMM), followed by May singles (away from breeding areas) at Dawlish Warren on 6th & 11th; Berry Head on 12th & 13th; Lundy on 13th; four at East Prawle on 18th followed by one there 20-22nd; and one at Abbotsham on 23rd (the only record from the N apart from Lundy). There was also a single at Prawle on 3 Jun.

Autumn. *Passage* was again slight, with the only records being singles in the Prawle area on 22, 23 & 26 Aug and 22 & 24 Sep; Bowling Green Marsh on 23 Aug (as last year perhaps from nearby breeding areas); East Soar on 9 Sep; and the *last* at Dawlish Warren on 29 Sep (IL).

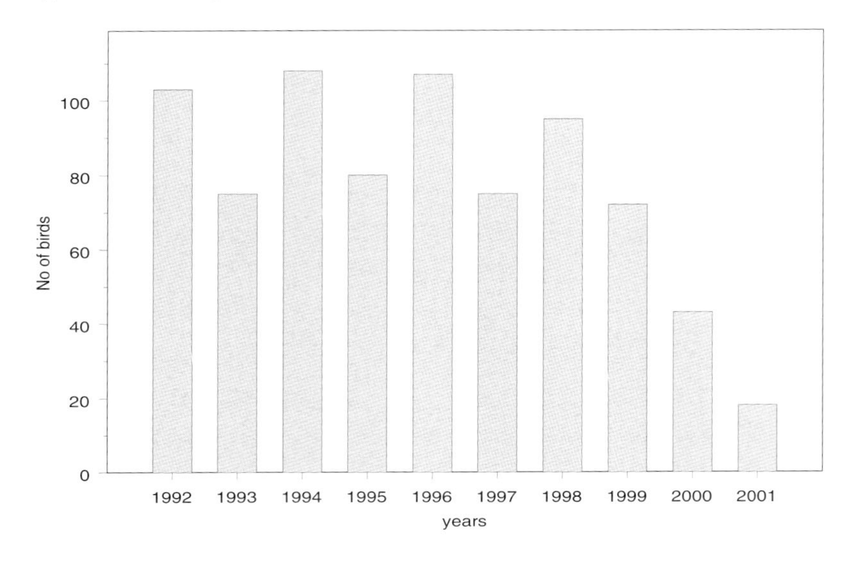

Fig. 23: Turtle Dove - Total number of birds recorded (excluding breeding pairs) in Devon 1992 - 2001 (Source: DBRs)

ROSE-RINGED PARAKEET *Psittacula krameri* **Ring-necked Parakeet**

Escapee resident

Again only one bird.

One at West Putford on 24 Jun "flew around calling for half an hour then flew off east"(AMJ).

CUCKOO *Cuculus canorus* — Common Cuckoo

Fairly numerous migrant breeder and passage visitor. BoCC (Amber).

A total of 107 mainland records from 44 sites (cf. 136 records from 97 sites in 2000).

Breeding. Records of juvs from Berry Hd, Burrator (fed by Yellowhammer – unusual host species) (RMcC), Cadover Bridge (fed by Meadow Pipit), Grimspound, Headland Warren (fed by Skylark – also unusual host species) (MGM) and Sherrill, nr Dartmeet.

Spring. The *first* was one on Lundy on 5 Apr (LFS), followed by singles at Sharkham Pt on 13 Apr, Dawlish Warren on 17th & 21st, and then a widespread arrival from 23rd (as last year). Passage on Lundy involved eight singles to 27 May.

Autumn. Only three birds recorded in Aug and none in Sep, with the *last* being a juv at Grimspound on 17 Aug (DR). There were no autumn records from Lundy.

BARN OWL *Tyto alba*

Scarce resident breeder. SPEC(3), BoCC (Amber), Devon BAP.

Reported from 50 widely distributed sites (cf. 82 in 2000). Access to farm sites will have been restricted by FMD; hence under-recorded this year.

Breeding. BOT data give a total of 72 yg reared from 24 sites, giving a productivity rate of 3 yg per site (BOT/KRG). Sites as follows: Bovey Valley (4 yg at one site); Budleigh Salterton (4 yg at one site); Exeter (9 yg from three sites within 10km radius); Holsworthy (6 yg at one site); Kingsbridge (12 yg from five sites within 10km of town); Okehampton (9 yg from three sites within 10km of town); Southleigh (2 broods of 3 & 5 yg from one site); Totnes (15 yg from six sites within 10km of town) and Yealmpton (5 yg at two sites within 5km of town).

Mortality and ringing recoveries. Birds were found killed on roads at Bridestowe, Clyst St George and Whiddon Down. Birds in 2001 were recovered up to six years after ringing, and up to 48km from the ringing site, all from the two established Barn Owl projects in Devon; see Ringing Report for more details.

LITTLE OWL *Athene noctua*

Not scarce resident breeder (naturalised introduction). SPEC(3).

Recorded from 27 sites (cf. 35 in 2000).

Breeding. Confirmed at Cox Tor (RJL), Culmstock (HWa), Higher Metcombe (GHG), Killerton (DJJ), Sandford (SGM), Soar area (EM,H&JH) and Tidwell, Landscove (JMW).

Distribution. All reported sites in areas defined in *DBR 2000*, with the bulk of records (65%) in a southern area between Netton in the W and Otterton in the E, and a typical smattering of records from the NE – Culmstock, Huntsham, Killerton, Morchard Bishop, Sandford, Uffculme and West Raddon. Further north there were records from Diddywell (Northam), Instow and South Molton, where a bird was found dead on 20 Jan. Noteworthy was a cluster of four prs in the Underdown (Haldon) area (JGa). There were also two records from usual sites on the western edge of Dartmoor (Cox Tor and Criptor Newtake), suggesting that the species may go unrecorded in W Devon.

TAWNY OWL *Strix aluco*

Fairly numerous resident breeder. SPEC(4).

89 records from 65 sites – better reported than in any recent year, probably reflecting a change in recording habits due to FMD as observers concentrated closer to home.

Breeding. Confirmed at Bellever Plantation, Castle Woods (Okehampton), Dalwood, Heatree Cross (nr Manaton), Higher Metcombe, Huccaby, Maidencombe, Slapton Ley and Stover CP. The highest count was of six birds calling in Torrington Cemetery during Apr.

SHORT-EARED OWL *Asio flammeus*

Scarce passage/winter visitor; former breeder (last bred 1943). SPEC(3), BoCC (Amber).

A healthy tally of 87 records from 14 sites (cf. eight in 2000 and 19 in 1999).

First winter-period/spring. A probable four individuals, with singles at Bursdon Moor on 27 Jan

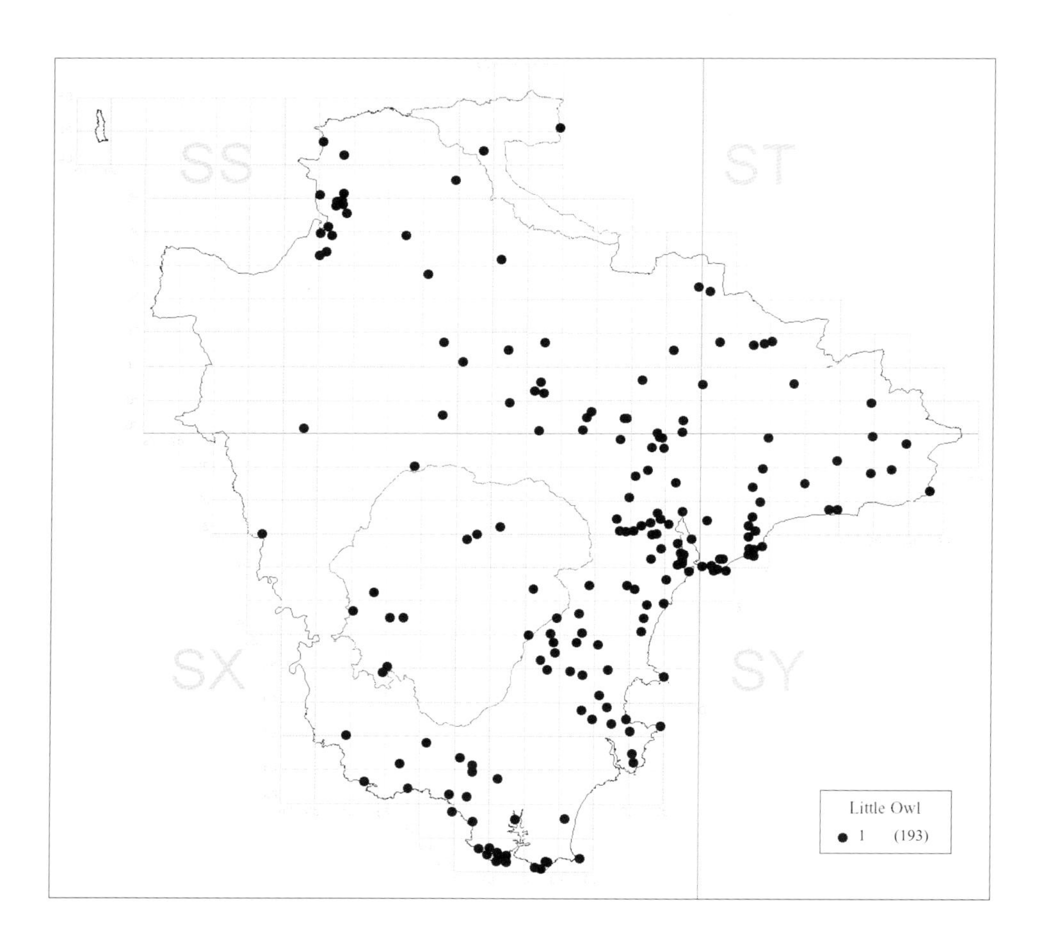

Map 7: Little Owl.
Distribution of all records for all months. Most records referred to single pairs /territory.
Data: Based upon a ten year period 1992-2001. Source: DBRs) (see Overview on page 25)

Map 8: Nightjar.
Distribution of churring males. Abundance is based on maximum count at each location. The species appears to have extended its range westwards during the past decade. A repeat national breeding survey is planned in the next couple of years, this should yield some interesting results. The maximum counts for the two main sites on Dartmoor (28) and Haldon (96) are shown directly on the map.
Data: Based upon a ten year period 1992-2001. Source: DBRs) (see Overview on page 25)

(RKT,MSS) and 7 & 12 Apr (FHCK,RMHa), one flying west, mobbed by gulls, at East Prawle on 29 Apr (SM) and one on Lundy on 23 May (LFS).

Summer. One at Witheridge Moor on 29 Aug (RJJ).

Autumn/second winter-period. An impressive series of records at *Exminster Marshes* throughout the period began with one on 2 Oct and peaked at five on 11 Nov, with two or three seen on a number of dates to 11 Dec (RSPB *et al.*). *Dawlish Warren* also saw a good run of records, perhaps involving some overlap with the Exminster birds, with two in off the sea on 20 Oct (JEF) followed by two on 12 Nov (JEF,KRy) and one on 11 Dec (LC). *Bursdon Moor* held birds almost continuously 2 Nov-9 Dec, followed by a year-end straggler on 31 Dec. The peak there was three on 15-16 Nov and 8 Dec (FHCK,IM *et al.*). *Elsewhere* two over Staddon Pt, Bovisand (SGe) and one in the Venford Res area (WJD) on 21 Oct (the first on Dartmoor since 1999); two in the Soar area on 22 Oct (PSa); up to two on Lundy on eight dates 22 Oct-18 Nov (LFS); singles at Rippon Tor (possibly same as Venford bird) on 28 Oct (SMRY); Skern on 31 Oct (RGM)*(see col plate)*; Coleton Fishacre on 1 Nov (JMW); Prawle area on 4 Nov (TM) and 10 Nov (JLt); in off the sea at Hartland Point on 11 Nov (MD,AR); Assycombe Hill, nr Fernworthy on 28 Dec (RS) and Braunton Burrows on 31 Dec (JTi).

Nightjar - *Tom Whiley*

NIGHTJAR *Caprimulgus europaeus* — European Nightjar

Not scarce migrant breeder. SPEC(2), BoCC (Red), UK BAP, SW BAP, Devon BAP, Exmoor BAP.

Haldon and SW Devon breeding sites. At the principal site, *Haldon Ridge*, 71 nests were located, with an additional 25 churring ♂♂ on newly felled areas (RK). This total includes Little Haldon, where ten churring ♂♂ were recorded during June (GGT). *Elsewhere in SW Devon* there were records (single prs/churring birds unless stated otherwise) from Cann/Great Shaugh Woods (ten churring ♂♂, *cf.* seven in 2000) (GGT); Chudleigh Knighton Heath (DS); Heathfield (PP); Stover CP (pr with two broods, three yg fledged) (JDA)*(see col plate)*; and Teignmouth ("one heard churring in council estate") (JEF).

E Devon Commons breeding sites. These sites were better covered than last year, with records as follows: Colaton Raleigh Common (five prs with six additional churring ♂♂) (DJJ); Bicton Common (pr + three other churring ♂♂) (DJJ); Venn Ottery Common (3 ♂♂) (GHG); East Budleigh Common (2 ♂♂) (GV); and single birds at Aylesbeare (BSh,CyS) and Lympstone (DJJ) Commons and Squabmoor Res (AWo). Breeding was confirmed nr Higher Metcombe (GHG). Elsewhere in *E Devon* Trinity Hill NR held five birds (RBo), Ashclyst Forest two prs (DJJ) and Sampford Common one ♂ (DJG).

Dartmoor, Mid and N Devon breeding sites. On *Dartmoor* the highest count was 11 churring ♂♂ in Bellever Plantation on 21 Jun (DSGp), though FMD precluded any records from Fernworthy, another important site. Soussons Plantation produced a low count of five ♂♂ in poor conditions on 28 Jun (DSGp). There were no records from the Burrator area. Records from non forestry sites around the edge of Dartmoor: at least three birds at Roborough Down (RS,PAS,PFG,PMM); at least two displaying ♂♂ at Trendlebere Down (DR); and one ♂ at Wooston Castle, nr Drewsteignton (possibly a new site) (StH). One nr Cornwood in Sep (see below) was probably a migrant but was in suitable breeding habitat. Just *N of Dartmoor* two prs bred at Abbeyford Woods, Okehampton (GAV,JWV); one flying over the A3079 at Cookworthy Moor on 27 Jul (BCM) was the only other record from *Mid- or North Devon*.

Arrivals and departures. The *first arrival* was one churring at Heathfield on the early date of 27 Apr (PP); the only spring *migrants* away from breeding sites were singles on Lundy on 13 May (TWW,ML,PM) and Dawlish Warren on 18 May (DWNNR) (both sites unusual). The last record from breeding sites was two displaying at Trendlebere Down on 31 Jul (DR). The *latest* record was a single at Hawns and Dendles, nr Cornwood on 6 Sep (NB); there were no other records involving potential autumn migrants.

SWIFT *Apus apus* — Common Swift

Fairly numerous breeder and passage visitor.

Breeding. Breeding confirmed at: Aveton Gifford (5 prs); Peverell, Plymouth (one pr fledged two yg; first breeding here for 38 years - see separate report); Prawle; Princetown; Stoke Canon; Sidborough NR, Morchard Bishop; Tiverton; Westcombe; and Bideford (3 prs).

Spring. The *first* records were single at Jennett Pt NR (MG) and Slapton Ley (DEk) on 17 Apr. These were followed by singles at Shebbear on 19th and again at Slapton Ley on 22nd; thereafter there were daily records from 24th; the first double figure count was ten at Exminster Marshes on 26 Apr. Numbers were generally low until late May, with the peak Apr count being 40+ at Slapton Ley on 28th. The only count >100 in early May was 298 at Dawlish Warren on 10th; the next was *c.*2,000 at Slapton Ley on 27th, the highest count of the period here.

Late summer/autumn. Counts ≥100, all in Jul: 111 W over Prawle area on 1st; 250 at Starcross on 10th; 300 at Slapton Ley on 11th; 100 at Aveton Gifford on 22nd; 480 at Prawle on 28th, with 127 next day. The highest count in Aug was of 50 at Slapton Ley on 12th. The *last* record was one at Torcross on 29 Aug (DJa).

ALPINE SWIFT *Apus melba*

Rare vagrant

A dramatic record of an individual with a missing tail feather at Berry Hd on 1 Oct, 'in off the sea' pursued by an Arctic Skua before soaring over the headland on two occasions during the morning (ML,MD,AR), was the 30th Devon record and the first since one in 1999, also at Berry Hd. (One of six in 2001 accepted by BBRC).

KINGFISHER *Alcedo atthis* — Common Kingfisher

Not scarce resident breeder, with local winter movement towards coast. SPEC(3), BoCC(Amber).

388 records from approximately 130 locations (cf. 213 records at 95 locations in 2000).

Table 84: Monthly distribution of all passage records:

	J	F	M	A	M	J	J	A	S	O	N	D
Sum max from all sites	3	22	8	1	8	13	38	37	68	44	69	56
WeBS county total '00	11	8	6	1	1	3	5	8	19	18	13	15
WeBS county total '01	8	7	1	1	-	1	3	3	18	14	19	19

Breeding Confirmed only at Aveton Gifford (2 prs), Beam Weir (R Torridge) and Grand Western Canal.

Non-breeding. Records from every *estuary* in the county, with the great majority around the Taw/ Torridge Ests, followed by the Exe. *Inland* sites with multiple records included (nos of records in brackets): Burrator Res (7); Grand Western Canal (21); Fernworthy Res (3); Huccaby (W Dart)(5); R Carey (5); Roadford Res (6); Uffculme (4). An interesting record was one in the quarry at Berry Hd on 14 Oct.

HOOPOE *Upupa epops*

Very scarce passage visitor, mainly in spring.

A better year than 2000, with five spring records and one in summer.

Spring. One at Sidmouth 19-21 Mar (*per* ML) was followed by singles at Higher Metcombe 26-29 Mar (GHG), Budleigh Salterton on 4 Apr (JAr), Netherton (Teign Est) (MKn) and Wiggaton (Ottery St Mary) (GHG) on 10 Apr.

Summer: One at West Hoe, Plymouth on 9 Jun (MTd).

WRYNECK *Jynx torquilla*

Very scarce passage visitor, mainly in autumn. SPEC(3), BoCC (Red), UK BAP.

A good year, with one spring bird (as every year since 1997) and nine in the autumn.

Spring. One at Hatherleigh Moor on 10 May (WHT).

Autumn. All singles: in a garden at Clayhidon on 25-26 Aug (MRe); at Aylesbeare Common on 1 Sep (MSW); Dawlish Warren 29 Sep-5 Oct (thought killed by Kestrel) (SMRY,JRD,JEF,LC *et al.*); Pig's Nose Valley on 29 Sep (PMM,TM); Soar Mill Cove 11-12 Oct (PHA); Lundy (LFS) and Start Point (SRT,VRT) next day; and East Allington (PBo) and Prawle (PMM) both on 16 Oct.

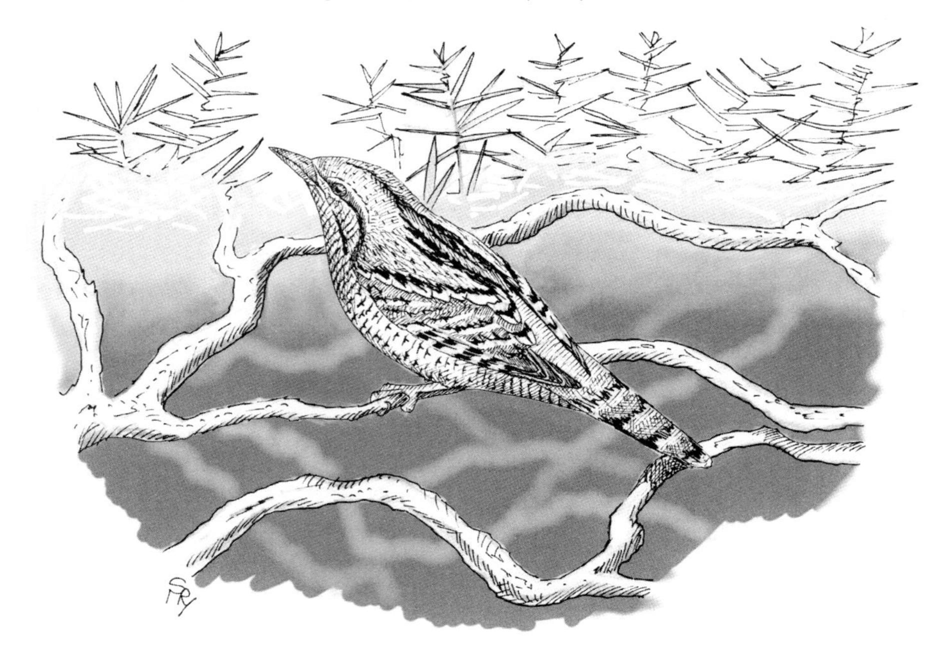

Wryneck - (Dawlish Warren) - *Steve Young*

GREEN WOODPECKER *Picus viridis*

Fairly numerous resident breeder. SPEC(2), BoCC (Amber).

A total of 230 records (cf. 150 in 2000) from 100 sites (cf. 99 in 2000).

Breeding. Confirmed at Blackborough, Burrator Res, Dartmeet, Dawlish Warren, Exeter University, Galmpton, Honiton, Kerswell Down/Wood, Maidencombe, Manaton, Meldon Res and Woods, Okehampton, Rackerhayes, South Brent and Tawstock.

Unusual observations. These included one at Pottington on 16 Jun 3km from the nearest wooded area where the species is usually seen; one on open moorland at Holne Moor on 29 Dec; and an ad female feeding on fallen apples at Hooe, Plymouth 17-19 Dec ("in *BWP* [Green Woodpecker] is described as the most insectivorous and specialised of the true woodpeckers and this behaviour is noted as very unusual") (SRT,VRT).

GREAT SPOTTED WOODPECKER *Dendrocopos major*

Fairly numerous resident breeder.

279 records from 115 sites (cf. 167 records from 93 sites in 2000).

Breeding. Confirmed at the following sites: Berrynarbor, Burrator Res, Combestone (Dartmeet), Dart Valley, East Anstey, Grand Western Canal, Higher Metcombe, Langston (nr Kingston), Meldon Woods, Okehampton, Sourton Cross, Sticklepath, Stover CP, Thornworthy (Fernworthy), Tiverton, Virginstow, West Putford and Wistlandpound Res.

Autumn. Probable *migrants* included one at Prawle on 17 Sep, followed by a further ten records of singles in Oct and early Nov that might have involved birds on the move, as might singles in the Soar area on 29 Sep, 20 Oct and 3 Nov, and two over Berry Hd on 20 Oct. A series of records at Staddon Pt, Bovisand over the period perhaps also involved migrants.

LESSER SPOTTED WOODPECKER *Dendrocopos minor*

Not scarce resident breeder. BoCC (Red).

A total of 18 records from 14 sites (cf. 20 records from 13 sites in 2000 and 32 records from 20 sites in 1999). No records from the Crediton, Exmouth or Plymouth areas. Breeding proved at two sites.

E Dartmoor. Records (singles except where stated) from E Dartmoor and surrounding core area: Combestone, Dartmeet (RCh); Fingle Bridge, Drewsteignton (FHCK); Dunsford Wood (JMW); Fernworthy Res (MGM); Finlake (RK); Hapstead, Buckfastleigh (JMW); Hembury Woods (pr fledged two yg) (SMRY,BB,GV,PMM); Stover CP (pr feeding yg)(JDA); Woodash, Bovey Valley (DR) and Yarner Wood DJSu).

Exeter area. Ebford (first record by observer here) (AHk) and Pinhoe (PWE).

N Devon. Chittlehamholt (DHWM) and Yeo Vale (RGM).

2000 addition: One at Otter Est on 20 May (RJJ).

WOODLARK *Lullula arborea* — Wood Lark

Scarce resident breeder; migrant status uncertain. SPEC(2), BoCC (Red), UK BAP.

Breeding season records understandably scant and few winter records; five autumn migrants.

Breeding. The Feb Dunchideock records (see below) were followed by "one or two all year" (*per* JEF). Records of single singing birds Apr-Jun, all from the key areas: Chudleigh Knighton (RK); Exminster (IM,RNK); Holloway Barton Farm, Dunchideock (SMRY) (possibly one of above birds); Peamore, Alphington (BDW) and Whitcombe, Kenn (KRG).

First winter-period/spring. (some records will refer to birds on territory): three (two singing) at Finlake on 6 Jan (DS); six in stubble at Cockington on 14 & 20 Jan (ML) and one singing there on 23 Feb (with one other), 2 & 8 Mar (J&HR); two at Exminster on 24 & 28 Jan (MSW,BBH) and 19 & 24 Feb (DC,MAS,PMM) and one in stubble there on 12 Apr (RMHa); four singing at Dunchideock in Feb (*per* JEF); one at Prawle Pt on 30 Apr (PBo).

Autumn. Two at Pig's Nose Valley on 13 Oct (JCN); one S over Huccaby on 17 Oct (RHi); one in the Soar area on 21 Oct (PSa) and one W over Staddon Pt on 11 Nov (SGe). There were no records for the *second-winter period*.

SKYLARK *Alauda arvensis* — Sky Lark

Numerous resident breeder; fairly numerous passage migrant and winter visitor. SPEC(3), BoCC (Red), UK BAP.

A massive 516 records from 96 sites (cf. 196 records from 98 sites in 2000).

Table 85: Monthly distribution of flocks > 50 and maximum counts:

	J	F	M	A	M	J	J	A	S	O	N	D
Number of flocks ≥ 50	4	-	-	-	-	-	-	-	1	15	5	5
Maximum count	200	40	6	14	20	12	30	13	50	300	350	300

Breeding. The only confirmed records: Braunton (nests/yg possibly destroyed by cutting of grass for silage), Huntsham (two nests) and Prawle. Peak site counts of singing ♂♂/territories: 14 at Saunton Sands, 11 at Huxham, 10-12 at Bursdon Moor, eight at Stoke Pt, seven at Netton, six at Dawlish Warren and five on set-aside at Nether Exe. The only notable record from upland areas was 30 birds at Rival Tor on 15 Jul.

First winter-period/spring. Flocks of ≥50: 62 at East Charleton on 7 Jan; 70 at Soar area same day; 80 at Huxham on 14 Jan; 200 at Exminster Marshes on 20 Jan.

Autumn/second winter-period. *S Coast* counts of ≥50: *c.*625 over Staddon Pt 15 Sep-30 Nov including 207 on 1 Nov; 117 at Dawlish Warren on 9 Oct, with 110 on 21st, 65 on 27th and 72 on 29th; 50 at Soar on 22 Sep followed by 200 on 9 Oct, 300 on 20 Oct, 350 on 3 Nov, 100 there next day and 75 on 20 Nov; 57 in Prawle area on 13 Oct, 60 there on 26 Oct, 53 next day and 60 on 17 Nov; 75 at Berry Hd on 20 Oct with 60 there next day; 247 W at Weston on 21 Oct; and 150 at Otter Cliffs on 30 Dec. *Elsewhere*, the only counts >50 were: 55 at Fernworthy Res on 27 Oct; 60 at Kerswell Cross on 20 Dec; and 300 at Higher Rixdale Farm, Ideford on 2 Dec (NCW).

Table 86: HIRUNDINES - Monthly maximum counts:

	J	F	M	A	M	J	J	A	S	O	N	D
Sand Martin	-	-	200	1000	200	190	57	500	75	3	-	-
Swallow	-	-	30	1800	300	13	60	700	5000	1000	4	1
House Martin	-	-	8	40	300	50	250	600	1000	142	-	-

SAND MARTIN *Riparia riparia*

Not scarce migrant breeder and fairy numerous passage visitor. SPEC(3), BoCC (Amber).

Breeding records from several sites (including one particularly unusual habitat), with the usual concentration in upper Culm valley sand quarries, and overall numbers similar to those in recent years. Away from the breeding sites, there were higher peak counts than in 2000 in both spring and autumn.

Breeding. Colonies were at: *Burlescombe area* (Burlescombe Quarry, 20 active holes on 4 Jun; Redhill Quarry, 190 holes on 17 Jun; and Town Farm, a new site, ten holes on 4 Jun); *Uffculme area* (Broadpath, 50 holes in May deserted by 13 Jun; Houndaller, 240 holes on 4 Jun; and Hillhead, a few scattered holes); *Cadover Bridge* (max 24 birds on 12 Jun); *Cullompton* (34 at holes on 21 Jun); *Knapp Mill*, Aveton Gifford (two prs probably bred); *Lee Moor* China Clay works (21 active holes) and nearby *Torycombe Valley* (30 active holes, a new colony thought to be linked to the Lee Moor decline); *Lifton* (10-20 birds at holes on 22 Jun); and *Stoke Canon* (one pr still feeding yg in hole on 27 Aug). A new site, and a most unusual record, involved several prs nesting in hay-ricks in W Devon *(see col plate)*. There were no birds at the usual site near Dury Farm, *Postbridge* on 5 May. Apparent territorial activity over drainage holes in a retaining wall observed on *Lundy* in Apr (LFS), but no subsequent reports of breeding. (N.B. Colonies in the Burlescombe and Uffculme areas can be very mobile, even within a breeding season; the birds prefer newly exposed sand faces, and at times are affected by heavy rain (PC)).

Spring. *First* four at Ilfracombe on 8 Mar (NST,IM), followed by eight at Bowling Green Marsh on 11th; thereafter more or less daily. Of 48 counts ≥50, 42 were in Apr, the *highest* being at Slapton Ley (*c.*1,000 on 22 Apr, 500 on 21st and 300 on 18th) and Sherpa Marsh (300 on 25 Apr). Other spring counts >100 were from Bowling Green Marsh (seven dates, 2-26 Apr), Lundy (8 & 29 May), Roadford Res (3 Apr) and Slapton (24 & 25 Mar and 3,10 & 20 Apr).

Autumn. Counts of 50 in late Jun at Bowling Green Marsh perhaps heralded the start of autumn migration; of eight subsequent counts ≥50, two were in Jul, five in Aug and two Sep, the *highest* being at Prawle (500 on 21 Aug and 300 on 26th), Slapton Ley (200 on 31 Aug and 100 on 29th) and Beesands (100 on 19 Aug). On 6 Jul at Dawlish Warren, a flock of 41 feeding over a calm sea was an unusual sight. The *last* birds were two at Start Farm on 21 Oct and one at Northam Burrows on 23rd (MAS).

SWALLOW *Hirundo rustica* **Barn Swallow**

Numerous migrant breeder and passage visitor. SPEC(3), BoCC (Amber).

Later main arrival compared to 2000, and lower peak counts in both spring and autumn. First Dec records since 1997 and some evidence of a good breeding year.

Breeding. The very few breeding reports suggest a productive season; at Bremridge, Ashburton, six nests had 22-26 yg in Jun, and four had second broods in Sep; at Owley, South Brent, it was "a very successful year for productivity from the six nests"; and at the Grand Western Canal a pr were feeding five yg in Jul. Also a probable breeding record for central Plymouth (Central Park allotments – the first seen there in eleven years (SGe)).

Spring. The *first* was at Slapton Ley on 13 Feb (*per* ML), with no other records until mid-Mar when singles at Slapton on 15th and Seaton Marsh on 17th and four at Dawlish on 18th. Daily from 21 Mar (about a week later than in 2000), with the first double-figure count being 11 over Berry Hd on 23rd. Apart from 30+ over New Bridge, R Dart, on 31 March, the first large flock was of 150 at Yeo Vale on 10 Apr. The *largest spring groups* were on Lundy (1,800 on 26 Apr, 1,000 on 29th, and 300 on 8 May), and although up to 200 were recorded from the mainland in late April, there were only 18 counts ≥50 in the spring, nine occurring 26 Apr – 2 May.

Autumn. Of 20 flocks ≥200 recorded in autumn, 16 were in Sep, four in Aug and two in Oct. *Substantial passage* was from late Aug to mid-Oct. The largest flocks were 5,000 on Lundy on 18 Sep, *c.*1,000 at Lee Moor on 21 Sep, 1,000+ at both Soar and Thurlestone next day and at Prawle on 9 Oct. The largest number at Dawlish was 550 on 19 Sep, and the peak passage at Staddon Pt was 15-25 Sep with the highest daily passage being 700+ E in *c.*40 mins on both 23 & 24 Sep. In Nov, almost daily from various S coast sites (max four at Prawle on 4th) until 14th, then none until the *last birds*, singles at Exeter on 1 Dec (RWk) and at Dawlish Warren on 13th (KRy).

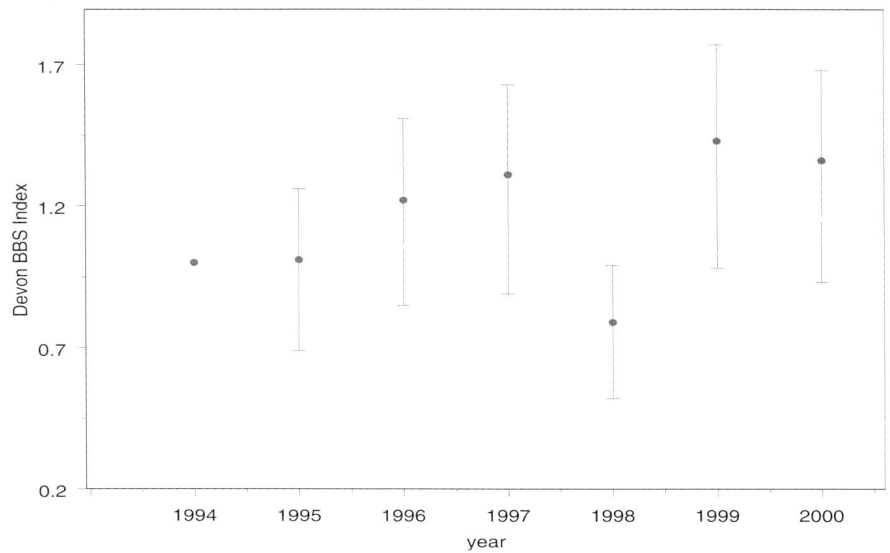

Fig. 24: Swallow **-** Devon BBS Indices 1994 - 2000, showing means with 95% confidence intervals (Source: BTO; see seperate report)

RED-RUMPED SWALLOW *Hirundo daurica*

Rare vagrant.

Recorded in the county for the fifth consecutive year.

One at Lundy 26 Oct (M Ferris, M James) (Accepted by BBRC).

2000 addition. One at Hallsands on 30 Apr 2000 (PJ Barden), the year's second. (Accepted by BBRC).

HOUSE MARTIN *Delichon urbica*

Numerous migrant breeder and passage visitor. BoCC (Amber).

Later first arrivals, lower spring peak and fewer large flocks than in 2000; perhaps also a slight decrease in overall breeding numbers.

Breeding. Few records submitted: 149 occupied nests in Aveton Gifford parish (including 100 in village) was the same as in 2000; 40 prs in Clennon Valley; 25+ prs at Hunter's Court; and 25 occupied nests at Sherracombe (The Poltimore Arms) on Exmoor. The Devon *House Martin Population Survey*, incorporating some of the above data, recorded 542 occupied nests, compared to 713 in 2000, but there were three fewer counters, and if only the 19 study areas counted in both years are included, then the decrease is one of 8%, from 561 to 514 (RCT; see also *The Harrier* No.69).

Spring. *First arrivals* were singles on 18 Mar at Staverton (PE) and at Slapton the next day. Then almost daily from 24 Mar, but the only flocks ≥50 were in May: 75 at West Putford and 80 on Lundy on 8th; 300 at Slapton Ley on 28th; and 100 on Lundy on 30th.

Autumn. Of 14 flocks ≥200, one was in Jul, five in Aug and eight in Sep. Several large flocks reported *25-26 Aug* (600 at Prawle and Elberry Cove on 25th; 396 at Dawlish Warren and 300 at Prawle on 26th) and *17-25 Sep* (450 W at Exmouth on 17th, the same day that the max count of 137 occurred at Staddon Pt and 50 were over Roadford Res; a Lundy max of 400 on 18th; thousands along the E Devon coast on 19th; 1,000+ passing Soar on 22nd; 500 sheltering from strong winds in the Church Valley at Wembury on 23rd; 188 at Dawlish Warren on 24th; and 600 at Elberry Cove on 25th). There were then no records between 25 Sep and 1 Oct (two at Holne Bridge). Most subsequent records were *9-13 Oct*; the highest counts were on 9th with 142 at Dawlish Warren, 60 nr. Kingswear and 100 in the Prawle area; and 12 were recorded from both Coleton Fishacre and E Prawle on 12th. A late high count of 50 at Berry Hd on 22nd preceded the *last birds*, two at Prawle on 27 Oct (PMM).

SYNCHRONISED COUNTING OF HOUSE MARTIN MIGRATION

A bold attempt to synchronise counts of House Martin movements was carried out on 19 Sep when eleven observers from The Axe Vale & District Conservation Society positioned themselves at eight cliff sites along 37km of coast between Cain's Folly (Dorset) and Orcombe Pt. In a firm NW wind, it was estimated that the observers may have had a total of 5,000 House Martins and Swallows in their view at 08.00-09.00h. If it is assumed that each observer could count over a distance of 0.5km, and that birds were distributed all along the coast, it was also calculated that as many as 46,000 House Martins and 4,000 Swallows could have been present in the study area at that time. The highest number actually observed was at Branscombe Mouth ("thousands") whereas none were seen at Axmouth Harbour and Seaton Marshes. Birds were variously described as circling, coming in off the sea or drifting east. *All information supplied by DEDC.*

RICHARD'S PIPIT *Anthus novaeseelandiae*

Rare vagrant.

Another three autumn records of this now annual visitor.

One over Ugborough Moor on 22 Sep (MD), one all day on the headland at Berry Hd on 18 Oct (ML *et al.*) and one in Tillage Field, Lundy 23-25 Oct (M Ferris & M James).

TREE PIPIT *Anthus trivialis*

Fairly numerous migrant breeder, scarce passage visitor. BoCC (Amber).

Similar distribution and number of records to 2000, but breeding comparisons affected by FMD access restrictions.

Breeding. Breeding confirmed only at Cann Wood and nr Venford Res (food-carrying ads), but likely at 25 other sites. *Strongholds* are around Dartmoor, Culm Measures in the NW, and across mid-Devon to the East Devon Commons. *Sites with four or more prs* were: Aylesbeare Common (six); Bellever

Plantation (six); Burrator area (four at Combshead Tor and six at Deancombe); Huccaby (four); and Warren House area (four). Only two prs were recorded for Haldon Forest (*cf.* 50 prs five years ago and four in 2000).

Spring. *First arrival* was one at Yarner Wood on 28 Mar (PP); then two displaying at Witherdon Wood on 12 Apr before becoming more widespread from 20th. Migrants included two at East Prawle on 20 Apr, then singles at Slapton on 21 Apr, at Prawle Pt on 30 Apr and 3 May, and at Dawlish Warren on 8 & 12 May.

Autumn. The first migrant was at West Putford on 6-7 Jul (AMJ), but not noted at coastal sites until late Aug with a single at Start on 23rd, followed by six at both Soar on 25th and Berry Hd on 26th. At several coastal sites by end of Aug, with a max of 16 at Start Pt on 26th. Passage continued through Sep, with a max count of six past Staddon Pt on both 15 & 17 Sep during the 15-23 Sep peak period, but there were only three Oct records, the max being yet another six on 13 Oct. The last birds were two NW past Staddon Pt on 21 Oct (SGe).

Richard's Pipit - *Mike Langman*

MEADOW PIPIT *Anthus pratensis*

Numerous resident breeder, passage and winter visitor. SPEC(4), BoCC (Amber).

Compared to 2000, counts lower in the first winter-period but similar in the second.

Table 87: ***Monthly distribution of flocks ≥25 and maximum counts:***

	J	F	M	A	M	J	J	A	S	O	N	D
Number of flocks ≥25	4	2	7	-	-	-	1	1	34	45	7	3
Maximum counts	50	28	76	24	17	6	30	30	300	1300	100	50

Breeding. Clearly under-recorded, particularly because of FMD access restrictions, as the only records came from Bursdon Moor, Exminster Marshes (nine prs), Prawle area, Start Pt (six prs) and Yar Tor (nest with yg).

First winter-period/spring. The max winter count was 50+ at Slapton Ley on 28 Jan, and other counts ≥25 were from Waddeton, Exminster Marshes, Isley Marsh and Great Western Canal; in addition 29 were recorded from 1km^2 at SX8478 during the BTO Winter Farmland Bird Survey. The largest spring counts were from Dawlish Warren during Mar and Apr, peaking at 76 on 15 Mar.

Autumn/second winter-period. Detailed observations at Staddon Pt (SGe and see separate report) indicated that the main passage was 15 Sep – 13 Oct with a max of 310 on *10 Oct* (*c*.200 per hour). The highest count at Dawlish Warren was also on 10 Oct (IL), with 1,217 in four hours giving a passage rate of *c*.300 per hour. *Passage rates* of just under 200 per hour also occurred on *4 October* at both Dawlish (total of 751) and Staddon Pt (total of 187), but the highest count of the autumn was 1,300 in the Soar area on 9 Oct, a day which also produced 424 past Dawlish and 300 at Berry Hd. *Other autumn days* with high counts included: 26 Sep (300 at Start Pt); 29 Sep (300 in Prawle area); 8 Oct (the Lundy max of 500); 12 Oct (300 at Start Pt, 200 at Soar); and 13 Oct (250 at Start Pt, 222 past Staddon Pt and 210 at Prawle Pt). Much smaller numbers occurred from then on, with peaks in late Oct of only 90 (at Dawlish Warren on 29th), in Nov of 100 (at Prawle on 10th), and in Dec of 50 (at West Charleton Marsh on 22nd and Exeter Peamore on 28th). Many of the autumn birds were counted in flight, but birds on the ground included 70 roosting at Mardon Down on 18 Sep, and 60 feeding on craneflies at Beer Hd on 21 Sep.

ROCK PIPIT *Anthus petrosus*

A. p. petrosus: fairly numerous resident breeder, passage and winter visitor.

Recorded from about 30 sites (similar to 2000) with a slightly higher max count at the main site. Lower numbers in the first winter-period compared to 2000.

Table 88: Monthly distribution of flocks ≥10 and maximum counts

	J	F	M	A	M	J	J	A	S	O	N	D
Number of flocks ≥10	1	1	-	-	-	1	1	4	4	4	3	5
Maximum counts	20	29	7	2	5	13	13	28	21	33	15	76

Breeding. Only recorded from Budleigh Salterton, Dawlish Warren, Prawle Pt and Sidmouth, though counts of 13 at Wembury in Jun and Jul indicate that this important site is not just a wintering area.

First winter-period. Very few records and the only counts in excess of ten were at Wembury with 20 on 20 Jan and 29 on 28 Feb (*cf. c*.50 there in both months in 2000).

Autumn/second winter-period. Most records from Wembury, where the highest counts were all in Dec (76 on 1st, 63 on 6th and 50+ on 26th). Otherwise, the only counts ≥10 were in the Prawle area (28 on 25 Aug, 40 on 9 Dec and 21 on 29th).

WATER PIPIT *Anthus spinoletta*

Scarce passage and winter visitor.

Similar to other recent years, with 32 records from 19 sites (cf. 37 records from 17 in 2000); almost all on the south coast and estuaries and mostly in Jan and late Dec.

Table 89: ***Monthly distribution of all records:***

	J	F	M	A	M	J	J	A	S	O	N	D
Sites	8	2	2	3	-	-	-	-	-	1	2	7
Birds	16	4	2	3	-	-	-	-	-	1	2	11

First winter-period/spring. In *Jan* recorded from Axminster Marsh (one on 1 Jan (DWH)); Otter Est (max three on 7 Jan (BBH)); Teign Est (one at Passage House on 6 Jan (DS)); Thurlestone Marsh (four on 15 Jan (PJR)); Topsham (one on 19 Jan (BG)); Blaxton Marsh, Tavy Est (three on 25 Jan (NST)); Totnes (one on 27 Jan (JLb)); and Isley Marsh (one on 29 Jan (JEW,RSPB)). In *Feb*, one at Topsham 4-28 Feb

(MKn *et al.*) and up to three on Otter Est on 27th (THS). *Thereafter*, singles only at Totnes on 1 Mar and 1 Apr (JLb), Exeter Riverside CP 20-24 Mar (JRD), Hallsands on 7 Apr (AKS) and the last, in summer plumage at Exminster Marshes on 17-18 Apr (RD,TAS).

Second winter-period. *Early records* were from the Taw/Torridge Est (singles at Velator on 29 Oct and 20 Nov and at Braunton Burrows on 21 Nov (RJ)) and Stoke Pt (one on 13 Nov (TF)), but in *Dec* came from the Axe, Exe, Kingsbridge and Otter Ests, Totnes and Prawle as follows: one on the Starcross golf course on 1 & 26 Dec (DJG); two on the Otter Est on 16 Dec (GV); at least one at West Charleton Marsh 14-31 Dec and two in Bowcombe Creek on 27 Dec (PBo,PSa *et al.*); one at Horsley Cove, Prawle on 29 Dec (JCN); one at Colyford on 30 Dec (IJW); and one at Totnes on 31 Dec (JLb).

YELLOW WAGTAIL *Motacilla flava*

M. f. flavissima. Scarce passage visitor; casual breeder (last bred 1986 and probably 1994). BoCC (Amber).

Fewer records than in 2000 and smaller flocks; as usual; mostly in the autumn, with 25-27 Aug being key days in the 14 Jul – 27 Oct passage period.

Table 90: Monthly distribution of all mainland records:

	J	F	M	A	M	J	J	A	S	O	N	D
Sites	-	-	-	6	9	-	1	13	17	6	-	-
Birds	-	-	-	18	37	-	1	113	86	10	-	-
Maximum counts	-	-	-	3	5	-	1	26	6	-	-	-

Spring. The *first bird* was at Bowling Green Marsh on 14 Apr and the last of spring at nearby Exminster Marshes on 23 May (RSPB); the max Apr count of three on 22nd was also at the latter site, whereas the max count for May, and for the whole spring passage period, was the five at Dawlish Warren on 2 May. Apart from these Exe Est sites, ones and twos were recorded from Aveton Gifford, Berry Hd, Prawle, South Huish Marsh, and the Teign Est on the S coast, and from Northam Burrows and Velator in the north. On Lundy, there was a max of three on 13 May.

Autumn. As in 2000, the *first* to return was at Dawlish Warren, a single on 14 Jul (*cf.* 31 Jul in 2000). The next flew over Ashburton on 1 Aug (the only *inland* record apart from one at Roadford Res on 19 Sep). There were only two other records in the first three weeks of Aug, but from 20 Aug to end of Sep, records came from 25 sites, though with a max count of only 26 and only four *double-figure flocks*. Three of the latter occurred on 26 Aug (22 past Exmouth seafront (ARB); 14 at Exmouth seafront (GS); and 26 at Start Pt (PSa)) and the other was 20+ in the Soar area on 25 Aug (SRT,VRT). Smaller numbers were also recorded from Axmouth, Dawlish, Prawle, South Huish and Start at this time, but *counts >five* came only from Bolberry Down (max six on 18 Sep), Budleigh Salterton (max six on 2 Sep), Dawlish Warren (max eight on 27 Aug), Prawle area (max six on 27 Aug and 12 Sep) and Start farm (max seven on 23 Aug). The max *Oct* count was of three at Prawle on 10th; then singles at Prawle and Start on 13th were followed by one at Berry Hd on 18th and the *last of the year* at Alphington on 27th (CSt).

M. f. flava ('Blue-headed Wagtail'). Rare passage visitor.

At least two mainland ♂♂: one at Velator on 4 Apr (JEW); a summering individual in the Prawle area 7 May – 11 Aug (PMM *et al.*), though no records for it in Jun; and one at Slapton on 28 May (PSa).

GREY WAGTAIL *Motacilla cinerea*

Fairly numerous resident breeder, not scarce passage/winter visitor. BoCC (Amber).

No evidence for any change in numbers or distribution. Widely recorded especially during autumn, with some useful daily counts of migrants.

Breeding. The only records came from: Beam Weir, Bridgerule, Bow, Buckfastleigh (three prs), Dawlish, Dewerstone, Farley Water (three prs), R Heddon, Honiton Bottom, Huccaby, Hudley Mill, Ideford Mill, Jennet Pt NNR, Lustleigh, Lydford Gorge, Orford Mill, Piles Copse, Plymbridge, Shipley Bridge, Taw Est, Tiverton (a factory leat), Totnes and Yeo Vale.

Non-breeding. A marked contrast between the highest count in the first winter-period of five at Kilbury on 22 Feb and the six counts ≥10 in *Sep* (though mostly birds moving along the coast): max daily count past Exmouth seafront was 22 on 10 Sep out of a total of 115 going W 25 Aug – 22 Sep (ARB); at Staddon Pt a total of 62 going W in autumn (peak period 15-29 Sep) included 10 on 15 Sep (inc. one group of four), 12 on 16th, 10 on 17th and 10 on 20th (SGe); at Dawlish there were ten on 15 Sep and 12 on 22nd. In *Oct*, the max coastal count was of six at Dawlish Warren on 10th, but nine were recorded from the R Culm at Culmstock. On Lundy, one on 20 Sep conforms to the mainland passage period.

PIED WAGTAIL *Motacilla alba* — White Wagtail

M.a.yarrellii. Fairly numerous resident breeder and passage/winter visitor.

Smaller and fewer large counts compared to 2000.

Table 91: Monthly distribution of flocks ≥25 and maximum counts

	J	F	M	A	M	J	J	A	S	O	N	D
Number of flocks ≥25	2	2	-	-	-	-	-	8	17	19	4	2
Maximum counts	60	50	21	11	8	16	16	60	135	292	300	45

Breeding. Only recorded from Barton Pines, Dartmeet, Dawlish Warren, Exminster Marshes, Exmouth, Grand Western Canal (Crownhill Bridge), Higher Metcombe, Huccaby, Prawle, Stokenham, Thurlestone and West Putford.

First winter-period/spring. None of the large roosts reported in 2000, the maxima being 60, on a catamaran in the Avon Est on 2 Jan, up to 50 at Exeter Peamore on 16 Feb and 30 at Passage House Inn on 6 Jan.

Autumn/second winter-period. The first large gatherings were in Aug with 50-100 at South Molton and 60 at Slapton Ley. Peak *passage* at Staddon Pt was *20 Sep – 13 Oct* with a max daily count of 35 passing 07.05-08.35h on 10 Oct (SGe). The highest Dawlish Warren count of 292 was also on 10 Oct (see also under Meadow Pipit), with 151 there on 9 Oct and 135 on 29 Sep (IL). Other high coastal counts included *c.*100 feeding around pig-pens at Otterton cliffs on 4 Sep, and a similar number by the R Exe on 23 Nov, at Soar on 9 Oct and at Prawle on 12 and 14 Oct. *Roost counts* included: the highest count of the year – *c.* 350 at Lee Moor China Clay works on 21 Oct, 300 at Fleet Mill, R Dart on 9 Nov; 120 in a Plane tree on The Plains at Totnes on 13 Nov; *c.*100 in a pre-roost gathering at Sourton Cross on 30 Oct; and 90 at Hennock Res on 20 Sep. Not all records specify activity or habitat, but the largest gathering among several reported from ploughed fields was 50 at Galmpton on 4 Oct.

***M. a. alba* ('White Wagtail').**

Not scarce passage visitor; bred 1956.

An average year, with records from ten sites in spring and 14 in autumn, but more birds in both spring and autumn compared to 2000.

Table 91: Monthly distribution of all records:

	J	F	M	A	M	J	J	A	S	O	N	D
Sites	-	-	1	7	4	1	-	4	12	4	1	1
Birds	-	-	3	40	8	2	-	14	84	39	1	1
Maximum counts	-	-	2	26	5	2	-	4	22	28	1	1

Spring. Late in arriving compared to 2000, the *first* were up to two at Dawlish Warren 22-30 Mar (IL). No other records until the spring peak of 26 at Dawlish on 10 Apr (LC), and then more widespread but with a late Apr max of only three (at Sherpa Marsh, RJ *et al.*); singles occurred at Bowling Green & Exminster Marshes, East Prawle and South Huish. Passage continued until 13 May when there were five on Lundy (ML), but the only mainland May records came from Soar, S Huish and Velator. An isolated occurrence was of two on Goodrington seafront on 5 Jun (DHa).

Autumn. Three at Dawlish Warren on 19 Aug, peaking at four on 27th, (IL) were the *first* of the autumn passage, which also included one at Brixham on 22nd (RJJ), three at South Huish Marsh on 24th (NPR) and two in Pig's Nose Valley, Prawle on 28th (NPR). In *Sep* recorded on most days, the max being 22 at Pig's Nose Valley on 8th. Six to eight occurred in the Prawle area on 1st, 9th and 17th (JCN,PMM), at Dawlish on 18th (JEF,IL) and at Start on 26th (ML); otherwise one-three only from Berry Hd, Instow, Lundy, Slapton, Soar and Staddon Pt. *Inland* records included one at Upper Tamar L on 8 Sep (BMc) and five at West Putford on 15th (AMJ). Only six records in *Oct* included high counts from

Mt Batten, Plymouth Sound (28 W on 28 Oct (TF)) and 10+ at Prawle Pt on 10th (G&BJ); otherwise singles at Broadsands on 6th (BMc) and at Slapton Sands on 9th (BW). In *Nov*, the only record was of one at Dawlish Warren on 6 Nov (KRy), this being followed by a *very late bird* at Chelston, Torbay on 24 Dec (CJP).

DIPPER *Cinclus cinclus* — White-throated Dipper

Not scarce resident breeder.

Recorded from just 61 sites (cf. 60 in 2000, 75 in 1999), distributed over 20 rivers, but generally under-recorded, with few records from the main moorland strongholds. Total included 26 sites during the breeding season (Feb–Jun), probably referring to c. 28 breeding prs (cf. 35 in 2000, 57 in 1999, 42 in 1998, 64 in 1997 and an estimated county total of 500–700 prs in the Tetrad Atlas 1977–85), with breeding confirmed at just 13 sites.

Breeding. Records (single prs unless stated otherwise) from: *R Torridge*, Halsdon DWTNR; *R Exe* at Tiverton, Ludwell Valley, Sowton Br (ad feeding juv); *R Gissage*, Honiton Bottom (ad feeding 2 yg); *R Otter*, Higher Metcombe (juv); *R Carey*; *R Kenn*, Babel's Br (feeding yg); *R Bovey*, Houndtor Wood (fly-catching); *R Teign* at Trenchford Res (singing), Fingle Br, Dunsford (juvs); *R Hems*, Littlehempston (collecting nest material); *Gara brook*, Gara Br; *R Dart* at Newbridge, Dartbridge (singing), Buckfastleigh (ads feeding juv), Holne Br, Ashburton, Dartington (yg in May), *R Avon*, Shipley Br; at Portworthy (in a working claypit); on Devon Exmoor, *E Lyn River*, Watersmeet (nestbuilding); and elsewhere along *N Coast*, *R Sterridge*, Berrynarbor (ad feeding juv). No breeding at two regular sites on *R Lemon* at Newton Abbot and *R Avon* at Hatch Br.

Non-breeding. Found at most of above locations, but also *R Culm* at Uffculme and Culmstock; *R Coly*; *Farley Water*; Efford Marsh, Plymouth; *R Yealm*, Puslinch Br; Spekes Mill, Hartland; and Heddon's Mouth. Noteworthy was one flying along centre line of B3231 nr Braunton on 23 Oct, possibly mistaking the very wet road surface for a watercourse (P&RM), and one flying over moorland at Merripit Hill at dusk on 25 Nov (RS).

WREN *Troglodytes troglodytes* — Winter Wren

Abundant resident breeder.

Breeding. Confirmed breeding at 11 sites, and well distributed along Grand Western Canal. Notable counts: up to nine singing at Clennon Valley, Paignton; six at Seaton Undercliff, and five at Kenwith NR. Present at another 28 sites during breeding season, including the only Dartmoor records at Meldon and Soussons Plantation. Undoubtedly under-recorded.

Non-breeding. Notable *higher counts*: 21 on 5 km circular walk nr Stockland on 14 Jan; second winter period max 20 at Prawle area on 27 Oct and 10 Nov (JCN); 14 at Witheridge Moor on 23 Dec.

DUNNOCK *Prunella modularis* — Hedge Accentor

Abundant resident breeder. SPEC(4), BoCC (Amber).

Recorded from 66 widespread localities across the County.

Breeding. Few records, but breeding proven at 11 sites with 22 territories. Notable counts: 14-16 at Prawle area in May; five prs at Netton, nr Noss Mayo on 23 May; three sang at Clennon Valley, Paignton, on 12 May, where two prs raised seven yg; two prs and four fledglings at Manor Gardens, Exmouth; at Okehampton pr bred but yg taken by cat; on Dartmoor juvs at Huccaby. At least 16 individuals in Prawle area Apr-May (*cf.* 21 in 2000, 20-34 in 1999).

Non-breeding. Peak counts: 26 on 5 km circular walk nr Stockland on 13 & 14 Jan; ten at Dawlish Warren on 2 Dec; 50 at Prawle area on 20 Oct with 30 there on 10 Nov.

ROBIN *Erithacus rubecula* — European Robin

Abundant resident breeder; probably not scarce passage/winter visitor. SPEC(4).

Breeding. Confirmed at only 14 sites, with breeding season (Mar-Jul) records from another 29; a very early incubating bird in a nestbox inside a factory at Tiverton on 2 Feb (RJJ); four singing at Kenwith NR on 12 May; three prs and no successful fledging until May at Higher Metcombe; two prs fledged five yg at Manor Gardens, Exmouth; successful broods also at Clennon Valley (Paignton), Lydford Gorge, Northam, Prawle area, Stokenham, Thurlestone, and on Dartmoor at Huccaby and Piles Copse.

First winter-period. Max counts included 37 along 5 km walk nr Stockland on 14 Jan, seven at Stover CP on 2 Feb and 19 along 4 km of the Grand Western Canal (Greenway Br-Ayshford Br) on 10 Feb.

Autumn/second winter-period. Perhaps two influxes of migrants: late Sep in Plymouth area with 17 at Mt Gould hospital grounds on 21st (unprecedented numbers here), 20+ at Jennycliff on 25th and, just along coast at Staddon Pt, 10+ feeding in a loose flock on same date; at Prawle area counts of 20+ from 25 Aug-10 Nov, max 110 on 20 Oct and 100+ on 27 Oct, when other high counts at this time included 47 at Berry Hd on 20 Oct, 20+ at Staddon Pt (see also Black Redstart). Also 14 continental types around the lighthouse at Start Pt on 14 Oct. On Lundy, there were ten counts in the range 40-65 birds between 8 Oct-5 Nov, suggesting significant passage (A Taylor). Late autumn/winter counts included at least 20 between Shipley Br and Avon Dam on 19 Nov and 30 at Prawle area on 31 Dec. One was well out on the open moor, amongst rock clitter at Wild Tor Well, Dartmoor, on 15 Dec.

NIGHTINGALE *Luscinia megarhynchos* — Common Nightingale

Very scarce/rare migrant breeder and passage visitor. BoCC (Amber), SPEC(4).

Three spring records, including one probable breeder.

One singing at Duryard CP, Exeter, 26 Apr-30 June, was present for the fifth consecutive year, and probably indicative of a breeding pr (RAda *et al.*). Another sang at Hope's Nose 5 Apr-22 May (WJD *et al.*) but no evidence of breeding, and a migrant was seen in scrub at Pottington on 26 Apr but did not sing (RJ).

BLACK REDSTART *Phoenicurus ochrurus*

Scarce passage visitor and very scarce winter visitor; bred 1942 and 1949. BoCC (Amber).

A typical year with unremarkable numbers in either winter-period; few spring records and fewer autumn migrants than recent years, with numbers dwindling into Dec. Monthly presence fairly typical apart from one particularly early migrant. Recorded at 64 mainland sites (cf. 67 in 2000, 87 in 1999, 57 in 1998), with, as usual, a mainly southerly distribution: 77% of records in S of county; 9% inland including two on high ground of Dartmoor; and 14% N Coast. A probable total of 124 birds (allowing for duplication of long-stayers) (cf. 114 in 2000, 159 in 1999, 101 in 1998).

Table 93: Monthly distribution of all records (long-staying birds counted only once in month):

	J	F	M	A	M	J	J	A	S	O	N	D
Sites	24	6	12	2	2	-	-	-	1	23	14	21
Birds	29	8	12	2	1	-	-	-	1	58	18	25

First winter-period/spring. On the mainland in *Jan-Feb*, 28 birds (including nine ad ♂♂) most along *S coast*, except singles at Abbotsham Cliffs, Baggy Pt, Heddon's Mouth, Ilfracombe, and inland at Buckfastleigh. Max of three at both Ladram Bay and Thurlestone. *Spring migrants* included singles at Kingston (♀) on 5 Apr, Dawlish Warren (♂ in moult) on 11 Apr and Lundy on 13 May.

Autumn. The first of autumn was ♀/1w bird at Buckfastleigh on 28-29 Sep, followed by a ♂ and two imm/♀♀ on 4 Oct at Devonport Dockyard (see also wintering), with main influx from mid-Oct beginning with singles at Prawle Pt on 13th & 18th and Berry Hd on 14th. Significant arrivals from 19 Oct, with six in Prawle area and four at Start Pt; followed on 20-22 Oct by another nine at Prawle area and widespread records of ones and twos elsewhere along the S Coast from Budleigh Salterton to Staddon Pt, Bovisand; also at Dartmeet and Mel Tor (Dartmoor). A single was also on Lundy on these dates, but maximum day count here was 14 on 30 Oct. Singles then noted through into Nov with three in Prawle area on 10 Nov. Described as the smallest autumn passage for some time at Kingston, nr Erme Est.

Second winter-period. Wintering birds present from 4 Oct at Devonport (where a ♂ returned to same locality as previous two winters), and from 17 Nov at Ayrmer Cove, Ringmore. In Dec, an estimated 28 wintering birds (including eight ♂♂) at 22 sites (*cf.* 26 at 22 sites in 2000, 19 birds at 16 sites in 1999, 41 birds at 29 sites in 1998). Most on *S coast* with at least seven in *Plymouth area* and four together in the car park at South Huish on 16 Dec. None *inland* this year. On *N coast* singles at Appledore, Baggy

Above: Bobolink at Prawle, Paul Nunn, October 2001

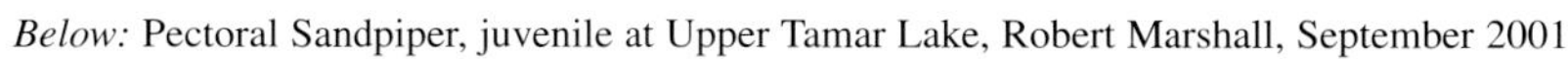

Below: Pectoral Sandpiper, juvenile at Upper Tamar Lake, Robert Marshall, September 2001

Top left: 1st Winter Franklin's Gull (centre) and adult Ring-billed Gull (left) among Common Gulls, Teign Estuary, April 2001. - Bob Normand

top right: Male Black Duck, Slapton Ley, Feb 2001 - Chris Proctor
above left: Little Ringed Plover, Upper Tamar Lake, Sep 2001 - Robert Marshall
above right: 1st summer Night Heron, Stover CP, Apr 2001 - Ron Champion

below: Iceland Gull, Teign Estuary, Jan 2001 - Bob Normand

Above: Bobolink at Prawle, October 2001, another view - Bob Normand

Below: Juv. Curlew Sandpiper, Upper Tamar Lake, Sep 2001 - Robert Marshall

This page: above: Yellow-browned Warbler, Lundy, 1 October 2001 - A. M. Taylor
middle & bottom left: 1st Winter Franklin's Gull, Hope's Nose, Mar 2001 (with Black-headed Gull) - Chris Proctor
below right: A rare picture of a Short-eared Owl in flight, Skern, Oct 2001 - Robert Marshall
bottom right: Adult Long-tailed Skua, Hope's Nose, Jul 2001 - Chris Proctor

Opposite page: top left: Sand Martins nesting in hay-rick, W. Devon *(see systematic list)* - Chris Robbins
top right: Greenfinch showing partial xanthrochroism in plumage, Dunsford, spring 2001 - John Woodland
upper middle left & right: nestling Nightjar about to fledge and adult male Nightjar, Stover CP. - Ron Champion
middle left: Male Snow Bunting, Ilfracombe, Mar 2001 - Robert Marshall
middle right: Juv. Arctic Tern, Westward Ho! Oct 2001 - Robert Marshall
bottom: Scarlet Rosefinch, Lundy, Oct 2001 - Richard Patient

87J

'his page:
ıbove: Little Auk, Broadsands, Nov 2001 - Chris
'roctor
ıiddle: Rose-breasted Grosbeak, Lundy, Oct
!001 - Richard Patient
ottom: Grey Phalarope, Upper Tamar Lake, Oct
!001 - Robert Marshall

Opposite page:
top: Juv. Mediterranean Gull, Westward Ho! Sep
2001 - Robert Marshall
bottom left: Raven, Lundy, May 2001 - Robert
Marshall
bottom right: Green Woodpecker, Yeo Vale, May
2001 - Robert Marshall

Above: Kentish Plover, Dawlish Warren, Mar 2001 *(see systematic list for details)* - Bob Normand
Below: Chiffchaff *(tristis / abientinus),* this individual shows many of the characteristics of the race *tristis,* Lundy 29 Oct 2001 - A. M. Taylor

Map 9: Black Redstart - Migrants
Maximum count of migrants (Sep-Nov & Mar-May) at each location.
Data: Based upon a ten year period 1992-2001. Source: DBRs) (see Overview on page 25)

Map 10: Black Redstart - Wintering
Maximum count of wintering birds at each location.
Note the importance of urban areas, the dispersed occurence of migrants, and the highly coastal distribution of the wintering population. Records of wintering inland are scarce and particularly noteworthy.
Data: Based upon a ten year period 1992-2001. Source: DBRs) (see Overview on page 25)

Pt, Barnstaple, Hartland Quay and Woolacombe.

REDSTART *Phoenicurus phoenicurus* — Common Redstart

Fairly numerous migrant breeder and passage visitor. SPEC(2), BoCC (Amber).

Another below average mainland passage involving just 58 birds.

Table 94: Monthly distribution of all records (away from breeding areas):

	J	F	M	A	M	J	J	A	S	O	N	D
Sites	-	-	1	9	3	-	-	3	8	6	1	-
Birds	-	-	2	15	5	-	-	5	17	14	1	-

Breeding (Apr–Jul). Casual records of an estimated 40 prs/singing ♂♂ from just 30 sites on and around *Dartmoor* (*cf.* 36 sites in 2000, 30 in 1999, 32 in 1998, 47 in 1997, 25 in 1996, 42 in 1995 and 41 in 1994). *Casual records*: first was at Huccaby on 17 Apr, where three sang to 30 Jun; four at Piles Copse (3 prs feeding yg on 18 Jun), three at Combestone; two prs/♂♂ at Bellever Plantation, Burrator, Cadover Br, Criptor Newtake (juv noted), Dury Fm, Fernworthy, Hound Tor, Postbridge; and single prs/♂s recorded at a further 25 sites. On *Exmoor*, the usual few records, this year from Farley Water, a pr bred at Martinhoe village hall, and a pr fledged six from a nestbox at East Anstey (1 May-30 Jun); elsewhere along *N Coast* at least three juvs noted at Woody Bay on 16 Jun.

Spring. The *first* was a ♂ at Dawlish Warren on 29 Mar, with the same or another there the next day. Next were singles at Exminster Marshes on 8 Apr and Sharkham Pt on 13th. First multiple sightings were three at Berry Hd and ♂ & ♀ at Dawlish Warren on 14 Apr, when also a ♂ in an urban garden in Heavitree, Exeter. Then mainland migrants recorded on 11 further dates with a late ♀ at Budleigh Salterton on 25 May.

Autumn. The *first* migrants were singles at: Dawlish Warren on 25 & 26 Aug, Start Pt on 26th and East Soar on 27th. Ones or twos from many S Coast headlands in Sep, and also singles at Haldon Underdown and Buckfastleigh. The last record on Dartmoor (ad with a juv) again came from Fernworthy on 1 Sep. Twelve singles from 11 coastal sites during Oct (with two at Start Pt on 20 Oct), and the *last*, also at Start Pt, on 21 Oct.

WHINCHAT *Saxicola rubetra*

Not scarce migrant breeder and passage visitor. SPEC(4).

Similar numbers to past ten years during migration periods, but a dearth of breeding records continued a rather 'stable underestimate' of breeding numbers on Dartmoor. Follow-up survey work to the 2000 Moorland Survey may help clarify its breeding status.

Table 95: Monthly distribution of all records (away from breeding areas):

	J	F	M	A	M	J	J	A	S	O	N	D
Sites	-	-	-	4	10	-	-	10	12	6	-	-
Birds	-	-	-	5	17	-	-	16	52	32	-	-

Breeding. *Casual records* (May–Jul) from a meagre 14 *Dartmoor* sites, relating to a minimum 21 prs/singing ♂♂ (*cf.* 36 prs at 25 sites in 2000, 24 prs at 13 sites in 1999), including at least five at Headland Warren (and several family parties noted in late Jul); exact number in Warren House area supersite difficult to ascertain due to multiple site names and lack of grid references accompanying records; four-six at Ditsworthy Warren area and two at Grimspound, Piles Copse (feeding fledglings in bracken), Rival Tor, Soussons area and Venford Res. Single prs at a further 13 sites: Avon Dam, Burrator, Cadover Br, Challacombe Fm, Cuckoo Rock, Holne Moor, Holwell Tor, Horse Ford, Meldon, Newleycombe Lake, Ringleshute Mine, Sourton Cross, Statts Br. *Devon Exmoor* reverted to its usual under-watched state with nine prs at Farley Water and a ♀ at Martinhoe Common being the only records.

Spring. The *first* were single ♂♂ at Berry Hd on 24 Apr, Dawlish Warren and Exminster Marshes on 29 Apr, and at the latter site and Slapton Ley on 30 Apr. During May, a further 17 migrants reported from ten sites, including at least five at Dawlish Warren, with latest singles at Prawle on 21st and Berry Hd on 27th. First on *Dartmoor* was a ♀ at Sourton Cross on 8 May.

Autumn. The *first* on mainland: singles at Prawle area and Orcombe Pt on 19 Aug (same date as 2000); South Milton Ley on 22 Aug; then daily from 25 Aug at many S Coast localities, including four at

Axmouth Marsh on 26 Aug, three at Haldon Forest and two at Woodbury Common on 3 Sep. A total of 52 birds at 12 sites during Sep (*cf.* 74 at 18 sites in 2000), with max six at Prawle area on 29th, from where again the majority of records emanated: 20 birds in Sep, 18 in Oct. Elsewhere, ones-twos at 17 sites. *Last* on *Dartmoor* was at Mardon Down on 3 Oct, whereas the last mainland coastal migrants were three at Start area on 21 Oct and one at Prawle area on 27 Oct. On Lundy, a report of a late one on 2 Nov.

STONECHAT *Saxicola torquata*

S. t. hibernans: not scarce resident breeder, passage and winter visitor. SPEC(3), BoCC(Amber).
Continues to benefit from the series of mild winters. Well reported from across the county, although fewer breeding records this year due to FMD access restrictions. Breeding records from inland sites, away from the main moorlands, would be useful.

Breeding. A min of 183 mainland prs from *c.*85 sites Mar–Jul, a decline on previous years almost certainly due to FMD access restrictions (*cf.* 267 prs at 129 sites in 2000, 196 prs at 97 sites in 1999, 210 prs at 84 sites in 1998). On *Dartmoor*, remarkably, *casual* records obtained from roadside observation gave a similar total to last year despite FMD restrictions, with a min of 83 prs at 34 sites (*cf.* 86 prs at 49 sites in 2000, 82 prs at 42 sites in 1999, 70 prs at 40 sites in 1998), including (prs/singing ♂♂): 18 between Heatree Cross and Blackslade Water (including five at Bonehill Down, Blackslade Mire and three at Holwell Down); eight at Holne Moor, five or six family parties in Aish Tor area; five between Venford Res and Saddle Br; three at Corndon Down, Challacombe Cross-Firth Br and Wigford Down; and one-two at remaining sites. Late summer observations suggested a good breeding season on Dartmoor: e.g. 20+ juvs at Ingra Tor/Routrundle area on 17 Aug (RS), and a loose flock feeding in a field at Huccaby peaking at 14 on 9 Aug (RHi). Again poorly reported from *Exmoor*, with three prs at Farley Water. *Elsewhere* on *N coast*: only five records of single prs at Forda (nr Croyde), Foreland Pt, Hartland Pt, Woollacombe Down, Woody Bay (♀+2 juvs) and Yelland. On *S coast* at: Berry Hd (three); Dawlish Warren (two or three; yg from 17 May); Start Pt (three + eight fledglings); and Wembury (two). Casual records from *East Devon Commons* at Aylesbeare (two prs, juvs from late Jun-Jul) and East Budleigh (three including ♂ with juvs). *Culm grasslands*: min four prs at Witheridge Moor. *Other inland sites*: Haldon Forest (14 prs on heathland and young conifer plantation), Bovey Heath, Stoke Canon, Trinity Hill, and Upper Tamar L.

First winter-period/spring. A large increase in records, with at least 110 birds reported from 44 mainland sites (*cf.* 66 from 32 in 2000, 72 from 33 in 1999, 40 from 22 in 1998). Distribution mainly coastal; *inland* sites included 17 birds at ten sites on *Dartmoor* which was fewer than last year, probably due to less observer coverage. Difficult to distinguish migrants but a ♀ at Hendom Cross, Huntsham, on 29 Mar was unusual at that location, and nine at Dawlish Warren on 16 Mar may have included migrants. On Lundy, just one, on 4 Apr.

Autumn/second winter-period. Max counts, all at *coastal sites*: Prawle area 27 on 25 Aug, 59 on 24 Sep, 33 on 20 Oct, declining to just three on 29 Dec; Soar Mill Cove 40+ on 19 Sep but only one pr in whole Soar area by 20 Nov; Berry Hd 13 on 2 Sep and 20 Oct; Dawlish Warren 20+ mostly ♂♂ on 9 Oct; Start Pt 22 on 11 Sep (considered migrants) and 17 on 12 Oct; Wembury 20 on 18 Sep. On *Dartmoor* few autumn records, but *wintering* birds (one–three prs) at Dunnabridge, Fernworthy, Hew Down, Holne Moor, Mardle Head, Scorhill Down, Venford Res, Wild Tor Well. On *Exmoor* at Great Black Hill and Molland Common. On *Culm grasslands* at Witheridge Moor and Bursdon Moor. *Elsewhere wintering* at 22 lowland sites including Grand Western Canal, Hartland Pt, Isley Marsh, Kenton, Otterton Cliffs, Seaton Marshes, Stoke Canon, and Trinity Hill. On Lundy, the maximum day count was 20 on 21 Oct.

S. t. maura* (‘Siberian’ Stonechat) *(rare vagrant)*: *previously unrecorded in Devon.
Amazingly two records of this eastern race – the first in the county.

A ♀ or 1w was at Start Pt on 26 Sep (ML,BMc); see separate Report. This was followed by one at Lannacombe on 13 Oct (DS,ARHS). (These were among 14 records for 2001 accepted by BBRC, mostly in late Sep in NE England.)

WHEATEAR *Oenanthe oenanthe* — Northern Wheatear

O. o. oenanthe: fairly numerous migrant breeder and passage visitor.

Main part of text refers to O. o. oenanthe or those not racially identified (a proportion of passage records refer to O. o. leucorhoa). FMD access restrictions limited records of spring arrivals and breeding birds on Dartmoor. Conversely, a high number of coastal migrants were noted in both spring and autumn.

Table 96: Monthly distribution of numbers of birds at well-watched sites, and all records (away from breeding areas) :

	J	F	M	A	M	J	J	A	S	O	N	D
Berry Head	-	-	52	37	13	1	-	22	21	11	2	-
Dawlish Warren	-	-	145	58	24	3	7	83	43	21	2	-
Prawle area	-	-	-	62	41	1	3	76	110	25	-	-
Slapton Sands	-	-	55	28	3	-	-	6	-	-	-	-
Start Point	-	-	-	26	-	-	-	33	15	8	-	-
Sites	-	-	23	23	10	3	4	20	30	21	4	1
Birds	-	-	345	244	113	6	12	313	360	119	6	1

Breeding. On *Dartmoor, casual* records from roadside observations equated to min 41 prs at 31 sites (*cf.* 82 prs at 46 sites in 2000, 98 prs at 33 sites in 1999, 77 prs at 36 sites in 1998); all single prs, with juvs noted at Aish Tor, Burrator, Challacombe Down, Dewerstone, Grimspound, Haytor, Headland Warren, Vitifer, and Warren House Inn. *Elsewhere*: on *S coast* perhaps two prs bred at Start area (PE), a family party was seen here on 17 Jun (JHB); possible breeders or late migrants were one at Prawle Pt on 3 June and on *N coast* two at Skern on 29 May and a ♀ at Foreland Pt on 29 Jun. On *Exmoor*, a ♂ at Farley Water on 30 Jun. The coastal breeding population is surely underestimated, particularly on the *N Coast*; the Tetrad Atlas estimated 'breeding' in over 40 coastal tetrads. The current Devon Atlas Project should provide clarification.

Spring. The *first* arrivals were two at Frogmore on 14 Mar, then daily from 16 Mar when 12 at Dawlish Warren, eight at Slapton Sands, and singles at Berry Hd and Exeter Riverside CP. First on *Dartmoor* was one at Cadover Br on 19 Mar, followed by singles at Dunnabridge and Prince Hall on 25th, then none until a ♂ at Blackdown, Mary Tavy on 25 Apr and Merrivale the next day. Peak spring passage late Mar, with a smaller peak in late Apr. Double-figure counts: ten at Berry Hd and 20 at Dawlish Warren on 17 Mar, and the total of 145 (42% of county total) there in Mar reflecting greater observer effort of IL *et al.*; counts elsewhere were 12 at Beesands, 30 at Slapton Sands and 17 at Riverside CP on 24 Mar (when 100 at Portland, Dorset); 20 at Berry Hd on 27 Mar. During Apr, 28 at Prawle area on 16th and 17 on 30th, 15 at Berry Hd on 9th. Low single figure counts of migrants continued until early Jun with one at Dawlish Warren on 1st and 2nd and two at Berry Hd on 11th. Just one record on Lundy: seven on 13 May.

Autumn. The *first* migrants were singles at Prawle area on 22 Jul and Dawlish Warren (juv) on 23 Jul. More regular at *S coast* sites from 27 Jul. Majority of migrants at *Prawle area* with seven double-figure counts in Sep (max 17 on 9 Sep). Slightly fewer at *Dawlish Warren* where mostly single-figure counts except ten on 19 & 24 Aug. Elsewhere: in *Soar area,* max 20 on 18 Aug and 19 Sep; at *Start Pt*, 12 on 23 Aug, 11 on 26 Aug and 10 on 27 Aug; 18 (mostly juvs) at Haldon racecourse on 3 Sep; 11 at Thurlestone GC and at Kingswear on 23 Sep. The *last Dartmoor* record was one at Cox Tor on 10 Nov (RJL); and the last at *mainland coastal sites* were singles at Langerstone Pt on 17 Nov, and at Velator on 4 Dec (DC). On Lundy, there were two on 2 Nov.

O. o. leucorhoa* ('Greenland Wheatear') *(probably not scarce passage visitor).

Several probable records. In *spring*: at Huxham (Exeter) on 29 Apr (DJJ), and at Dawlish Warren (♂) on 19 May (GS) with two there on 29 May (IL). In *autumn*: two ♂♂ at South Milton Ley on 21 Sep (SRT,VRT), three (ad ♂, 2 ♀♀) at Northam Burrows on 2 Oct (IM), one at Start Pt on 12 Oct (DJJ), and singles at Berry Hd (G&BJ) and Beesands (ML) on 15 Oct.

RING OUZEL *Turdus torquatus*

Scarce migrant breeder (on Dartmoor and Exmoor) and passage migrant. SPEC(4), BoCC (Red), Dartmoor BAP.

Breeding records not surprisingly few, but a fairly good autumn passage both inland and on the south coast.

Table 97: Monthly distribution of all records:

	J	F	M	A	M	J	J	A	S	O	N	D
Sites	-	-	-	1	1	1	1	2	3	9	3	-
Birds	-	-	-	1	3	1	2	2	9	24	16	-

Breeding. The only records were three territories located in the Warren House area during the season, with five other records of singles or a pr from the same area. The other main *Dartmoor* stronghold, Tavy Cleave, went unvisited and there were no records from *Exmoor*. A late bird at Fur Tor on 20 Aug was probably a lingering local breeder.

Spring. The only recorded passage bird was a ♀ at Kingswear on 19 Apr (MI).

Autumn. By contrast, there were at least 22 migrants on *Dartmoor* (which saw an excellent autumn berry crop) 6 Sep – 3 Nov, with groups of three at Ugborough Beacon on 22 Sep, five at Shipley Bridge on 28 Sep and at Dartmeet on 11 Oct, and six at Meldon Res on 1 Nov. There were also Oct singles at Dartmeet on 1st, Watern Hill on 12th and Challacombe Down on 15th. The only *other inland* record was of a juv on the Haldon Ridge on the early date of 29 Aug. On the *coast*, passage began with one at Soar on 29 Sep, followed by at least 13 further birds here, including eight on 3 Nov and six next day. The Prawle area had at least six birds from 13 Oct, including four in Pig's Nose Valley on 21st. The *last* were two on Lundy on 7 Nov (LFS).

BLACKBIRD *Turdus merula* — Common Blackbird

Abundant resident breeder. Passage/winter status uncertain. SPEC(4).

In the absence of any survey data relating to breeding, difficult to determine anything meaningful from a number of casual records.

Breeding. Breeding noted at 14 sites. The only records specifying counts of prs or territories were 16 ♂♂ at Dawlish Warren on 12 May and three prs at Higher Metcombe. First song at Holne was on 9 Feb; the first fledgling at Broadsands was on 21 Mar; the first fully fledged juv at Alphington was on 10 Apr; and two nesting attempts by a pr in a Derriford garden failed due to predation. A series of spring counts in the Prawle area peaked at 23 birds on 13 May.

First winter-period. The peak count at Dawlish Warren was 30 on 5 Jan. Other notable counts: 53 on a 5km route near Stockland on 14 Jan and 25 along the Grand Western Canal on 10 Feb.

Autumn/second winter-period. *Passage*: ten birds on the cliffs at Berry Hd on 21 Oct were considered migrants, and there were ten continental 1cy ♂♂ at Start Pt same day; 30+ at Dawlish Warren on 22 Oct; 30+ at Stoke Pt 27 Oct; at Staddon Point a max day count of 31 on 29 Oct; four counts >30 in the Prawle area from the end of Oct, including 61 on 10 Nov; 60+ including migrant 1cy ♂♂ at Soar on 20 Nov. *Inland counts* >20: 59 along the Grand Western Canal on 16 Dec, 34 at Beer on 29 Dec and 30 at Lipson, Plymouth on 27 Oct.

Ringing recoveries. A male caught at Roborough, Plymouth in Jan, had been ringed all of 2km away at Derriford in 1994, and another male, ringed in Crediton in 1995, was found dead there in Nov 2000, apparently not having moved at all; further details in the Ringing Report.

FIELDFARE *Turdus pilaris*

Numerous winter and passage visitor. SPEC(4^{w}), BoCC (Amber).

Some large numbers on Dartmoor and in NE Devon during the autumn.

Table 98: Maximum flock size and number of flocks >50

	J	F	M	A	M	J	J	A	S	O	N	D
Max flock size	300	250	250	9	-	-	-	-	1	500	1000	600
No. of flocks >50	9	8	4	-	-	-	-	-	-	4	24	11

First winter-period. Eight flocks of >100, with a max of 300 at Nether Exe on 28 Jan; there were also eight other counts of >50. Higher counts were again predominantly from the Dartmoor and Exeter-Tiverton areas. The *last* of four Apr records was two in the Huntsham area on 7th (RGr).

Second winter-period. The *first* was one at Trinity Hill on 28 Sep (RBo); the next in the county were also there, but considerably later, on 16 Oct. The first records from Dartmoor and Huntsham were on 22-23 Oct, with the main arrival picking up over the next week. At Staddon Pt, the peak passage period was

28 Oct – 5 Nov, with a max day count of 860 on 1 Nov. Approx 1236 birds moved through there during the autumn (SGe - see separate report). The first triple figure flock elsewhere was 500 at Huccaby on 30 Oct, building to *c.*1,000 there on 2 Nov, heralding a period of high numbers on *Dartmoor*, including 400 at Criptor Newtake on 1 Nov, 250 at Willsworthy, Peter Tavy on 10 Nov, 200 at Burrator Res on 1 Dec and ≥100 at Widecombe 27-28 Oct, Ashburton 29 Oct, Western Beacon 1 Nov, Ringmoor Down 8 Nov, Huccaby again on 23 & 28 Nov, Harford Moor 24 Nov and Princetown 16 Dec. *Elsewhere*, counts ≥100: 100 at Soar on 28 Oct; 200 at Shute Hill Woods 28-29 Oct; 800 at Grand Western Canal on 3 Nov, with 100 there on 25 Dec; 130 at Hendom Cross on 17 Nov; 500 there on 29 Nov; 15 at Huxham on 16 Dec; 600 at Kerswell Cross on 26 Dec.

SONG THRUSH *Turdus philomelos*

Numerous resident breeder, passage and winter visitor. SPEC(4), BoCC (Red), UK BAP.

Autumn passage well recorded on the south coast.

Breeding. Noted at 13 sites: Dawlish Warren, Dunsford Woods, East Budleigh, Higher Metcombe, Huccaby, Ideford Mill, Kenwith NR, Langston, Lipson, Okehampton, Sticklepath, Stokenham and Tiverton. Counts referring to singing ♂♂/prs included five at Huccaby, Lee Abbey and Witheridge Moor, three at Prawle and two at Clennon Valley, Huntsham Lake, Kenwith NR, Martinhoe and Netton.

First winter-period/spring. The maximum was a remarkable 70 at Isley Marsh on 19 Jan (IM). Other counts >10: 15 in the Prawle area on 7 Jan; 32 at Darracott on 13 Jan; 12 singing in the Huntsham area on 3 Feb; 16 at Huccaby on 1 Mar; 21 at Exeter Canal on 4 May.

Autumn/second winter-period. Counts >10: 183 migrants at Staddon Pt over the period, beginning on 17 Sep, with peak passage 28 Oct – 5 Nov and a max day count of 54 on 29 Oct; 41 in small flocks going S at Dawlish Warren on 14 Oct followed by 10+ on 30th; 11 at Berry Hd on 13 & 20 Oct; 30 at Start Pt on 14 Oct; and nine double figure counts in the Prawle area from mid Oct peaking at 20 on 29 Dec.

REDWING *Turdus iliacus*

Numerous/abundant winter and passage visitor. SPEC(4^{w}), BoCC (Amber).

A considerable number of high counts in both winter periods.

Table 99: Maximum flock size and number of flocks >50

	J	F	M	A	M	J	J	A	S	O	N	D
Max flock size	100	88	100	16	-	-	-	-	2	350	500	900
No. of flocks >50	12	2	3	-	-	-	-	-	-	7	11	12

First winter-period/spring. Twenty widely distributed flocks of 50 or more, with a max of 350 at Higher Metcombe on 5 Mar. Other counts: 100 at Stoke Canon on 6 Jan; 91 at West Park, North Molton on 10 Jan; 89 at Exminster Marshes on 11 Jan; 50 at Black-a-tor Copse on 13 Jan; 100 at Kitterford Cross, Ugborough on 13 Jan; 50 at Babel's Bridge on 14 Jan; 100 at Lower Coly same day; 80 at Morwelldown Plantation, Tavistock on 16 Jan; 70 at Grenofen on 18 Jan; 85 at Belstone Common on 24 Jan; 60 at West Putford on 28 Jan; 50 at Flete Estate on 29 Jan; 65 at Stover CP on 29 Jan; 60 at Totnes on 3 Feb; 50 in the Huntsham area on 6 Feb; 88 at Peamore, Exeter on 13 Feb; 60 at Huccaby on 6 Mar; 100 at Puslinch Creek on 7 Mar; 60 at Hendom Cross on 10 Mar. Only two Apr records, on Lundy and at Rackerhayes, were followed by singles on Lundy on 5 May and the *last*, at Berry Hd on the very late date of 13 May (RBe) (the last in 2000, on 11 May, was also very late).

Autumn/second winter-period. The *first* were two at Huntsham on 29 Sep (RGr); the next were five at Huccaby on 9 Oct. Thereafter there were daily records from 12 Oct, with the first double figure count being 22 at Efford Marsh, Plymouth on 14 Oct. *Peak passage* and counts ≥50 commenced with hundreds passing overhead at night at Buckfastleigh on 28 Oct. Next day there were 100 at Exminster Marshes and 70 at Higher Chieflowman, Tiverton. The max day count at Staddon Pt was 1,480 on 1 Nov, and a total of 2,529 passed through there over the period, with a peak 21 Oct – 9 Nov (SGe – see separate report). *All other counts* ≥50: 500 at Grand Western Canal on 3 Nov with 900 on 16 Dec and 200 on 25 Dec; 88 at Wembury on 6 Nov; 50 at Kenwith NR on 10 Nov; 80 at Witheridge Moor on 11 Nov; 120 at Exminster Marshes on 16 Nov; 50 at Huccaby on 22 Nov, with 65 there on 28th; 82 at Westacott, Barnstaple on 28 Nov; 50 at Saltram Wood on 1 Dec; 110 at West Putford on 2 Dec; 100 at Burrington and 50 at Lower

Coly on 23 Dec; 50 at Combe Martin on 23 Dec; 200 at Witheridge Moor on 23 Dec; 124 at Holne on 24 Dec; 600 at Kerswell Cross on 26 Dec; 100 at Peamore on 28 Dec; 120 at Pig's Nose Valley on 31 Dec. Despite these high counts elsewhere, numbers were generally low on Dartmoor throughout the period (DSGp).

MISTLE THRUSH *Turdus viscivorus*

Numerous resident breeder, passage and winter visitor. SPEC(4), BoCC (Amber).

Breeding. Confirmed at only six sites: Braunton, Challacombe, Higher Metcombe, Huccaby (three prs), Slapton Ley and Stokenham.

First winter-period. No notable counts but for three singing along a half-mile length of path at Yealmpton on 17 Jan. The max count was again only four, at Chudleigh Knighton on 6 Jan and Morwelldown Plantation, Tavistock on 16 Jan. Feb was the peak month for song at Huccaby, with birds heard on 16 dates.

Post-breeding/autumn flocks and passage. A fairly good year, with 14 *double figure counts* as follows: 65 at Killerton on 25 Aug; 40 at Huccaby on 21 Sep, with 13 on 19 Jul and 12 on 20 Sep; 24 at Andrew's Wood, Loddiswell on 5 Sep; 22 at Wigford Down on 11 Jun, with at least ten on 22 Jul; 20 at Piles Brook on 27 Jun and at Huntsham on 17 Oct; 17 at Crediton Golf Course on 23 Aug; 12 at Willesleigh Farm, Barnstaple on 16 Aug; 11 at Bigbury on 16 Aug and at Wrangaton on 11 Oct; ten at Two Bridges on 7 Aug. *Potential migrants* at S coast sites: one W at Prawle Pt on 24 Sep; one flying around lighthouse at Start Pt on 13 Oct; eight at Berry Hd on 28 Oct; 12 NW over Staddon Pt on 10 Nov.

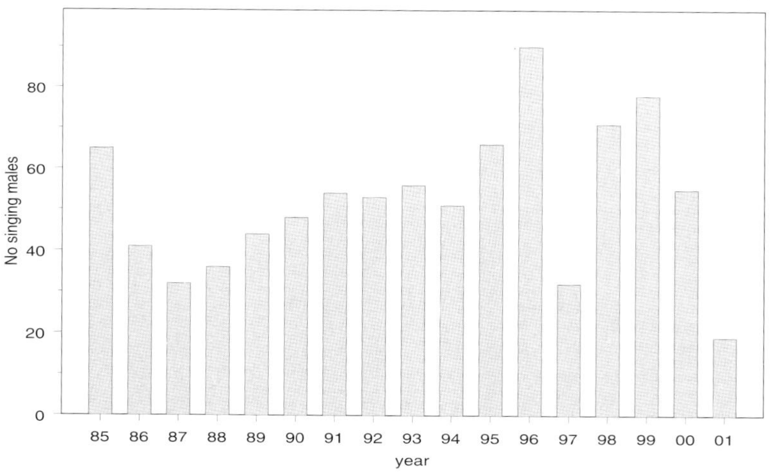

Fig. 25: Cetti's Warbler - Number of singing males recorded in Devon 1985 - 2001. (Source: DBRs, including summary table in DBR 1995)

CETTI'S WARBLER *Cettia cetti*

Not scarce, but very localised resident breeder.

Recorded from ten sites.

Breeding. A minimum of 19 birds recorded Apr–Jul at six sites: Exeter Canal (7 birds); Slapton Ley (6); Bere Ferrers (1) 4–19 Apr (PFG); Exeter Riverside CP (1) 3 Jun (DS,NPi); Exminster Village (3) 26 Apr (DC); Hackney Marshes (1) 20-30 Apr (EM). FMD access restrictions have undoubtedly drastically reduced the number of breeding records submitted. No full count was completed at Slapton Ley where last year 32-33 singing birds were recorded (BW).

Non-breeding. Away from usual sites, eight records from seven sites including: singles at Clennon Valley 7-9 Sep (ML); Hallsands 28 Sep (PHA); Prawle Pt DBWPS reserve 9–11 Oct (AKS,JCN); and

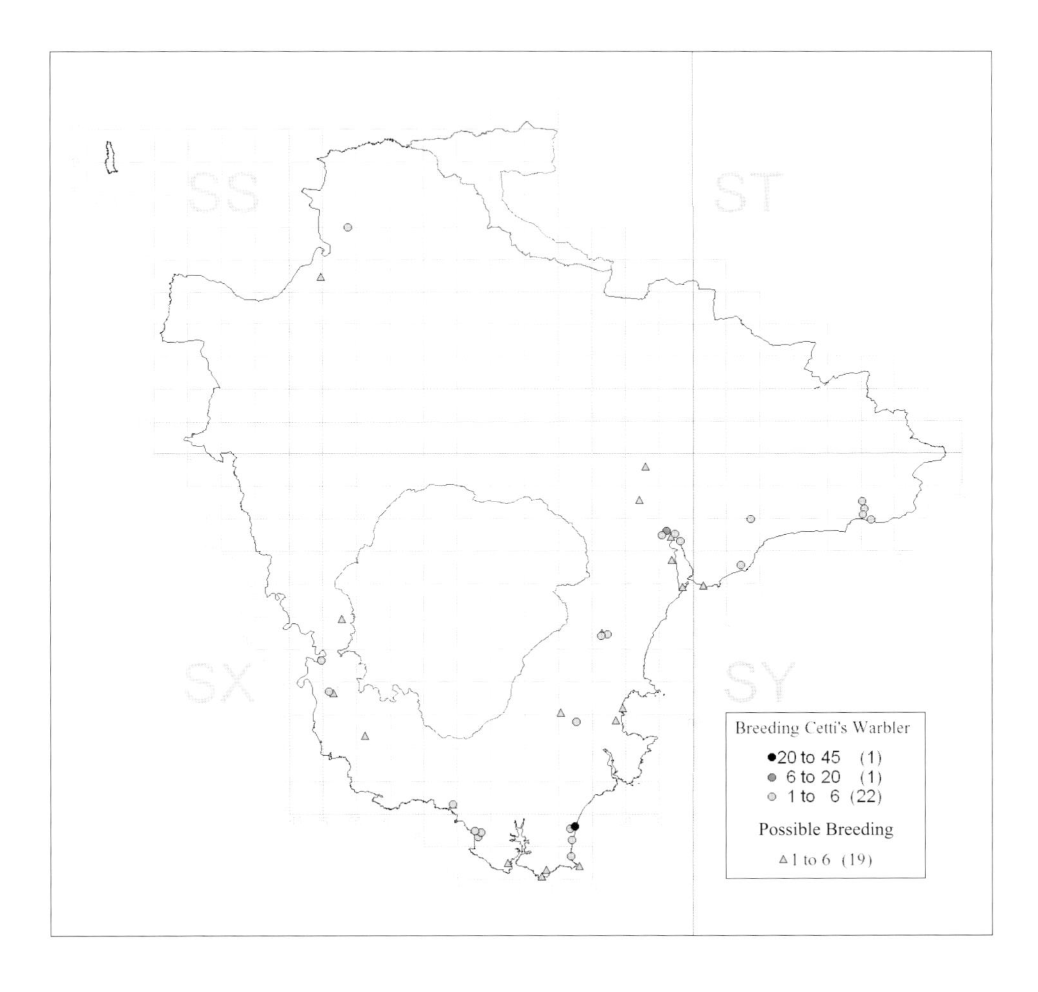

Map 11: Cetti's Warbler.
Distribution of records with obvious migrants excluded. Abundance is based on the maximum count of singing males /territories at each location. Slapton Ley is the county stronghold, whilst Exeter Canal and Exminster Marshes (Countess Wear to Turf Locks) is the second most important area. Other locations appear to hold fewer territories than would be expected for their size, e.g. South Milton Ley where habitat improvement for this species is in progress. Habitat creation for this species is also in progress at DBWPS' South Huish Reserve.
Data: Based upon a ten year period 1992-2001. Source: DBRs) (see Overview on page 25)

Sherpa Marsh, (singing) 8 Dec (ALa). Two birds were recorded at South Milton Ley on 29 Jan, and two also there on 21 Sep (RBu,SRT,VRT).

GRASSHOPPER WARBLER *Locustella naevia* — Common G. Warbler

Not scarce or scarce migrant breeder and passage visitor.SPEC(4), BoCC (Red).

Table 100: Monthly distribution of all passage records:

	J	F	M	A	M	J	J	A	S	O	N	D
Sites	-	-	-	17	13	4	4	2	2	2	-	-
Birds	-	-	-	30	16	4	4	2	2	2	-	-

Breeding. Reeling males between mid-May and the end of Jul, indicating potential breeding sites, recorded at: Aveton Gifford, Bursdon Moor (2), Cann Wood, Dunnabridge, Exminster Marshes, Hartland Forest (2), Laughter Tor, Mardon Down, Melbury Plantation (2), Northam Burrows, Prawle, Roadford Res, Routrundle, Statts Bridge, Thornworthy Common, Upper Tamar L and Westward Ho!. The only *confirmed* breeding record was one feeding yg in the nest on Roborough Down 11 Jun (RS).

Spring. Passage started on 13 Apr with *arrivals* at six sites between Plymouth and Torbay, including three at Sharkham Pt (ML,BMc). New arrivals were then noted almost daily throughout Apr with numbers tailing off notably during first half of May. Max counts: Slapton Ley, four on 14 Apr; and Berry Hd, three on 29 Apr. There was just one report from Lundy, on 13 May.

Autumn. Single passage birds at: Bridford on 24 Aug; nr Avon Dam on 10 Sep; Soar on 29 Sep; Prawle on 9 Oct; and *lastly* Lannacombe 15 Oct (PMM).

SEDGE WARBLER *Acrocephalus schoenobaenus*

Not scarce migrant breeder and passage visitor. SPEC(4).

Breeding. As last year, recorded from at least 20 scattered sites mid-May – mid-Jul. Max counts of singing males: Aveton Gifford (6), Yagland (5), Exminster Marshes (5), Northam Burrows (10), Kenwith (8) and Prawle (4). No birds were recorded breeding at Stover CP (JDA).

Spring. The *first* arrival was noted at Exminster Marshes on 11 Apr (RSPB). Widely scattered arrivals recorded almost daily from then on with the first N coast record at Kenwith on 16 Apr. Max counts of up to six at Northam Burrows 25 Apr, 20 on Lundy on 2 May, 15 Slapton Ley 4 May and 25 Exminster Marshes 7 May. One, presumably a late arrival, was singing in the northern fort at Berry Hd on 17 Jun.

Autumn. Passage underway from early Aug with small numbers throughout the month. More movement in Sep with largest numbers apparently moving in the first three weeks. Most records were from the S coast. Max counts of ten at Soar in a small area of crop field on 25 Aug (SRT,VRT), 24 at Torcross on 3 Sep (AKS) and 35 at Torcross on 22 Sep (PHA) with six recorded there on 2 Oct and just one on 11 Oct. The *last* was one at Dawlish Warren on 13 Oct (IL).

REED WARBLER *Acrocephalus scirpaceus* — Eurasian R. Warbler

Not scarce, but very local, migrant breeder and passage visitor. SPEC(4).

Breeding. Records from 18 sites mid-May - mid-Jul, with the following max counts of singing males: Beesands (10), Broadsands (4), Clennon Valley (4), Coombe Cellars (6), Dawlish Warren (7), Kenwith (8), Lower Otter Valley (3), Sherpa Marsh (11), Slapton Ley (6) and West Charleton Marsh (4). A bird seen carrying food at Velator Res on 19 Jul was the only *confirmed* breeding record.

Spring. The *first* was at Slapton Ley on 3 Apr (BW). The main arrival started on 11 Apr with three at Exminster Marshes, and thereafter daily, the highest counts being: seven at Slapton Ley 16 Apr; four Dawlish Warren 23 Apr, and nine on 13 May; six Sherpa Marsh 1 May; and eight Otter Est 9 May and nine Dawlish Warren 13 May. The first ever spring record for Stover CP was a single on 20 May (JDA).

Autumn. Numbers at breeding localities were swelled with juvs during late Jul and Aug, with, for example, fourteen recorded at Dawlish Warren on 20 Aug. Birds continued to be seen in ones or twos throughout Sep with four still present at Clennon Valley on 6 Sep. Singles were recorded at a number of sites throughout Oct and two were at Slapton Ley on 20 Oct. The *last* records were of singles at Prawle Pt and Lundy on 1 Nov, and one at Soar Mill Cove on 5th (AJL).

2000 correction. Delete the record of one at Prawle Pt on 14 Mar.

MELODIOUS WARBLER *Hippolais polyglotta*

Very scarce or rare passage visitor.

Two records, as in 2000, on typical dates.

One at Pig's Nose Valley, Prawle on 28 Aug (NPR); and one at Soar Mill Cove on 1 Sep (JGS,PJS,PSm).

2000 correction. Finders of the Pig's Nose Valley bird 17-19 Sep 2000 were MKn,BMc and TWW.

DARTFORD WARBLER *Sylvia undata*

Not scarce resident breeder. SPEC(2), BoCC (Red).

For the third consecutive year there was a significant decline in breeding numbers with a 30% reduction in the number of territories (cf. 10% in 1999; 21% in 2000). A similar fall, of 32%, was recorded on E Devon Commons (cf. 9% in 2000) but numbers were down by 44% on the Aylesbeare/Harpford Commons RSPB Reserve (cf. 3% increase in 2000). Away from E Devon, declines of 43% in the Dartmoor NP and 83% on the S coast were recorded, but there was no change on the N coast.

Breeding. A minimum 137 territories (98 prs) (*per* RCT *et al.*) (*cf.* 196 territories in 2000 and 249 in 1999). The stronghold is still the *E Devon Commons* with 92 territories (67 prs) (*cf.* 135 territories in 2000 and 149 in 1999); in the remainder of *E Devon*, seven territories were recorded (*cf.* six in 2000 and seven in 1999). The *S coast* apparently held only one territory (*cf.* six in 2000); *Dartmoor NP* held 13 territories (*cf.* 23 in 2000 and 29 in 1999); *N coast* held six territories (same as 2000); and elsewhere in Devon, 18 territories were identified (*cf.* 20 in 2000).

Changes in status. With winter mortality rates likely to have been low and with much apparently suitable habitat remaining unoccupied, a significant part of the recorded *decline in numbers* could have been the result of incomplete coverage due to FMD access restrictions during the early part of the breeding season. However, although recording effort was reduced on the E Devon Commons, all site visits were at the most productive time, so it was likely that there was a genuine reduction in population this year but perhaps not to the extent of nearly a third. Similarly, FMD restrictions make it difficult to assess any real *change in distribution* although there may have been some losses along the S coast. On the E Devon Commons, distribution remained similar to last year, although within the Dartmoor NP some new sites were identified. No information was received from Exmoor where the population on the Somerset side has been expanding in recent years.

Non-breeding. Autumn/winter records away from breeding sites: the *S Coast* regularly holds dispersing birds with up to five seen in the Soar area 29 Sep-4 Nov (mo) and singles at Prawle Pt 15 Oct (BG), Start Point 21 Oct (BMc) and Dawlish Warren 29 Oct (JEF). During the winter, singles were noted at Trinity Hill 8-9 Dec (RBo,DJCo), Slapton Ley 22 Dec (BW) ad Dawlish Warren 5,8 & 28 Jan and 2 Feb (IL). Unusually, one was seen with Stonechat on farmland at Westclyst, Nr Exeter on 14 Jan (RCT) perhaps indicating further association between the two species away from their breeding grounds. Along the *N. coast*, singles were seen at Baggy Pt on 17-18 Feb (RJ,CGy). Without the significant altitudinal movement shown by some continental birds breeding at high altitudes, a number were noted on *Dartmoor* during the coldest months: two on Brent Moor 6 Jan (PMM); two at Brat Tor 3-4 Feb (RAJJ); and singles at Cold East Cross 20 Jan (RS) and 17 Feb (RHi), Trendlebere Down on 9 Feb (RHi), Holne Moor on 15 Feb (RHi), Criptor Cross on 1 Nov (RJL), Riddon Ridge on 10 Nov (RHi), Avon Dam area (three singles in different locations on 19 Nov) (RS) and Pew Tor on 31 Dec (RS).

Record request. *To aid interpretation of breeding records, observers are asked to supply grid references with all records and the species co-ordinator (RCT) particularly welcomes details of breeding records away from the E Devon Commons.*

SUBALPINE WARBLER *Sylvia cantillans*

Rare vagrant. SPEC(4).

A ♂ present briefly in a Wembury Garden on 8 Apr (C&R Rhodes), is the 14th Devon record (but only the fourth for the mainland) and the first since 1997. (One of seventeen in Britain in 2001, accepted by BBRC).

1997 addition. A ♀ on Lundy on 2 May 1997 was trapped, ringed and photographed (D Clifton, C McShane *et al.*); the ninth record for Lundy (the other 1997 record, a ♂ on 15-16 Jun as documented in *DBR* 1998, now becomes the tenth). (Accepted by BBRC).

BARRED WARBLER *Sylvia nisoria*

Rare passage visitor. SPEC(4).

Three records, but only one on the mainland (cf. two, both on mainland, in 2000).

Singles on Lundy on 13 & 27 Oct (LFS); and a very late individual at Dawlish Warren 6–9 Nov (JEF *et al.*), the only other Nov record in Devon being at Slapton in 1978.

LESSER WHITETHROAT *Sylvia curruca*

Not scarce migrant breeder and passage visitor.

Table 101: Monthly distribution of all passage records:

	J	F	M	A	M	J	J	A	S	O	N	D
Sites	1	-	1	11	22	7	5	7	3	1	-	-
Birds	1	-	1	16	28	11	5	11	4	1	-	-

Breeding. An apparent improvement on last year, with records of singing males from ten sites between late May and mid-July: a pr seen throughout this period at Dawlish Warren and an ad plus juv on 19 Aug; one was collecting food at Prawle Pt on 23 Jun; and an adult was seen with two young at Berry Hd NNR on 11 Jun. Other sites included: Aveton Gifford, Bickington, Littlecombe Shoot, Sharkham Pt and West Charleton Marsh. Away from the South Hams, singles were recorded at Neadon Cleave, Manaton 4-12 May, Buckfastleigh on 19 May (unusual around Dartmoor), Roadford Res on 5 Jul and the Grand Western Canal on 6 Jul.

First winter-period. Overwintering birds were noted in a private garden in Ilfracombe 20-25 Jan (MAP) and Braunton, where one was trapped and ringed at a private site on 11 Mar (BG). Previous wintering records in Devon, typically in gardens, have been in 1999, 1997/98, 1995, 1989/90 and 1985/86.

Spring. The first was at Start Pt on 21 Apr (PSa), and passage began in earnest just a day later than last year, on 24 Apr; the bulk of birds arrived 24 Apr - 5 May in mainly S coast localities, though there were singles at both Penhill Marsh and Pottington on the Taw Estuary on 26 Apr. Highest counts were three at Berry Hd on 29 Apr and four at Bowling Green Marsh on 8 May.

Autumn. Records in the first half of Aug came mostly from probable breeding sites, but likely migrants were: three Pig's Nose Valley on 22 Aug, with the last two here on 29 Aug; three Dawlish Warren on 25 Aug, with the last here on 27 Aug; one Axmouth on 26 Aug; one South Milton Ley on 2 Sep one Berry Hd singing on 7 Sep, with two there on 9 Sep; and one Soar on 22 Sep. The *last* autumn record was of a late bird at Prawle on 17-18 Oct (PMM).

WHITETHROAT *Sylvia communis* — Common Whitethroat

Fairly numerous breeder and passage visitor. SPEC(4).

Breeding. Records of birds from numerous sites across the county, with highest counts: seven singing ♂♂/prs E Dartmoor along road between Heatree Cross and Cockingford Br on 20 May, five prs Starehole Valley,Bolt Hd on 29 May; 26 birds in the Prawle area on 30 Jun; six singing ♂♂s on Hartland Pt on 23 Jun; six singing ♂♂s at Yelland on 25 Jun; and eight at mid-Soar on 1 Jul. On 14 Jul a count of 36 birds in the Prawle area presumably included juvenile birds.

Spring. The *first* arrival was a single bird at Berry Hd on 7 Apr (RBe). Then seen mainly along the S coast almost daily, in ones and twos until 23 Apr when there were six at Prawle; on 25 Apr, nine were counted at Berry Hd, ten at Dawlish Warren and ten at Hope's Nose. The first N coast record was of one at Woolacombe on 19 Apr. Another wave of arrivals started in the first week of May with singing males recorded: 30 on Lundy on 3 May; 20 in Ilfracombe and 16 at Prawle on 4th; 20 at Jennycliff on 6th; 13 at Dawlish Warren on 8th; 20 at Portlemouth Down on 12th; and 33 in the Prawle area on 13th.

Autumn. Migrants and breeders are difficult to separate in Aug, but likely migrants were 30+ in the Soar area on 25 Aug, 16 at Dawlish Warren on 20 Aug and 25 in the Prawle area same day. Small numbers were seen at coastal sites throughout Sep with highest counts: ten in Starehole Valley on 4 Sep; ten at Prawle on 9 Sep; and nine at Soar on 22 Sep. The latest records were of a bird in subsong at Soar Mill Cove on 12 Oct, a single in Pig's Nose Valley on 16th (JCN) and one on Lundy on 28th (LFS).

GARDEN WARBLER *Sylvia borin*

Fairly numerous/numerous breeder and passage migrant. SPEC(4).

Breeding. Recorded from 26 sites 15 May-31 July, mostly singles except: at least five territories at Huccaby, four singing ♂♂s at Sourton Cross, two singing ♂♂s at Cann Wood, Haldon Forest and Coombe Cellars; and two prs at both Andrew's Wood, Loddiswell and West Putford. None recorded at Stover CP.

Spring. The *first* was at Passage House Inn on 20 Apr (EM), and on 27 Apr singles were noted at Galmpton, Kingswear and Prawle. There followed a steady trickle of mostly singles across the county through the first half of May, with the highest count by far being *c.*30 on Lundy on 8 May (IM).

Autumn. Passage evident from the very beginning of Aug, peaking second half of month and trailing off through Sep. Highest counts: three at Prawle on 4 Aug; four on 19 Aug rising to seven the next day at Dawlish Warren, with five there on 25 Aug; and five at Soar also on 25 Aug. Oct records: one Berry Hd 2 Oct; at least one at Prawle 9-20 Oct; Start Pt on 14 and 19 Oct; and two on Lundy and one at Soar on 20 Oct. The last were on 1 Nov, on Lundy (LFS) and at Prawle (JLt,PMM).

BLACKCAP *Sylvia atricapilla*

Numerous migrant breeder and passage visitor; and scarce winter visitor. SPEC(4).

Breeding. Poorly recorded throughout May and Jun, with highest counts of singing ♂♂ being eight at Clennon Valley and six at Seaton Undercliff. In the first half of Jul, juvs were noted at Lipson, Burrator Res, Galmpton and Wistlandpound Res.

First winter-period/spring. A minimum 78 *wintering* birds reported from at least 43 sites 1 Jan - 15 March, with records predominantly from the S coast. As expected most records came from gardens, and the highest counts were in Exeter, with five in both Alphington and Wonford. *Spring passage* was very slow through the second half of March with a max count of just five at Berry Hd on 21st. A heavier passage started in early Apr when five ♂♂ were singing at Slapton on 8th, and on 13th, 20 were at Lannacombe and 12 at Jennycliff, Plymouth. Low numbers continued to be seen through the second half of April with highest counts of seven at Dawlish Warren and eleven at Berry Hd on 29 Apr suggesting a small influx.

Autumn/second winter-period. *Autumn passage* started in early Aug with low numbers (though six at Prawle on 4th) around S coast headlands, and a modest passage later in the month with eight at Soar on 25 Aug being the monthly max. More movement in mid-Sep, with max count of 44 birds on 17 Sep at Prawle, and 28 there a week later; also 12 at Froward Pt on 23 Sep. Small numbers continued to be recorded throughout Sep and early Oct, then: 18 at Berry Hd and 40 on Lundy on 20 Oct; 75 on Lundy, 20 in Pig's Nose Valley and 18 at Start Pt on 21st; and 20 ♂♂ in one bush at Dawlish Warren on 22nd. Nine still at Prawle on 1 Nov, 12-15 on Lundy on 1st-4th and six at Dawlish Warren on 3rd, but scarce elsewhere by early Nov. Mostly singles recorded mid-Nov to end Dec at 44 sites, with many *wintering* birds as usual in gardens (max of three in Clennon Valley on 19 Nov and Peverell, Plymouth on 12 Dec).

Ringing recovery. A bird ringed at Kenn, Exeter in Jul 2000, was found dead in Morocco in Mar 2001; further details in the Ringing Report.

YELLOW-BROWED WARBLER *Phylloscopus inornatus*

Very scarce passage visitor; rare winter visitor.

A good year, with twelve birds recorded (cf. six in 2000), the highest number since 1996 when up to fourteen were recorded.

One at Berry Hd 12-13 Oct (ML,SJA *et al.*) was quickly followed by an impressive arrival of at least five, all on 13 Oct: one Gammon Hd (JCN); one Lannacombe (AKS); and a remarkable three on Lundy (SL & A Cooper, RM Patient) *(see col plate)*. Singles were also at Start Pt on 14-17 Oct (PSa *et al.*), Soar Mill Cove on 20th (MKn *et al.*) and on Lundy on 29-30 Oct (two) and 1-2 Nov (LFS). Finally an individual first seen at Slapton Ley on 18 Nov (SRT,VRT) near the sewage works, was relocated on 1 Dec and remained until 11 Dec (BW *et al.*).

WOOD WARBLER *Phylloscopus sibilatrix*

Fairly numerous migrant breeder, very scarce passage visitor. SPEC(4). BoCC (Amber)

Breeding. Recorded from 23 sites with highest counts: four at Yarner Woods NNR; five singing ♂♂ at Fingle Bridge; three prs at Hunter's Inn; three in Meldon Woods; two ♂♂ and a ♀ at Dunsford Woods; six at Castle Drogo; four singing ♂♂ at Stoke woods and Steps Bridge; and three at Dewerstone.

Spring. The *first* arrival was at Yarner Wood NNR on 24 Apr (PP), with a N Devon record shortly after on 26 Apr at Holywell, Bratton Fleming (AW). Away from breeding sites: ♂ singing at Slapton Ley on 28 Apr (MKn); one at Mt Gould, Plymouth on 3 May (RWBu); singles on Lundy on 2 & 9 May (LFS); and one singing all morning in a Chillington garden on 10 May (DEk,BW).

Autumn. As usual, no autumn records of birds on passage; the *last* bird was in the Dart Valley on 28 Aug (RHi).

COMMON CHIFFCHAFF *Phylloscopus collybita*

P. c. collybita: numerous migrant breeder and passage visitor; scarce winter visitor.

Breeding. Poorly recorded, with highest counts: six Cann Wood on 12 May; ten at Chaddlewood, Plympton on 17 May; eight at Hardwick Wood on 28 May; and ten in the Prawle area on 30 Jun.

First winter-period/spring. A minimum 61 *wintering birds* at 33 sites Jan to early Mar, with most typically in damp areas with sallows. Highest counts were eight at Galmpton Sewage Works on 7 Jan, with twelve there on 3 Mar, and five between Slapton Bridge and Deer Bridge on 2 Feb. The *earliest singing* bird was in Plymouth (Buddleia bush in a pub carpark) on 14 Feb. *Spring passage* birds started arriving along the S coast mid-Mar, with high counts including: 12 at Dawlish Warren on 17 Mar, rising to 22 on 22nd; 12 at Exeter Riverside CP on 20th; 30+ at Hallsands on 22nd; 11 at Berry Hd on 23rd; and 14 at Lannacombe on 24th. A large fall occurred at Moorgate, S Dartmoor on 30 Mar. Modest counts of no more than a dozen at any one site continued through Apr and May: 80 at Lannacombe on 13 Apr (AKS) and 20 at Prawle on 5 May were the highest in this period.

Autumn/second winter-period. Slightly increased numbers noted from mid-Jul were presumably in part due to the presence of local juvs, but 20 at Dawlish Warren on 4 Aug and 23 at Berry Hd on 5 Aug surely indicated *passage*. Small numbers were recorded steadily through Aug picking up towards the end of the month and into Sep. Highest counts were: 20 at Start Pt on 26 Aug; 16 Clennon Valley on 7 Sep; 35 Berry Hd and 15 E Soar on 9 Sep; and 43 Prawle area on 17 Sep. A good count of 75+ at S Milton Ley on 21 Sep was followed by 30+ at Berry Hd the next day; 20 at Barton Pines on 24 Sep; 39 Prawle area on 25 Sep; 32 Start Pt on 26 Sep; and 32 Dawlish Warren on 29 Sep. Numbers dropped steadily through the first half of October with highs of 13 at Berry Hd and 29 in Prawle area on 9 Oct, and 20 at Start Pt on 14th. After this, just single-figure counts at coastal localities into Nov, except on Lundy where 35 were counted on 20 Oct and there were still up to 13 in first week of Nov. The highest counts of *wintering* birds were: seven in Clennon Valley and six at Efford Marsh,Plymouth on 5 Dec; five at Slapton Ley on 13 Dec; and ten plus at Countess Wear Sludge Beds on 11 Dec.

Ringing recoveries. A bird ringed at Slapton Ley in Jul 2000, was recaptured in Spain in Mar 2001, and one ringed at Prawle in Oct 2000 was recaptured in Portugal in Jan 2001; further details in the Ringing Report.

***P. c. abietinus/tristis*: ('Scandinavian' and 'Siberian' Chiffchaffs)**

Scarce passage visitors; rare winter visitors.

(All records refer to 'grey' Chiffchaffs that could not be assigned to a particular race. The records committee is still reviewing claims of P. c. tristis, and all observers are reminded that a full description is required.)

At least 11 birds, a similar number to last year, and typically in late autumn: one at Dawlish Warren 21 Oct – 17 Nov, with two there on 24 Oct (IL) and three on 17 Nov(GS); two on Lundy 29 Oct – 6 Nov (LFS); one at Broadsands 2–31 Dec (ML); at least four at Countess Wear Old Sludge Beds on 11 Dec (MKn); and singles at Coombe Cellars on 21 Dec (MKn,TWW), Kilmington on 30th (IJW) and Kenwith NR on 31st (DC).

2000 correction. Delete records specifically relating to *P.c. tristis*.

WILLOW WARBLER *Phylloscopus trochilus*

Abundant migrant breeder and passage visitor. BoCC (Amber).

Breeding. Poorly recorded from late May – Jul, with highest counts of nine at Witheridge Moor on 20 May, seven on Aylesbeare Common on 16 Jun and seven at Trinity Hill NR on 21 Jun.

Spring. Passage began on 22 Mar with the *first* at Hallsands (PBo), then one-three birds daily over the next week at S coast localities; the first N Devon record was one at Braunton on 30 Mar. Numbers picked up briefly with nine in Radford Park, Plymouth on 31 Mar, seven at Dawlish Warren and 12 at Slapton Ley on 1 Apr, but no more large counts until mid-Apr when: 20 at Lannacombe on 13 Apr; 12 at Berry Hd on 14th; 11 at Dawlish Warren on 16th; 15+ at Jennycliff on 17th; ten at Witheridge Moor on 21st; 140 on Lundy on 25th; and 16 at Berry Hd on 29th. Numbers moderated over May with the highest count being just ten in the Prawle area on 3rd.

Autumn. Passage appeared to be underway from the beginning of Aug, with 20-25 at Moorwell pond, East Prawle and at least ten at Dawlish Warren on 4 Aug; 15 present on morning of 6 Aug at Dawlish had departed by the afternoon. Later in Aug were 12 at Dawlish Warren on 16 & 20 Aug, and 12 at Prawle and 17 at Start Pt on 26th. Twenty at Moorwell Pond on 1 Sep indicated 'the autumn's second influx' (AKS) but after this, numbers throughout the region dropped considerably. Four were on Lundy on 20 Sep. Mostly singles were seen at S coast headlands in Oct with one in subsong at Prawle on 10th, a late individual along the Grand Western Canal on 20th and four in Pig's Nose Valley on 21st. The *last* was one in Clennon Valley on 12 Nov (ML).

GOLDCREST *Regulus regulus*

Numerous resident breeder, passage and winter visitor. SPEC(4). BoCC (Amber).

Breeding. Very poorly recorded, with records from just 14 sites between Apr and Jul.

First winter-period. Mostly low numbers recorded, with highest Jan counts being six at Dawlish Warren on 1 Jan, seven at Squabmoor Res on 13th, six at Efford Marsh LNR Plymouth on 15th and ten at Slapton Ley on 20th. Ten were at Galmpton Sewage Farm and six at Dawlish Warren on 3 Mar, and evidence of spring passage came from Berry Hd with eight on 17 Mar, Dawlish Warren with 16 on 23 Mar and Lundy with a max of five on 11 Apr.

Autumn/second winter-period. *Passage* was very poor with several observers noting lower than average counts. Main passage was during the latter half of Oct, trailing off into Nov, but earlier high counts were 15 in the Prawle area on 25 Sep, nine at Berry Hd on 12 Oct, ten plus at Start Pt on 13th and 20 at Lannacombe on 14th. Sixteen were at Prawle on 20 Oct, twelve at Soar and 30 on Lundy on 21st, ten plus at Dawlish Warren on 27th & 28th and 15-20 at Lannacombe on 27th; otherwise mostly single-figure counts. The highest Nov count was fifteen between Shipley Bridge and Avon Dam on 19th, and in Dec there were 15-20 at Hallsands and 20 at Lannacombe.

FIRECREST *Regulus ignicapillus*

Scarce (mainly autumn) passage migrant and winter visitor; possibly breeds. SPEC(4), BoCC (Amber).

Table 102: Monthly distribution of all passage and wintering records:

	J	F	M	A	M	J	J	A	S	O	N	D
Sites	11	5	5	1	-	-	-	-	7	30	26	19
Birds	11	8	8	1	-	-	-	-	8	142	41	30

First winter-period/spring. *Wintering* singles at 12 S coast sites Jan – mid-Mar, with three birds at Slapton Ley on 17 Feb, two at Kitley Pond on 24 Feb, and one inland at Buckfastleigh on 3 Jan. *Spring passage*: singles on several dates at Dawlish Warren in the latter half of March with two on 23rd; one at Berry Head on 24th; two at Kingswear on 25th; and in Apr, only singles on Lundy on 23rd and at Orcombe Pt on 29th. No May records.

Autumn passage/second winter-period. Much lower Sep counts than last year, due to less favourable weather conditions. The first of the *autumn* was at Staddon Pt on 17 Sep (SGe), two days earlier than in 2000. The next arrivals were singles at Berry Hd in the south and Berrynarbor in the north on 27 Sep, two at Wembury on 28th and singles at Dawlish Warren and Start Pt on 29th. Reports from 28 sites in *Oct*,

Map 12: Firecrest Migrant.

Maximum count of migrants (Sep-Nov & Mar-May).
Note the more dispersed occurance of migrants compared to the highly coastal distribution of wintering birds shown on Map 13. (as is the case with Black Redstart, see maps 10 & 11))
Data: Based upon a ten year period 1992-2001. Source: DBRs) (see Overview on page 25)

Map 13: Firecrest Wintering
Please refer to previous Map 12 for notes.
Data: Based upon a ten year period 1992-2001. Source: DBRs) (see Overview on page 25)

of small numbers on several dates primarily at well-watched S coast headlands and Lundy, with apparent influxes around 10 & 20 Oct. There were also three *inland* at Huccaby on 18 Oct, with the only other inland birds singles also on Dartmoor at Fernworthy on 8 Oct, Venford Res on 28 Oct and South Brent next day. Highest counts were: 11 in the Prawle area on 20 Oct (PMM) and on 21st, seven on Lundy (LFS) and eight at Start Farm (BMc). The year's most interesting Firecrest landed on a birder's tripod during a seawatch at Berry Hd on 19 Oct (MD)! In *Nov* a minimum of 39 birds reported from 24 sites, with highest counts of three at Stoke Gabriel on 2 Nov, five at Clennon Valley on 12th and four at Slapton Ley on 18th. In the north, one bird was still present at Berrynarbor in mid-November. A minimum of 28 *wintering* birds at eighteen sites during Dec, mostly ones and twos but three at Slapton Ley, Lannacombe and Beer.

SPOTTED FLYCATCHER *Muscicapa striata*

Fairly numerous/numerous migrant breeder and passage migrant. SPEC(3), BoCC (Red), UK BAP.
Similar to 2000, but fewer breeding records, and later arrival and departure dates.

Breeding. Reported breeding from only 17 sites, fewer than in 2000: Bickington (in nestbox); Blackborough; Braunton; Buckfastleigh; Exeter (at both Pinhoe and Hamlin Rd playing fields); Georgeham; Grand Western Canal; Huccaby (three prs); Jennet Pt NR; Netton (two prs); Princetown; Stokenham; Throwleigh; Tiverton; Torrington; West Putford (three prs) and Whitchurch Down. At Haldon, no boxes occupied for second year running. Two at Soussons on 1 Jul were the observer's first for the area.

Spring. In contrast to 2000, no records before two on 2 May at Prawle (PMM,PSa), and during May only small numbers recorded at coastal sites, with maxima of one at Berry Hd, three at Dawlish Warren, ten on Lundy and two at Prawle; also occasional migrants still at Prawle and Dawlish Warren in early Jun. First reported arrival at inland breeding site was on 10 May at Huccaby.

Autumn. Six at Modbury on 29 Jul may have marked the beginning of autumn passage which extended through Aug, Sep and into early Oct. In *Aug*, the highest counts were: ten at Stoke Canon on 9th; seven at both Prawle Pt and Soar on 27th; six at Wistlandpound on 7th and at Trentishoe Down on 15th; five at Frogmore on 23rd; and a staggering 250 on Lundy on 24th. In *Sep*, recorded from 17 sites, but the highest counts were all from Prawle: 15 on 9th; ten on 8th; seven on 9th and 25th; and five on 24th. Only four *Oct* records, all of singles: Shobrooke Pk on 5th; Exmouth and Prawle on 8th; and the *last* at Slapton on 18th & 19th (PHA,BW).

RED-BREASTED FLYCATCHER *Ficedula parva*

Rare vagrant.
Records of three autumn singles – similar to 2000 (three) and 1999 (five).

Single birds recorded: at Prawle Pt on 10 Oct (PHA,PSa *et al.*); at Pig's Nose Valley, Prawle on 16 & 18 Oct (BW,PMM); and on Lundy on 20-24 Oct, trapped and ringed on 23rd (JGa).

PIED FLYCATCHER *Ficedula hypoleuca*

Not scarce migrant breeder, scarce passage visitor. SPEC(4).
Breeding data less complete than usual, but indicate a season at least as successful as 2000. Earlier arrival and later departure dates compared to 2000.

Table 103: Summary of Nest Box scheme data (sites with >9 boxes with eggs in bold)

	Boxes with eggs (B)	Success boxes (S)	Eggs (E)	Young (Y)	Fledged (F)	Mean clutch (E/B)	% fledged (100F/Y)	Mean prod -uctivity* (F/S)
Bovey Valley Woods (BT)	5	5	35	33	33	7.0	100	6.6
East Anstey (KHf)	3	3	21	19	15	7.0	79	5.0
Neadon Cleave (BT)	7	7	49	43	42	7.0	98	6.0
Okehampton (GAV)	**34**	**26**	**230**	**206**	**158**	**6.8**	**77**	**6.1**
Yarner Wood (BT)	**35**	**30**	**229**	**188**	**175**	**6.5**	**93**	**5.8**

**F/S = mean number fledged per successful box*

Breeding. Some nestbox sites not visited due to access restrictions (e.g. Dunsford Woods), and in others not all boxes visited (e.g. 81% coverage at Okehampton). The mean number fledged per successful box exceeded six at Okehampton for the first time since 1992 (GAV), and at 6.08 was certainly up on the 5.5 of 2000. In the other main site, Yarner Wood, there was also a small increase, though this site

BOX - *NEST BOX DATA*

Although only relevant to a few species, nest boxes are a major source of data on breeding productivity or success, and such data have been tabulated in *Devon Bird Reports* since 1990 for Pied Flycatcher, Blue Tit, Great Tit and Nuthatch. The few other species for which productivity data are *systematically* collected in Devon include Barn Owls (also mostly from nest boxes), Chiffchaffs and some raptors, seabirds and waterbirds. The advantages of nest boxes are that the nests are easy to find (!), the habitat does not have to be disturbed and the contents of the nest can easily be examined. There are about twenty small nest box schemes in Devon, some of which have been monitoring populations for decades (see *Devon Fieldwork in 2000* report in *DBR* 2000). The longest running is that in Yarner Wood which started in 1955, and the largest is at Okehampton with nearly 300 boxes. A great deal of information can be obtained from nest boxes (and in many of the schemes this is sent off to BTO on Nest Record Scheme cards), and of course the nestlings can also be weighed and ringed. Only a summary of all this information can be included in the *DBR* tables, but some additional detail is provided in the annual Okehampton Nestbox Reports, several articles written by GAV (e.g. in *Devon Birds* 54 (2)) and an article, soon to appear in *Devon Birds*, on the history of Pied Flycatchers in Devon by PFG.

As the nest box tables have been slightly modified for *DBR* 2001, it is an appropriate time to try to explain what the tables are showing. There are now eight columns, five of raw data and three of derived values, as follows:

- *Boxes with eggs* (B) – the number of boxes in which at least one egg was laid.
- *Success(ful) boxes* (S) – the number of boxes producing at least one fledgling.
- *Eggs* (E) – the total number of eggs laid.
- *Young* (Y) – the total number of young hatching.
- *Fledged* (F) – the total number of young fledged i.e. successfully leaving the box.
- *Mean clutch* (E/B) – the mean number of eggs per box (E/B), a measure of parental condition (itself affected by food and weather).
- *%fledged* (100F/Y) – the percentage of young hatching that result in a fledgling, a measure of survival rate of nestlings.
- *Mean productivity* – the mean number of young fledged per *successful* box (F/S), an overall measure of parental success.

Another change has been to delete the row giving county totals, which can be misleading as some schemes do not submit data, and the totals are therefore incomplete, and liable to annual fluctuations depending on which schemes have submitted data. In studying changes from year to year, it is better to look at each site separately. **Also, the main sites, those with eggs in at least ten boxes for a species and therefore likely to generate the most meaningful figures, are given in bold**.

***Record request**. Please submit data from nest box schemes every year, using the above column notation, so that we have as complete a picture as possible of the populations of nest box species throughout the county.*

(We are very grateful to GAV for his constructive comments on the above changes.)

continues to have relatively low fledging rates. Elsewhere, breeding season records (singing ♂♂/pairs) from: Burrator (1); Chittlehamholt (4); Combestone, Dartmeet (10, producing 63 fledged yg); Drewston Wood/Fingle Bridge (2), Dunsford Woods (12), Hembury and Burchetts Woods (2); Huccaby (one - 4yg hatched, but deserted) and Ivybridge (1). In the N of county, singing males were at Spreacombe, Braunton on 3 May, at Whiddon Valley, Barnstaple on 9 May and at Woody Bay on 11 Jun; and a pr with juv at Witheridge Moor on 24 Jun (RJJ) was the observer's first record for the site. At Stover CP a ♂ appeared again, and held territory 26 Apr – 3 Jun.

Spring. There was a big gap between the *first* bird, a very early male at Cockington on 26 Mar (SJA,MRAB) and the next at Lannacombe on 13 Apr. Very light passage, involving mostly singles at Prawle (also on Lundy, with a max of eight on 2 May) continued until 19 May, but birds were on territory from 15 Apr (at Okehampton).

Autumn. Two at Prawle Pt on 5 Jul must have been early migrants, as perhaps was the juv/♀ at Cadworthy Wood, Shaugh Prior on 22 Jul. In *Aug*, one at Start Pt on 3 Aug and one at Prawle on 12th, preceded the main passage involving almost daily records from coastal sites which began on 19 Aug; records came mostly from Prawle and Start, but also from Berry Hd, Frogmore, Soar and Torcross, the max count being 30 on Lundy on 24 Aug and four at Prawle on 26th. The last record on Dartmoor was one nr Dartmeet on 23 Aug. In *Sep*, surprisingly few records, with only singles at Prawle on 2nd, 4th and 22nd, and at Soar on 21st. In *Oct*, singles at: Plymouth (Mt Gould) on 2nd; Frogmore on 6th; Start Pt on 6th and 14th; and the *last* at Berry Hd on 12th and 14th (RBe,SMRY).

Ringing recoveries. Of various ringed birds recaptured, often at nestboxes, the greatest distance was 139km, between Jun 1999 and 2000 (ignoring the migration to Africa and back in between!), and two birds recaptured in 2001 had first been ringed in 1996; further details in the Ringing Report. In the Okehampton Woods, 16 of 35 ads caught had already been ringed, 14 of them at Okehampton (GAV).

LONG-TAILED TIT *Aegithalos caudatus*

Fairly numerous resident breeder.

Similar flock-sizes, monthly pattern and level of reporting as in 2000.

Table 104: Maximum flock size and number of flocks >10

	J	F	M	A	M	J	J	A	S	O	N	D
Max flock size	20	25	7	6	14	29	18	16	20	25	30	20
No. of flocks >10	4	2	-	-	1	2	2	3	6	12	13	10

Breeding. Breeding reported from only nine sites: Berrynarbor (pr feeding 5 yg on *home-made* fruit cake!); Bowling Green Marsh (pr+3yg); Clennon Valley (4 prs on territory); Cockington (pr with nest material in Mar); Dawlish Warren (pr+5yg); Higher Metcombe; Huccaby; Northam; and Tiverton (pr+2yg). A large count for the summer was of 29 at Derriford (Plymouth) on 12 Jun (PFG).

First winter-period. The only flocks ≥ 20 were 20 at Burrator on 14 Jan, and 25+ at Stover CP on 2 Feb. Feeding on garden nut-feeders was reported from Maidencombe in Jan (13 birds), Okehampton and Exeter (six birds) in Feb and South Brent in Mar; one also occurred on a Plymouth (Peverell) bird table in Jan.

Autumn/second winter-period. Seven flocks of ≥ 20 were at: Huccaby on 19 Sep (20+); Dawlish Warren on 10 Oct (25); Berry Hd on 28 Oct (20); East Prawle Wood on 2 Nov (20); Slapton on 9 Nov (30); Berry Hd on 16 Nov (30); and Man Sands on 20 Dec (20). At least some local movement is suggested by the autumn coastal flocks. The only record from garden nut-feeders in this period was from Berrynarbor, where small parties were regular in Dec.

MARSH TIT *Parus palustris*

Fairly numerous resident breeder. BoCC (Red).

A total of 98 records from 57 sites, up on 2000, but with the same max count of six.

Breeding. Interesting breeding records include the pr that attempted to breed in the roof of a house in Yeo Vale on 12 Apr but failed, possibly due to the presence of 144 Pipistrelle bats, and the pr nesting a metre above the river in an old Alder stump at Berrynarbor on 29 Apr. Otherwise reported (all in Jun) only from Ford Wood, Bideford, Higher Metcombe, Plymbridge Wood (pr+3yg) and Woody Bay (pr+5yg). A family group of 5+ was seen nr Manaton on 16 Jul.

Non-breeding. The highest counts were: six (at Halsdon DWTNR in Mar); and four (by Erme Est in Oct, at Stover CP in Nov and on Witheridge Moor in Dec (inc. two feeding in the middle of a road)).

WILLOW TIT *Parus montanus*

Not scarce, but localised, resident breeder. BoCC (Red).

A total of 36 records from 19 sites (mostly in N Devon), similar to 2000, but with very few in the breeding season.

Breeding. The only evidence of breeding came from: Fernworthy Res (up to two in Jun & Jul, also in Jan & Feb; MGM,IM); Torrington Common (Feb-Jul presence; WHT); West Putford (present throughout year, AMJ); Wistlandpound (heard regularly at four sites in summer; IT,MAP); and Witheridge Moor (carrying food in May; RJJ).

Non-breeding. In addition to the above, records of ones and twos scattered throughout the year from: Exeter Canal on 7 Jan (AHk); Hartland Forest on 18 Jan (IM); Hembury Woods on 12 Jan (MGM); Lower Tamar L in Jan, Feb & Dec (DIJ,FHCK,BCM); Meddon in Aug & Sep (WHT); Melbury Plantation on 1 Dec (IM); Odam Moor Plantation on 15 Feb (WHT); Plym Est (Blagdon's Boatyard) on 3 Apr (NST); Plymouth (Efford Marsh) on 12 Oct (TF); Prewley Moor on 5 Aug (WHT);Roadford Res in Jan & Nov (FHCK,MGM); Sourton Cross on 19 Oct (RHi); Trenchford Res in Feb, singing (SMRY); and Upper Tamar L, Aug-Dec (mo).

COAL TIT *Parus ater*

Numerous resident breeder; scarce autumn passage/irruptive visitor.

More coastal records and large flocks than in 2000 and two reports of a good breeding season.

Breeding. Breeding records from only seven sites, but a good breeding year was reported from Berrynarbor, and at Yarner Wood the two prs that raised 21yg in nestboxes constituted the second best year since 1973.

Non-breeding. The only notable features are the coastal records in Sep & Oct from Berry Hd, Dawlish Warren, Prawle, Soar and Start area, some of which included the largest flocks of the year. Double-figure flocks comprised: 29 at Prawle on 9 Oct; 20+ at Stover CP on 2 Feb; 17 at Woodbury Castle on 14 Jan; 17 at Berry Hd on 13 Oct (with 15 on 9 Oct, and 11 on 28 Sep & 20 Oct), and also 12 at nearby Sharkham Pt on 12 Oct; 11 at Dawlish Warren on 21 Sep; and ten in Torbay area on 5 Oct. Also one on Lundy, where rare, 9-16 Oct. No records submitted of the Continental subspecies *P.a.ater*, and some observers emphasised that their coastal birds were definitely British *P.a.britannicus*.

BLUE TIT *Parus caeruleus*

Abundant resident breeder. SPEC(4).

No real evidence of change, but fewer and smaller large counts than in 2000.

Table 105: Summary of Nest Box scheme data (sites with >9 boxes with eggs in bold)

	Boxes with eggs (B)	Success boxes (S)	Eggs (E)	Young (Y)	Fledged (F)	Mean clutch (E/B)	% fledged (100F/Y)	Mean prod - uctivity* (F/S)
Boreston, nr Halwell (VC)	6	5	50	46	35	8.3	76	7.0
Bovey Valley Wods NNR (BT)	**14**	**9**	**102**	**90**	**70**	**7.3**	**78**	**7.8**
Crowden Copse (VC)	5	3	37	26	19	7.4	73	6.3
Dartington Hill (VC)	8	4	56	43	21	7.0	49	5.3
Neadon Cleave (SH)	**10**	**8**	**73**	**72**	**66**	**7.3**	**92**	**8.3**
Okehampton (GAV)	**72**	**63**	**635**	**542**	**506**	**8.8**	**93**	**8.3**
Yarner Wood (BT)	**52**	**40**	**457**	**358**	**291**	**8.8**	**81**	**7.0**

**F/S = mean number fledged per successful box*

Breeding. Nestbox data in terms of boxes used and productivity were very similar to 2000, and in general, site differences were maintained. At Dunsford Woods, where access was limited, it appears that there was an increase in breeding prs, though they had only mediocre breeding success.

Non-breeding. The largest counts were 40+ at Burrator on 24 Sep, 27 at Huccaby on 19 Sep, 24 at Woodbury Castle on 14 Jan, 22 at Prawle on 20 Oct and 20 there on 9 June and 31 Dec. It would appear

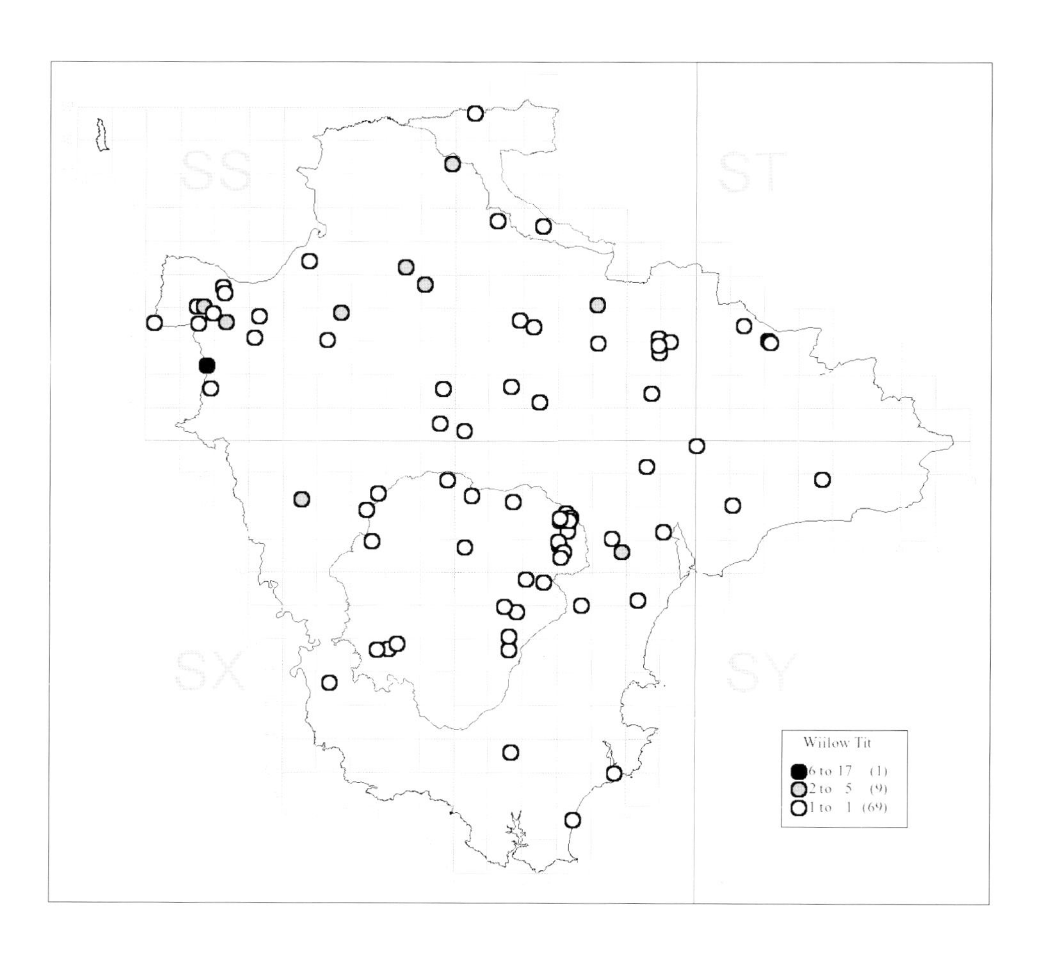

Map 14: Willow Tit.
Distribution of all records with obvious migrants excluded. Abundance is based on the maximum count of pairs at each location. The distribution closely follows the Culm Natural Area, whilst the importance of peripheral valley systems on Dartmoor is also evident.
Data: Based upon a ten year period 1992-2001. Source: DBRs) (see Overview on page 25)

from this that few observers send in even large counts of this species.

Ringing recoveries. One ringed at Bridford in Aug 1995 was recaptured there in Feb 2001, and two other birds had also survived at least three years but moved very little; details in the Ringing Report.

GREAT TIT *Parus major*

Abundant resident breeder.

No real evidence of change, but fewer and smaller large counts than in 2000.

Table 106: Summary of Nest Box scheme data (sites with >9 boxes with eggs in bold)

	Boxes with eggs (B)	Success boxes (S)	Eggs (E)	Young (Y)	Fledged (F)	Mean clutch (E/B)	% fledged (100F/Y)	Mean prod - uctivity* (F/S)
Boreston, nr Halwell (VC)	1	0	6	0	0	6.0	0	0.0
Bovey Valley Woods NNR (BT)	2	2	12	12	12	6.0	100	6.0
Crowden Copse (VC)	2	2	13	13	11	6.5	85	5.5
Dartington Hill (VC)	5	3	17	13	9	3.4	69	3.0
Neadon Cleave (BT)	3	3	20	20	19	6.7	95	6.3
Okehampton (GAV)	**46**	**40**	**297**	**270**	**248**	**6.5**	**92**	**6.2**
Yarner Wood NNR (BT)	**10**	**10**	**67**	**64**	**63**	**6.7**	**98**	**6.3**

**F/S = mean number fledged per successful box*

Breeding. At the main nestbox site at Okehampton, fewer boxes were used but productivity remained the same as in 2000. Although only ten boxes were used at Yarner Wood, this is the second largest site and here productivity increased from 5.4 to 6.3. Fifteen singing ♂♂ were recorded from Huccaby (in 9.3 hectares) but most other breeding reports referred to one-three pairs.

Non-breeding. The only counts ≥10 were: 12 at Dawlish Warren on 10 Oct; 11 at Woodbury Castle on 4 Jan; 11 at Prawle on 16 Oct; and 10+ at Stover CP on 2 Feb. One on Lundy, where rare, 25-27 Oct.

Ringing recoveries. Two birds, ringed at Sidborough and Derriford, were recovered after 1,858 and 272 days respectively (the latter terminated by a cat), but had gone nowhere; further details in the Ringing Report.

NUTHATCH *Sitta europaea* — Wood Nuthatch

Numerous resident breeder.

Recorded from 61 sites (cf. 37 in 2000 and 69 in 1999) but no reason to believe that this species is anything but stable.

Table 107: Summary of Nest Box scheme data (sites with >9 boxes with eggs in bold)

	Boxes with eggs (B)	Success boxes (S)	Eggs (E)	Young (Y)	Fledged (F)	Mean clutch (E/B)	% fledged (100F/Y)	Mean prod - uctivity* (F/S)
Boreston, nr Halwell (VC)	1	1	5	5	5	5.0	100	5.0
Bovey Valley Woods NNR (BT)	1	1	5	5	5	5.0	100	5.0
Okehampton (GAV)	**13**	**13**	**75**	**73**	**67**	**5.6**	**92**	**5.2**
Yarner Wood (BT)	2	1	15	6	6	6.0	100	6.0

**F/S = mean number fledged per successful box*

Breeding. Reported from only 12 sites, including four with nestboxes; for the 13 prs at the main one at Okehampton, productivity was slightly up on 2000. There were at least five prs in Avon Valley Woods, three breeding prs at Combestone, three ♂♂ singing at Hittisleigh, a pr with two juvs at Challacombe and four territories on Witheridge Moor.

Non-breeding. Counts of between three and six (though not necessarily involving birds seen together) were from Aveton Gifford tidal road, Halsdon DWTNR, Shipley Bridge, Stover CP and Watersmeet. Perhaps the only exciting Nuthatch of the year was at Dawlish Warren on 29 Sep and 11 & 12 Oct (ARo,JEF) as it was only the sixth site record since 1987, and the first in autumn.

TREECREEPER *Certhia familiaris* — Eurasian Treecreeper

Numerous resident breeder.

A total of 81 records from 51 sites is similar to other recent years.

Breeding. Hardly any breeding season reports, the only confirmed successful breeding being at

Dartmeet (family party of 4+ on 6 Jul) and Yarner Wood (pr+2yg in Jun).

Non-breeding. Max count was four at Shipley Bridge on 2 Jul, followed by threesomes at Exminster (church and hospital) on 3 Nov, Huccaby on 18 & 19 Nov and disputing territory at Lustleigh on 12 Dec. One at Lipson, Plymouth on 18 Feb was a first site record, and other rare occurrences were up to two at Berry Hd 5 Oct – 21 Nov and singles on Lundy 20 Sep & 21-24 Oct. *Unusual behaviour* included one on garden nut-feeder with Nuthatch at South Brent on 14 Mar, and another on ground, drinking water from a ditch at Witheridge Moor on 24 Jun.

GOLDEN ORIOLE *Oriolus oriolus* — Eurasian Golden Oriole

Very scarce passage visitor in spring, rare in summer and autumn (bred 1939). BoCC (Amber).

One mainland record as in 2000, but only one (cf. three) on Lundy.

A ♀ on Lundy 30-31 May (LFS), followed by a singing ♂ at Bowling Green Marsh on 3 Jun (MKn,RSPB).

ISABELLINE SHRIKE *Lanius isabellinus*

Rare Vagrant

A ♂, age uncertain, of the West-Central Asian race *(L. i. phoenicuroides),* was on Lundy on 28 Aug (PDv); only the fourth Devon record, the last in 1997, and previous birds have been 1w in Oct and Nov. (One of three in Britain in 2001 accepted by BBRC).

RED-BACKED SHRIKE *Lanius collurio*

Very scarce passage visitor; former breeder (last bred 1967). SPEC(3), BoCC (Red), UK BAP.

Similar to 2000, one in Jun, and five autumn records, four of these being in the Prawle area.

Summer. A ♀ at Grimspound on 9 Jun was mobbed by Whinchats and flew off N over Headland Warren (RS).

Autumn. The five records comprised an ad ♀ in the E Prawle area on 19 Aug (PMM *et al.*) and four 1w birds: at Prawle Pt on 9 Oct (PMM); at Sharkham Pt on 12 Oct (Colin Bath per BHNNR); at Langerstone Pt on 17-20 Oct (Jean Mullen *per* VRT,BW); and lastly at Ash Park Lane (Prawle area) on 11 Nov (PMM).

GREAT GREY SHRIKE *Lanius excubitor*

Very scarce winter visitor. SPEC(3).

Three only, as in 2000.

One at Dunsdon Farm NNR 1-16 Jan, present since 18 Dec 2000 (GPS), was also seen over the Cornish border; and one at Kilkhampton on 26 Feb (*per* ML). In the second winter-period, one at Fernworthy Res during Nov (RK).

JAY *Garrulus glandarius* — Eurasian Jay

Fairly numerous resident breeder, scarce and irregular passage visitor.

The usual wide scatter of reports of small numbers, but even fewer breeding reports than normal.

Breeding. Three reports only, at Higher Metcombe, Sticklepath and Stokenham.

Maximum counts. A few reports of double figure numbers, with Burrator a prime site, ten being reported on more than one occasion during the winter period; a similar number on Chudleigh Knighton Heath on 27 Jul.

Unusual records. One in a garden at Mt Gould (Plymouth) 27-28 Mar, normally only seen here in the autumn; four flying high and heading S over Kenwith NR on 6 Oct; two in a garden at Peverell (Plymouth) on 9 Oct; and two fighting a running battle with two Sparrowhawks at Dartmeet on 13 Nov.

MAGPIE *Pica pica* — Black-billed Magpie

Numerous resident breeder.

No large roost counts, but slightly more double-figure daytime counts than in 2000.

Breeding. Much under-reported as a breeding bird, with only seven locations reporting breeding: Derriford, Plymouth (in tree on roundabout island); Dousland; Higher Metcombe; Kenwith NR; Landcross; Northam; Prawle; and Stokenham.

Maximum counts. The only roosting party reported this year was of 27 at Lee Moor on 13 Sep. Larger daytime numbers were reported from various locations in autumn and winter: 24 at Wembury on 6 Jan; 16 at Braunton on 20th; 17 at Shaldon on 31st; 13 at Alphington on 14 Feb; 14 in Clennon Valley on 22nd; 14 at Berrynarbor on 2 Sep; 19 at Dawlish on 2 Oct, with 31 there on 10th; 15 at Burrator on 12 Dec; and ten in the Prawle area on 16 Dec, with 23 there on 31st.

CHOUGH *Pyrrhocorax pyrrhorocorax* — Red-billed Chough

Rare vagrant. BoCC (Amber).

Several records from the S coast at a time when also reported from Cornwall and Dorset. After one on Lundy in 2000, the first on the mainland since 1997.

One seen on farmland at the mouth of the R Dart on 1 Apr (MI) was also noted flying S at Slapton on the same date (T Reid); at Beesands on 11 Apr, it was then frequently reported from the Prawle area 11 Apr-12 May (KEM et al.) and back at the mouth of the R Dart on 15 May (MI).

Chough - *Steve Young*

JACKDAW *Corvus monedula* — Eurasian Jackdaw

Numerous resident breeder, irregular passage/irruptive visitor. SPEC(4).

Widely reported from rural areas, coastal locations and villages.

Table 108: Maximum flock size and number of flocks >25

	J	F	M	A	M	J	J	A	S	O	N	D
Max flock size	70	12	41	30	19	10	40	60	200	50	24	35
No. of flocks >25	3	1	1	1	-	-	5	5	3	2	-	1

Breeding. Reported from the following sites (prs given in brackets): Aveton Gifford (38), Baggy Pt (no count), Huccaby (5), Kingston (7), Princetown (2), Stoke Canon (7) and Virginstow (2). At Stoke Canon, five monitored breeding prs fledged an average of 1.8 yg per pr, a figure that corresponds with those quoted in *BWP* (DJJ).

Maximum counts. The largest were: 260 on Exminster Marshes on 6 Jan; 300 at Hope Cove on 6 Feb; 250 going to roost at Stoke Canon on 6 Jun; 200 going to roost at Moorgate on 26 Jul; 300 going to roost at Stoke Canon on 11 Aug with 400 going to roost at Burlescombe on 31 Aug; 200 at Ashburton on 17 Nov; 200 at Sourton Cross on 6 Dec; and 300 in one tree adjacent to Kingsbridge Est on 19 Dec. Morning counts at Stoke Canon in the latter half of Dec showed up to 700 dispersing from roost. Birds on passage at Staddon Pt (Plymouth Sound) totalled some 2,100 between 1 Oct and 15 Nov, the peak counts being 460 on 21 Oct, 550 on 1 Nov, 750 on 2nd, 100 on 3rd and 180 on 5th (SGe – see separate report).

Aberrant plumage. A bird showing a white collar similar to that characteristic of eastern races was at Otterton on 27 Feb; western races can show this feature as an aberration.

ROOK *Corvus frugilegus*

Numerous resident breeder.

Despite access restrictions a reasonable number of rookery counts received since many are visible from public roads.

Breeding. A survey in 10km grid square SS53 centred on Barnstaple located 16 rookeries containing a total of 446 nests; the largest rookery contained 104 nests, the smallest five. In Aveton Gifford Parish 152 nests were located, although one rookery in that area remained uncounted due to access restrictions. Elsewhere, reported from a total of nine locations (nest numbers in brackets): Broadsands (26), Hooe (29), Port Hill, Northam (30) and West Putford (32). The other reports were for rookeries containing less than 20 nests each. The count for Hooe showed a large drop from the previous year when 70 nests were located.

Maximum counts (away from rookeries). Max counts: 200 at Hope Cove on 6 Feb; 108 at Netton on 23 May with 240 at Killerton on 27th; 150 at Ugborough on 18 Jun; 400 at Huxham on 8 Jul with 120 at Maidencombe on 21st, 200 at Bolberry Down on 22nd and 150 at Smallhanger Waste on 28th; 250 at Hendom Cross on 11 Aug; 130 at Wembury on 24 Aug and 20 Sep with 225 there on 8 Nov; 200 at Ashburton on 17th; 175 at Wembury on 3 Dec; and 150 at Huntsham on 13 Dec. Large *pre roost or roost counts*: 1,000 at Huxham on 14 Jan, 600 at Stoke Canon on 20 Jun, 900 at Thurlestone on 15 Jul and 500 at Huxham on 16 Dec.

Aberrant plumage. The brown-plumaged bird at Northam Burrows is now in its thirteenth year.

Rook - *Ren Hathway*

CARRION CROW *Corvus corone*

Numerous resident breeder.

Larger non breeding season flock sizes reported than in 2000, but few breeding records.

Maximum counts. The largest counts were: 377 on mudflats at Dawlish Warren on 2 Jan with 110 on Exminster Marshes on 6th; 137 on Taw Est on 18 Jun with 125 at Yelland on 26th; and 360 on Teign Est on 31 Aug. There were 14 double figure counts, several of these also being on estuary or seashore locations.

Aberrant plumage. As well as the usual plumage abnormalities of white primaries, a leucistic individual with a dark head and primaries was reported nr Marldon, paired with a normal bird (AWri).

HOODED CROW *Cornus cornix*

Previously Cornus corvus cornix. Rare winter/passage visitor.

One on Lundy 24-25 Apr (LFS); and one on a tip at Kingsteignton 2-10 Jan BHH *et al.*), present from 2000, and possibly the same as the bird at Goodrington on 21 Jan (ML).

RAVEN *Corvus corax* — Common Raven

Not scarce resident breeder.

Widely reported from rural locations, also seen passing over suburban parts of both Exeter and Plymouth and nesting close to a suburban area elsewhere.

Breeding. Few reports, due to FMD access restrictions and the species' tendency to nest early in remote locations. Successful breeding reported from Aveton Gifford (2 prs), Berry Head, Buckfastleigh, Cann Quarry, Clennon Valley (two prs in an area surrounded by suburbia), the Higher Metcombe area, Owley Bottom (the first breeding in the Glaze Valley during the last 60 years) and Saunton Down.

Maximum counts. The largest counts were: 11 at Dawlish Warren on 16 Aug; 29 at Mary Tavy on 5 Sep; at least 70 in the air together over woods near Dittisham on 6 Sep; 20 flying N at Haldon on 20 Sep; at least 20 in Pig's Nose Valley on 16-17 Oct; 30 around a sheep carcass nr Shipley Bridge on 30 Oct; 34 at Piles Copse Tor on 1 Nov; and 50 at Hoccombe Hill, Exmoor on 2 Nov.

STARLING *Sturnus vulgaris* — Common Starling

Abundant resident breeder, passage and winter visitor. BoCC (Red).

Mid-Devon continues to hold large roosts. Very few breeding records.

Table 109: Maximum flock size and number of flocks >1000

	J	F	M	A	M	J	J	A	S	O	N	D
Max flock size	200 K	15 K	3 K	2	100	150	80	102	1000 K	3 K	5 K	1000 K
No. of flocks ≥1000	9	6	4	-	-	-	-	-	3	3	7	4

Breeding. Obviously under-recorded: five prs located in Aveton Gifford, the same as in 2000, but none at Higher Metcombe.

First winter-period roosts. Thousands disturbed from Bideford Bridge after midnight due to New Year celebrations, with 15,000 counted there on 26 Feb; 800 at Dawlish Warren on 13 Jan with thousands over Abbeyford Woods on 17th presumably going to roost at Thorndon Cross; and a large roost in the Okehampton area 9-10 Mar. Other large flocks reported during this period included: 200,000 at Lower Tamar L on 6 Jan; a monthly max of 4,000 at Dawlish Warren on 13 Jan, with two separate flocks totalling 2,500 on Exminster Marshes on 19th; 1,200 at Huntsham on 27 Jan; 12,000 at West Putford on 17 Feb; a monthly max of 3,000 at Hendom Cross on 26 Feb; 3,000 at Huntsham on 10 Mar; and 3,000 at Berrynarbor on 13th.

Autumn/second winter-period roosts. *Visible movement* included small groups coming ashore in the Soar area on 12 Oct, and a total of 2,555 past Staddon Pt 21 Oct - 15 Nov, with a peak of 425 on the first date (SGe – see separate report). *Reported roost activity*: 1,500 at Topsham on 13 Sep with 3,000 there on 23rd and 1 million at Brimford Bridge on 29th (also on 4 Dec); 550 on Taw Est on 15 Oct and 500 on Axe Est on 16th; 1,000 at Slapton Ley on 5 Nov with 5,000 there on 10th (currently roost activity at this site only seen in early Nov after which it moves elsewhere); 3,000 at Sharpham Marsh (Totnes) on 8 Nov and 2,000 on 9th; 1,000 heading towards roost at Princetown on 14 Nov; and a major roost at Thorndon Cross on 12 Dec (in 2000 this roost was estimated to hold in excess of 2 million birds). Other flocks reported during this period include: 1,500 on Exminster Marshes on 24 Oct with 3,000 at Hartland Point on 28th; 2,000 on Exminster Marshes on 15 Nov with 1,000 at Waddeton on 16th; 4,500 on Exe Est on 2 Dec with 1,000 at Huntsham on 19th; and 1,200 in flight over Witheridge Moor on 28 Dec.

Aberrant plumage and ringing recoveries. A juv with a white rump and tail, white primaries and some white on the head was seen near Exeter in May. Ringed birds recovered in 2001 included one in

April 2,283km E of Tiverton where it had been ringed in 1995, and one shot at Hemyock, only 30km away from where it had been ringed in1998; further details in the Ringing Report.

ROSE-COLOURED STARLING *Sturnus roseus* **Rosy Starling**

Rare vagrant

A 1s ♂ in Paignton 11-17 Jun (Mr F Vagges, ML *et al.*) and an ad briefly at Lee Moor on 14 Jun (RJL). (These and another 55 in Britain in 2001 accepted by BBRC, the second summer invasion in succession; this species ceases to be a BBRC rarity from Jan 2002).

HOUSE SPARROW *Passer domesticus*

Abundant resident breeder. BoCC (Red).

Reported from 67 sites (cf. 48 in 2000), although difficult to assess this species, which is giving cause for concern, without more detailed information.

Breeding. Reported from 11 locations: Ashburton, Colyton, Dartmeet, Dousland, East Charleton, Exbourne (40 prs in an old barn (NCW)), Exmouth, Northam, Okehampton, Torcross, Virginstow and West Putford. Whilst the number of sites reporting birds and the number of breeding reports show a welcome increase, some observers report a continued decline.

Maximum counts. Double figure counts reported from 28 sites, a similar number to 2000: 70 at Stockland on 14 Jan; 80 at Exbourne on 7 Apr, with 30 at East Prawle on 16th; 40 at Huxham on 8 Jul, with 100 at East Prawle on 21st and 32 at Dawlish Warren on 23rd; 42 in a Clennon Valley garden on 11 Aug was a yearly max; 40 at Okehampton on 14 Aug; 60 at Wistlandpound Res on 16 Aug and up to 100 at Velator on 31st; 100 at Poltimore on 9 Sep and, the largest flock of the year, 200 in a cereal field alongside the Grand Western Canal on 15th (RJJ); and 50 at Appledore on 6 Oct. Regular surveys of a garden in urban Plymouth gave the following monthly maxima: 34, 40, 41 & 11 for Jan-Apr and 24, 27, 24 & 35 for Sep-Dec (RMcC).

Aberrant plumage. A leucistic individual present in an Upottery garden from 15 Nov – 31 Dec.

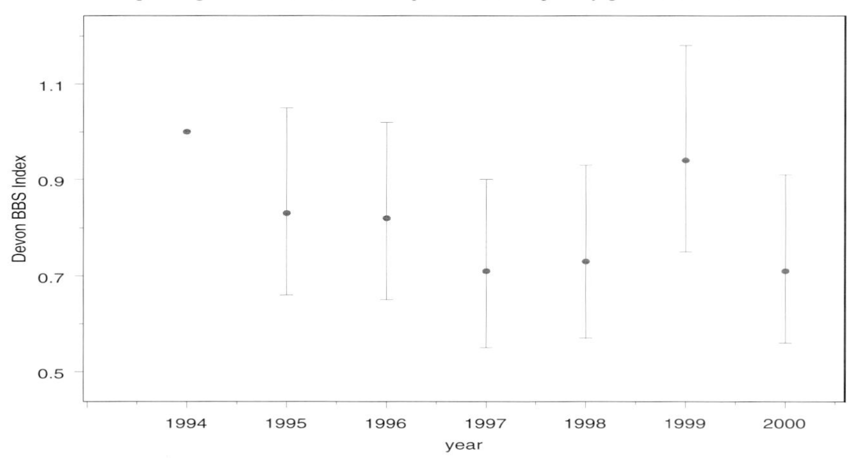

Fig. 26: House Sparrow - Devon BBS Indices 1994 - 2000, showing means with 95% confidence intervals (Source: BTO; see seperate report)

TREE SPARROW *Passer montanus* **Eurasian Tree Sparrow**

Rare or very scarce non-breeding visitor (last bred 1973). BoCC (Red).

Only three records, but yet another in the breeding season.

In the first winter-period, two overwintered in a garden at East Portlemouth (Mr R House), and there were two in an Okehampton garden on 31 Jan (GAV). In spring, one in the Prawle area on 7 May (PMM).

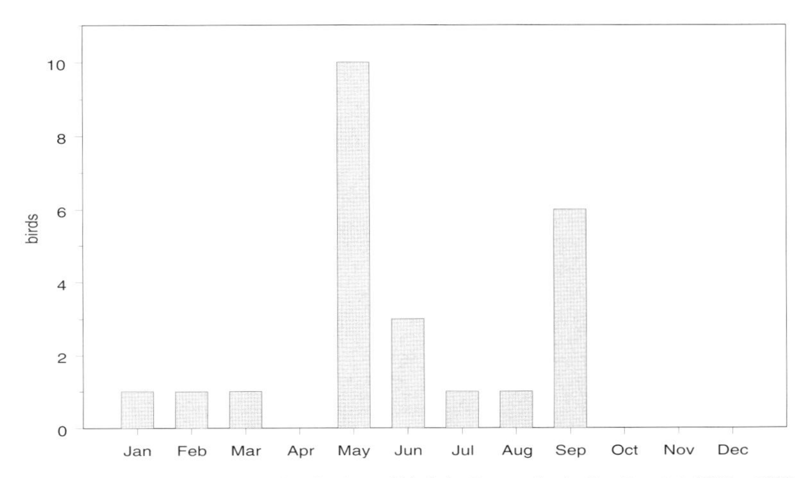

Fig. 27: Tree Sparrow - Monthly distribution of birds in Devon (including Lundy) 1992 - 2000. (Source: DBRs)

CHAFFINCH *Fringilla coelebs* **Common Chaffinch**

Abundant resident breeder, passage and winter visitor. SPEC(4).

Reported from 72 sites (cf. 97 in 1999), a return to lower reporting levels with fewer and smaller large flocks reported in either winter period.

Table 110: Maximum flock size and number of flocks >100

	J	F	M	A	M	J	J	A	S	O	N	D
Max flock size	120	120	350	6	7	5	2	2	100	50	100	250
No. of flocks ≥100	6	2	1	-	-	-	-	-	1	-	2	3

Breeding. The only counts were 10-12 pairs as usual at Higher Metcombe, and 11 singing ♂♂ at Huccaby.

First winter-period. Large counts: 100 roosting at Stover CP during Jan; 100 at Woodbury Castle on 2 Jan, 105 feeding in stubbles at East Charleton on 7th, 100 at both Crealy CP and Haldon on 14th and 120 at West Putford on 28th; *c*.350, the largest flock of the year at Fernworthy on 6 Feb (IM); 120 at Burrator on 24 Feb; and a regular flock of 150 at Huntsham during this period (last seen on 8 Mar). Several double figure flocks reported including some garden flocks.

Autumn/second winter-period. From 15 Sep to 30 Nov over 3,500 birds were recorded moving NW past Staddon Pt during early morning observations (SGe - see separate report); the highest daily rate was 553 in 1.5 hours on 9 Oct. Other high counts included: 100 in the Soar area on 20 Nov and 100 at Maidenford (Barnstaple) on 28th; 150 in Soar area on 10 Dec, 100 at Venford Res on 15th, 250 on the Taw Est at Barnstaple on 20th; and roosting by 100 birds at Stover CP during Dec.

Aberrant plumage and ringing recovery. Two unusually plumaged individuals reported: a sandy coloured bird with a silvery tail, faint wing bar and crown stripe 21-24 Mar visiting a feeder at Huntsham, and one resembling a ♂ Snow Bunting at Aveton Gifford on 8 Dec. A bird ringed and recovered at Sidborough had lived for at least seven years; further details in the Ringing Report.

BRAMBLING *Fringilla montifringilla*

Usually scarce passage and winter visitor, though not scarce in some years.

Few reported in first winter period following on from low numbers in late 2000; more reported in second winter period but still low in numbers with only 25 sites (cf. 38 in 2000).

Table 111: Maximum flock size and number of flocks >10

	J	F	M	A	M	J	J	A	S	O	N	D
Max flock size	6	1	several	-	-	-	-	-	-	20	9	4
No. of sites	5	2	1	-	-	-	-	-	-	7	17	12

First winter-period. Several at Burrator Res, four at Halwill Junction, two at Fernworthy and singles at Melbury and West Putford were the only records.

Second winter period. At Staddon Pt, 46 recorded on passage between 1 Oct and 15 Nov, with a daily peak of nine on 2 Nov and peak passage between 27 Oct and 15 Nov (SGe). *Elsewhere* on Lundy there was an autumn max of 13 on 3 Nov, and the first bird at other coastal migration sites was one at Prawle on 13 Oct followed by one at Start Pt on 18th and 20+ there on 21st; the first singles inland occurred at end of Oct. In Nov, mostly singles were reported from the coastal sites of Berry Hd, Bigbury, Dawlish Warren, Soar and Westward Ho! and even fewer inland sites. The max Dec count was of four feeding at a bird table in Chudleigh Knighton, whereas elsewhere there were two at Barnstaple, Otterton, Prawle and Venford and singles from four other sites.

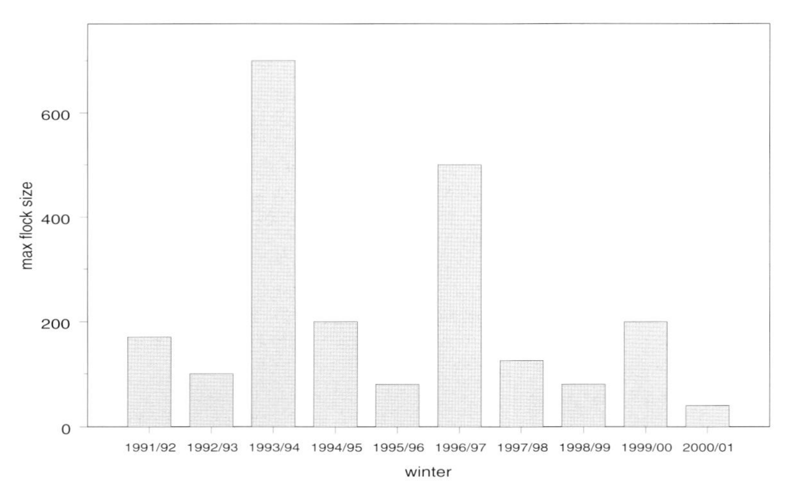

Fig. 28: Brambling - Max flock size recorded in Devon (Mainland only) 1991/92 - 2000/01 winters. (Source: DBRs)

SERIN *Serinus serinus* — European Serin

Rare passage visitor; casual breeder (last breed 1985). BoCC (Amber).

A welcome return for this species, not reported since 1998.

A ♂ in a finch flock in the Prawle area 24-26 Feb, noted as singing on the latter date (K E Vinicombe *et al.*).

GREENFINCH *Carduelis chloris* — European Greenfinch

Numerous resident breeder, passage and winter visitor. SPEC(4).

Relatively few reports in the first winter period; more reported in the autumn but again few reports in the second winter period. First reports of autumn passage at coastal sites later than in 2000.

Table 112: Maximum flock size and number of flocks >100

	J	F	M	A	M	J	J	A	S	O	N	D
Max flock size	150	20	28	20	18	16	10	20	80	250	200	100
No. of flocks >100	2	-	-	-	-	-	-	-	-	3	1	1

Breeding. Juvs noted at several sites. Considered to be a good breeding year at Berrynarbor.

First winter-period. The largest flocks were 150 roosting at Stover CP during Jan, 100 at Crealy CP on 14th and 45 at Slapton on 16th. Nine other locations reported low double-figure counts during this period.

Autumn/second winter-period. Largest flocks: 80+ at Dawlish Warren 22 Sep - 4 Oct rising to 150 on 9th and 250 by 14th; 50+ in the Soar area on 22 Sep; 75 at Berry Hd on 10 Oct rising to 120 on 12th; 50 on the Taw Est on 10 Oct, 60 at Slapton on 12th and 50+ at Northam Burrows on 16th; 50 at Slapton on 12 Nov and 200 at Maidenford (Barnstaple) on 28th; and a roost of 100 at Stover CP during Dec. Passage reported during this period: 450 past Staddon Pt between 15 Sep and 30 Nov with the peak period being 9 Oct - 30 Nov and a daily max of 65 on 18 Oct (SGe – see separate report); also 50 west past Weston on 21 Oct.

GOLDFINCH *Carduelis carduelis* — European Goldfinch

Fairly numerous migrant and resident breeder, passage and winter visitor.

Reported from 92 sites (cf. 79 in 2000) with flock sizes generally similar to recent years.

Table 113: Maximum flock size and number of flocks >50

	J	F	M	A	M	J	J	A	S	O	N	D
Max flock size	90	75	300	82	28	500	50	200	500	500	100	300
No. of flocks >50	2	1	2	4	-	1	1	3	12	11	3	2

Breeding. Reported only from Berrynarbor, Dunnabridge, Higher Metcombe, Huccaby, Lundy and Prawle. A flock of 500+ nr Thorverton on 2 Jun was large for the time of year; *BWP* suggests that summer flocks are composed of non-breeders, though the implications for breeding success are unclear.

First winter-period. Largest flocks: 75+ at Crealy CP on 14 Jan; 90 at Haldon on 22nd with 75 there on 7 Feb; 300 at Littleham on 10 Mar; 100 at East Portlemouth on 18 Mar, 80 at Bowling Green Marsh on 16 Apr and 82 at Prawle on 29th.

Autumn/second-winter period. Large flocks: 200+ in Soar area on 25 Aug; 110 on Exminster Marshes on 6 Sep with 135 there on 12th; 100 at Pottington on 19 Sep and 250+ in Soar area on 22nd with 500 at Wembury on same date; 250 at Kingswear on 25 Sep; 100 in Soar area on 9 Oct; 250 at Start Pt on 12 Oct (with 500 on 21st), 100 in Prawle area on 16th and 100 on Exminster Marshes on 31st; 100 at Waddeton on 17 Nov; and 300 at Braunton on 9 Dec with 250 there on 30th.

SISKIN *Carduelis spinus* — European Siskin

Not scarce, but local resident breeder; fairly numerous passage and winter visitor. SPEC(4).

Reported from 86 sites (cf. 111 in 2000), a typical level for normal years, 1997 and 2000 having been recent exceptions. There was a considerable movement in the autumn at coastal sites on the south coast but perhaps fewer inland records, posing a question as to the destination of those coasting birds.

Table 114: Maximum flock size, number of flocks >10 and number of sites

	J	F	M	A	M	J	J	A	S	O	N	D
Max flock size	50	40	35	25	6	2	20	-	50	140	13	60
No. of flocks >10	15	4	5	1	-	-	3	-	8	15	2	7

Breeding. Definite breeding activity only at Bellever (pr with yg) and Huccaby (three probable prs), but also present during breeding season at Burrator, Dart Valley, Fernworthy, Hennock, Newton St Cyres, Soussons and Stover CP. Also reported from Chudleigh (three on 1 Jul) and Peverell, Plymouth (two over on 18 Jun), both described as being unusual sightings for the time of year.

First winter-period. Fewer double figure counts than in 2000, the largest being: 20+ at Stover CP during Jan; 50 in Plymouth (Marsh Mills/Plym Est area) on 5 Jan; 20 at Burrator on 15 Jan and 50 at Kennick Res on 17th; 20+ at Haldon on 20 Feb and 40 at Fernworthy on 24th; and 35 at Blackborough on 7 Mar. Between 27 Jan-21 Mar, only ten were ringed in an Exeter garden (*cf.* 73 in 2000) (PWE).

Autumn/second winter-period. Larger than normal *coastal* movements: 563 past Staddon Pt 15 Sep-30 Nov with peak movement between 15 Sep and 21 Oct and max daily counts of 128 on 16 Sep and 127 on 17th (SGe – see separate report); 40 at Dawlish Warren on 24 Sep with 65 there on 2 Oct and 54 on 9th; up to 140 in Soar area on 9th with 50-70 in off the sea there on 12th; 28 at Berry Hd on 14th with 30 at Prawle on same date; 32 at Start Pt on 16th with 50 there on 17th and 60 on 18th; and 50 at Berry

Hd on 22nd. There were also many days with smaller counts from these coastal sites, and a Lundy autumn max of 40 on 28 Oct. *Inland* the largest counts were: 50 nr Dartmeet on 17 Sep; 13 at Fernworthy on 3 Nov; ten at Stover CP on 21 Nov; and in Dec, 20 at Huntsham on 8th, 20 at Lustleigh on 12th, 35 in Coly Valley on 22nd and 15 at Yarner Wood on 23rd.

Ringing recoveries. Birds ringed at Crediton and Exeter have been recovered in 2001 at distances up to 517km away, the latter involving a 1w ringed in Dec 1999 and recovered (dead) in Antrim in May 2001; further details in the Ringing Report.

LINNET *Carduelis cannabina* — Common Linnet

Fairly numerous or numerous resident breeder, passage and winter visitor. SPEC(4), BoCC (Red), UK BAP.

Similar levels of reporting to previous years, though possibly fewer large first winter-period flocks, and with the usual concentrations on the coasts in autumn which gradually reduce in size as birds move out to wintering areas, both in this country and on the near continent.

Table 115: Maximum flock size and number of flocks >100

	J	F	M	A	M	J	J	A	S	O	N	D
Max flock size	300	200	180	100	20	75	70	1000	1000	1000	650	2000
No. of flocks >100	4	1	1	1	-	-	-	4	6	12	5	4

Breeding. Widely reported during the breeding season, but few actual reports of breeding.

First winter-period. Largest flocks: 300 at Appledore on 9 Jan, 142 on Braunton Burrows on 19th and 300 at Bridford on 21st; 200 at Berrynarbor on 25 Feb; 180 at Huntsham on 31 Mar falling to 100 on 12 Apr.

Autum/second winter-period. Post-breeding flocks possibly gathering a week or so later than normal included 500 at Start 18 Aug - 1 Sep, with 200 there on 13 Oct and 1,000 on 21st; 1,000 at Soar on 25 Aug with similar sized flocks reported here during Sep falling to 400 in early Oct, 250 by mid Oct and 100 by early Nov; 600 at Bigbury on 29 Aug; 300 at Berrynarbor on 1 Sep; 400 at Wembury on 22 Sep; 200 at Skern on 3 Oct and 200 at Scabbacombe Head on 7th; 250 at East Charleton on 12 Oct, 240 at East Prawle and 100 at Kerswell Cross on 13th and 200 at Penhill Pt on 24th, 100 at Baggy Pt on 9 Nov and 120 at Velator on 10th; 350 on Roborough Down on 26 Nov and 650 at Maidenford (Barnstaple) on 28th; 200 in Bridford area during Dec and 200 on Taw Estuary at Barnstaple on 30th. *Autumn passage*: 918 past Staddon Pt 15 Sep-30 Nov with a peak day count of 111 on 20 Nov (SGe); 88 through Dawlish Warren on 10 Oct; and 200 at Weston on 21Oct.

Aberrant plumage. Two unusually plumaged birds reported, a paired ♀, basically white with brown speckling to the breast and black inner tail feathers on Witheridge Moor on 17 Jun, and one with a white head at Exmouth on 1 Nov.

TWITE *Carduelis flavirostris*

Rare winter and passage visitor. BoCC (Red).

One on Lundy on 2 Nov (JRD) was the first in Devon since one on 6 Oct 1991, also on Lundy.

LESSER REDPOLL *Carduelis cabaret*

Not scarce resident breeder, passage and winter visitor (other races may occur as winter visitors). BoCC (Amber).

No particularly large flocks in the first winter period, but rather larger numbers in the second winter period; no confirmed breeding.

Table 116: Maximum flock size and number of sites

	J	F	M	A	M	J	J	A	S	O	N	D
Max flock size	17	20	-	1	7	4	4	7	6	12	12	60
No. of sites	6	6	-	2	8	7	9	6	6	12	12	12

Breeding season. With restricted access to likely breeding sites, no confirmed records, but prs seen at

Bellever, Hartland Forest, Witheridge Moor (with juvs on 26 Aug) and Woodbury Common.

First winter-period. The largest groups were: 12 on the Exeter Canal on 7 Jan; 12 at Roadford Res on 14th and 17 at Haldon on 29th; 17 at Stover CP on 2 Feb and 20+ at Haldon on 20th. Other reports of smaller numbers, mostly from the above locations, with reports of singles from about five further locations.

Autumn/second winter-period. Coastal movement included 48 past Staddon Pt 9 Oct - 15 Nov with a peak of eight on 11 Nov (SGe); at Dawlish Warren, singles recorded on several days between Oct and Dec with two - three on some days at the end of Oct and into Nov and a peak of four on 10 Nov, thus corresponding to observations at Staddon Pt and suggesting similar movement pattern. On Lundy the autumn max was nine on 5 Nov. Double figure counts: 80+ roosting in the Haldon area Oct-Dec; 12+ at Stover CP on 19 Oct; 12 Roadford Res on 20 Nov; 15+ at Stover CP on 1 Dec; 60 at Aylesbeare Common on 4 Dec, 23 at New Bridge, Holne on 10th and 20+ at Yarner Wood on 23rd.

Aberrant plumage? A small bird showing characteristics of Common Redpoll *Carduelis flammea* was present at Topsham 4 Feb - 19 Mar (MKn *et al.*); for a full discussion of its identity, see separate Short Reports.

Redpoll species - (field notes-see separate report) - *Tom Whiley*

Crossbill - *Mike Langman*

CROSSBILL *Loxia curvirostra* — Common Crossbill

Scarce/not scarce breeder and irruptive visitor.

Few reports from the first winter period, but more widespread in the second winter period beginning with increasing numbers in Jul and passage noted in Jul and Aug including double-figure counts both during this period and later in the year. Family parties reported in May, Jul and Aug from two sites.

Table 117: Maximum flock size and number of sites

	J	F	M	A	M	J	J	A	S	O	N	D
Max flock size	9	25	-	-	20	12	31	40	2	15	14	5
No. of sites	3	2	-	-	1	2	4	7	2	4	7	1

Breeding. A pr with two juvs at Hennock Res on 27 May among a total of 20 birds present there, and small family parties at Haldon in Jul and Aug.

First winter-period. The few sightings included: eight at Bellever on 7 Jan with at least nine at Haldon on same date, one at Fernworthy on 13th and eight at Haldon on 29th; and 25-30 at Fernworthy during Feb and one at Haldon on 25th.

Summer, autumn and second winter-period. Numbers rose as early as *Jun*, with three at Fernworthy on 2nd and 17 at Trenchford Res on 24th; in *Jul*, 25 at Soussons and six at Stover CP on 12th, five at Fernworthy on 15th, 20 at Headland Warren on 23rd with 31 at Soussons on same date and one at Stover CP on 25th (and again on 7 Aug); in *Aug*, seven in flight over Mount Gould in Plymouth on 9th, seven in flight over Tawstock on 10th, 40 plus flying north and pausing briefly to feed at Burrator on 11th (with four there on 26th), five flying north over open moorland at Leedon Tor on 27th and one flying over Wistlandpound Res on 30th, with small number of other sites reporting singles at the end of the month. Largest *autumn* counts were 15 at Fernworthy on 6 Oct and 12 at Bellever on 27th; 10 at Bicton Common on 10 Nov, 14 at Haldon (obelisk) on 11th and 14 at Budleigh Salterton on 18th. The only report in Dec was five at Fernworthy on 3rd. Seven in a flock W over Staddon Pt on 9 Nov was the only obvious movement reported, although one flew E over the open moor at Gutter Tor on 26th.

SCARLET ROSEFINCH *Carpodacus erythrinus* — Common Rosefinch

Rare vagrant. BoCC (Amber).

Three autumn records from Lundy, probably involving three birds (cf. an absence in 2000).

One ♀/imm on Lundy on 8 Sep (NST), followed by an imm 9-13 Oct (BBH,MSS) and another, caught and ringed, on 26 Oct (LFS). *(see col plates)*

BULLFINCH *Pyrrhula pyrrhula* — Common Bullfinch

Fairly numerous or numerous resident breeder. BoCC (Red), UK BAP.

Reported from 90 sites throughout the year (cf. 80 in 2000) and 50 sites during the breeding season (cf 32 in 2000).

Breeding. More widely reported during the breeding season than in recent years with juvs seen at Berrynarbor, Braunton, Burrator Res, Dartmeet, Higher Metcombe, Northam, Sticklepath and Stokenham. In 2000 the report from Higher Metcombe indicated that they had not been seen there in the previous 30 years; in 2001 three or four pairs are thought to have bred in the area.

Maximum counts. As usual the majority of the records are of single birds or prs with the possibility of larger groups being seen during the winter periods; thus during the second winter-period there were eight on Saunton Down on 16 Nov, nine at Clennon Valley on 19th and seven on Exminster Marshes on 20th. Passage birds were recorded at coastal sites: at Dawlish Warren, six on 2 Oct; at Staddon Pt, out of a total of 22 over the peak period 18 Oct-14 Nov, six passed on 9 Nov with four on both 29 Oct and 14 Nov, while a group of six on 30 Oct were not moving; and at Berry Hd, nine on 21 Oct, including a group of six, and 15 on 28th, including a group of nine.

HAWFINCH *Coccothraustes coccothraustes*

Very scarce winter visitor; perhaps rare resident (breeding last confirmed 1973). BoCC (Amber).

Amazingly, 2001 data showed Devon as having the third largest known population of this species in Britain during the first winter period (behind New Forest and Forest of Dean).

First winter-period/spring. Recorded daily on a private shooting estate near *Longdown* until 24 Apr, feeding in an area of Hornbeam. Max count of 30 leaving woodland on 21 Jan, but with additional birds calling in wood, estimated total was *c*.40 birds. From mid-Feb, singing and gliding flight display regularly recorded. Also, three at *Doddiscombsleigh* on 11 Jan feeding on Hornbeam (RAda), and one also recorded here on 13 Feb (GV). The only spring records were on Lundy: one ♀ on 5 Apr (RGM), with one 4-28 May joined by another on 16th (LFS).

Summer. Records involved one adult ♀ and juv at *Southdown Farm, Soar*, on 10 Jul (CMi,PMM,PBo) and a bird perched high in a tree at *Chapel Wood* on 28 Jul (RJ).

Autumn/second winter-period. One at *Dawlish Warren* on 22 Oct (KRy) was only the second sighting there since 1954. Birds returned to the *Longdown* area on 1 Nov and remained until the year's end. Interestingly, despite a good Hornbeam seed crop, the birds switched to Field Maple as their food source. This tree is scattered over a wide area but in small numbers, and only 15 widely scattered birds were seen with no flocking (RAda). The only other records were of singles on Lundy on 4 Nov (LFS) and near *Castle Drogo* on 28 Dec (JTi).

2000 addition. One into roost at Haldon on 9 Jan (MRAB).

LAPLAND BUNTING *Calcarius lapponicus* — Lapland Longspur

Very scarce autumn passage visitor.

A poor year, with only two birds seen.

One at East Soar on 28 Oct (MKn,PAS), with one heard in the same area on 3 Nov (SRT,VRT); then one on Lundy 1 & 2 Nov (LFS).

Snow Bunting - *Ren Hathway*

SNOW BUNTING *Plectrophenax nivalis*

Very scarce passage and winter visitor. BoCC (Amber).

The usual scatter of winter records, mostly Feb, Oct & Nov and in the north of the county.

First winter-period. One flying over Dawlish Warren on 12 Feb (LC); and up to three, a ♂ and two ♀♀ at Capstone Hill, Ilfracombe 17-23 Feb *(see col plate),* with four on 29th (MAP *et al.*)

Autumn/second winter-period. The first were three at Berry Hd on 30 Sep (A Roe), then: one at Baggy Pt on 9 Oct (P&RM); three ad ♂♂ at Saunton Sands 17-23 Oct, then a pr there on 26th (RJ); up to three on Lundy 1-6 Nov (LFS); one on Saunton Down on 16 Nov heard calling before landing (RJ); and a ♂ on Cut Hill, Dartmoor, on 1 Dec (MLHS).

BLACK-FACED BUNTING *Emberiza spodocephala*

Rare vagrant.

The first Devon record of this E Asian vagrant, and only the third for Britain.

One on Lundy on 12 Oct (R M Patient *et al.*); see separate report. (Two in Britain in 2001 accepted by BBRC, the other being on Fair Isle 20-24 Oct.)

Black-faced Bunting - *Ren Hathway*

YELLOWHAMMER *Emberiza citrinella*

Numerous resident breeder, passage and winter status uncertain. BoCC (Red). SPEC(4).

Rather larger flock sizes reported in both winter periods than in 2000.

Table 118: Maximum flock size and number of flocks >25

	J	F	M	A	M	J	J	A	S	O	N	D
Max flock size	50	55	12	13	35	38	20	20	20	27	40	80
No. of flocks >25	2	5	-	-	1	1	-	-	-	1	1	2

Breeding season. Many reports during this period would suggest widespread breeding, with confirmation from the following localities (breeding pairs in brackets): Avon Dam; Braunton (4 prs); Bridford where a flock of 20 in Aug and Sep was composed mainly of juvs; Darracott (4 prs); Georgeham (3 prs); Lobb (2 prs); Higher Metcombe where juvs first seen in a garden flock of 38 during Jun; Pickwell Down (2 prs); and Putsborough (4 prs).

First winter-period/spring. Large counts: 50 at Aveton Gifford on 12 Jan with 37 at Finlake on 14th; 24 at Braunton on 5 Feb with 20 plus in a garden at Sandford on 11th and 55 at Prawle on 25th; 20 at Exbourne on 7 Apr and a monthly max of 17 at Higher Metcombe in Apr.

Autumn/second winter-period. Large counts: 14 at Wembury on 25 Sep; 40 near Slapton Village on 18 Nov; and 20 at Otterton on 2 Dec with 50 near Ideford on same date and up to 100 at Dunsford during the last two weeks of Dec. The monthly maxima at Prawle were 27 in Oct, 56 in Nov and 16 in Dec. Passage reported from Staddon Pt, with 12 between mid-Sep and end of Nov, the peak count being five on 3 Oct with the most active period being the last two weeks of Oct. A ♀ was seen to arrive in off the sea at Berry Hd on 2 Oct.

Cirl Bunting - *Mike Langman*

CIRL BUNTING *Emberiza cirlus*

Not scarce resident breeder, mostly in S Devon. SPEC (4). BoCC (Red), UK BAP, Devon BAP.

No RSPB monitoring or survey work was undertaken this year due to FMD access restrictions. Monitoring of selected sites is being undertaken during 2002, with a full 'national' survey to be carried out in 2003. Land under specific management for Cirl Buntings includes over 700 ha of winter stubbles, over 1,700 ha of low intensity grassland and 450 km of uncropped arable field margin; 45% of the population is now found on land in management agreements, with 95% of the population of this sedentary species being within 2 km of agreement land. Whilst the species has benefited from habitat management on farmland, it is now coming under increasing threat from residential development, as at Exminster.

Table 119: Maximum flock size, number of flocks >10 and number of sites

	J	F	M	A	M	J	J	A	S	O	N	D
Max flock size	14	15	2	6	3	2	3	12	15	20	16	24
No. flocks >10	2	3	-	-	-	-	-	3	4	6	1	3
No. of sites	12	11	6	7	11	11	7	10	11	13	13	11

Breeding season. Birds were recorded largely from the traditional coastal and estuary sites between Plymouth and Exeter. Despite only eight reports of proven breeding, 2001 appears to have been a good breeding season based on the number of young birds noted at the end of the season. Interesting records came from East Allington, Plymouth Jennycliff and Plymouth Hooe, with a pair breeding at Berry Hd. The population at Exminster continues to look vulnerable with a maximum of 3-4 pairs compared with 12 in 1998.

First winter-period. Flocks were reported from the usual areas, with double-figure counts including 14 at Snapes Pt on 18 Jan, 10 at Cockington on 20 Jan, 11 at Noss Mayo on 17 Feb and 15 at Prawle Pt on 25 Feb. Small numbers at Ludwell Valley CP again raised hopes, but did not stay on to breed.

Second winter-period. The only double-figure counts came from the Prawle area with a peak of 24 on 22 Dec. A flock using the stubble behind the beach at Broadsands peaked at eight on 2 Dec. There was a noticeable lack of records from Wembury during the year – does this reflect reduced coverage or a genuine population trend?

Record requests. *Records remain important to the ongoing conservation programme, and observers are encouraged to continue submitting sightings (including OS grid references) to the RSPB Cirl Bunting Project Officer (address on page iii) as well as to the County Recorder. Such records can help locate any birds outside known sites and are important for targeting future surveys and, more importantly, positive management work. Records from the Teign Valley are particularly sought.*

ORTOLAN BUNTING *Emberiza hortulana*

Rare vagrant. SPEC(2).

One at Hope's Nose on 2 Oct (A Roe), the first mainland bird since 1999 and a typical autumn record, though only the second in Devon for Oct.

LITTLE BUNTING *Emberiza pusilla*

Rare vagrant. SPEC(2).

One on Lundy on 1 Nov (C H & R J Rhodes).

REED BUNTING *Emberiza schoeniclus*

Fairly numerous resident breeder, passage and winter visitor. BoCC (Red), UK BAP.

As usual, most often reported singly or in very small numbers; some higher counts from areas though perhaps not always as a unified flock.

Table 120: Maximum flock size, number of flocks >10 and number of sites

	J	F	M	A	M	J	J	A	S	O	N	D
Max flock size	30	15	5	2	3	3	7	4	4	23	30	50
No. flocks >10	3	2	-	-	-	-	-	-	-	7	5	4
No. of sites	12	15	6	13	17	15	10	5	7	18	18	16

Breeding. Various reports show the species to be widespread and occupying many suitable habitats, with breeding confirmed from Aveton Gifford, Dawlish Warren, Kenwith NR, Kilmington and Witheridge Moor. Singing ♂♂ reported from Clennon Valley (breeding attempt failed), Colaton Raleigh, Combestone Tor, Countess Wear, Cripdon Down, Grendon Cot, Holne Moor, Piles Copse, Avon Head, Sherpa Marsh, Statts Bridge and Widecombe.

First winter-period: The largest counts were 12 at S Milton Ley on 9 Jan, 30 on Exminster Marshes on 20 Jan, 28 on Witheridge Moor on 29th and 10 at S Milton Ley on 25 Feb,

Second winter-period. Large counts: 15 at Sherpa Marsh on 12 Aug; 12 on Bursdon Moor and 10 on Witheridge Moor on 21 Oct with 30 on Witheridge Moor on 10 Nov; 23 at Dawlish Warren on 29 Oct; and 50 on Otter Cliffs on 30 Dec, the highest count of the year (BBH). Passage reported from Staddon Pt, with 14 between mid-Sep and end of Nov, the peak count being four on 29 Oct with the most active period being from third week of Oct to first week of Nov (SGe).

CORN BUNTING *Miliaria calandra*

Rare passage visitor; last breed 1980. BoCC (Red).

Four seen and photographed at Soar Mill Cove on 18 Oct (S Allen) were the first since 1999.

ROSE-BREASTED GROSBEAK *Pheucticus ludovicianus*

Rare vagrant.

The second Devon record of this N American vagran, the last in 1985.

A 1w ♂ on Lundy 6-9 Oct (S L Cooper and R M Patient); see separate report. (Accepted by BBRC,along with one on Scilly 13-14 Oct; these were the first in Britain since 1998.)

BOBOLINK *Dolichonyx oryzivorus*

Rare vagrant.

The third Devon record of this N American vagrant, and the second on the mainland, the last being in 1991.

A 1w in a set-aside field by the coast path east of Langerstone Point (Prawle area) 9-15 Oct (PMM *et al.*) was seen by many observers during its week's stay; although remaining faithful to one field, this bird could prove remarkably elusive. (Accepted by BBRC, the only other in 2001 being one in Yorkshire on 27 Oct.)

CATEGORY E SPECIES

These are species that have been recorded as introductions, transportees or escapes from captivity, and whose populations (if any) are thought not to be self-sustaining. They do not form part of the British List (BOU 1998). Species recorded in Devon that are on the E List, but *also* on the British List (A-C lists), are included in the main part of the Systematic List. Names and order of species in general follow those given by BOU on http://www.bou.org.uk/recbrlst3.html.

The main reason for including the records of E species is to alert observers to the presence of these species in the wild when confronted with potential genuine rarities. Some, such as Black Swan and Wood Duck, may also be in the process of establishing feral populations, and it is therefore particularly appropriate to monitor their numbers, distribution and breeding success, in relation to possible effects on other species.

BLACK-BELLIED WHISTLING DUCK *Dendrocygna autumnalis*

Southern North America, Central and South America

One at Bowling Green Marsh on 5 Jul.

BLACK SWAN *Cygnus atratus*

Australia and New Zealand

Most records came from various sites on the *Exe Est* Jun-Dec, with successful local breeding implicated for the second year running; the only other records of breeding in Devon were two failed attempts in 1993 at Clyst Marshes and Clennon Valley. Although initially recorded as six (pr with four juvs) at Starcross on 6 Jun (DJG), no more than five together were subsequently recorded, but four imms present at least until 9 Dec. *Elsewhere*, two at Slapton Ley on 14 Oct, and singles at: Taw Est on 2 Feb; Plymouth Cattewater on 26 Mar; Yealm Est on 13 Sep. It is worth noting that confirmed breeding was only recorded for three other pairs in the whole of the UK in 2000 (Ogilvie *et al.* 2002b).

SWAN GOOSE *Anser cygnoides*

East Asia

One on the Taw/Torridge Est in Jan, Aug, Oct and Nov.

BAR-HEADED GOOSE *Anser indicus*

Central Asia

All singles, often with Canada Geese, with most records in Aug; at least two individuals involved, perhaps one remaining in the north and another in the south (both seen on 29 Aug, at Upper Tamar L and Kingsbridge Est). In the *north*: on the Taw/Torridge Est area in Jan, Aug (recorded as a juv) and Oct; and at Upper Tamar L in Feb and Aug. In the *south*: on the Dart Est in Apr-May and at Dartington Oct-Dec; and in Aug at both Bowling Green Marsh and Kingsbridge Est.

EMPEROR GOOSE *Anser canagicus*

Siberia and Alaska

One in the Clennon Valley 9-28 Feb. A bird at Bowling Green Marsh on 31 May was considered to be a possible Emperor x Greylag Goose hybrid.

MUSCOVY DUCK *Cairina moschata*

Central and South America

Singles recorded from Exmouth, Slapton, Thorverton and Plymouth, but four (two breeding prs) at Okehampton on 1 Jun, with ten there on 29 Dec, and eight at Austin's Bridge, R Dart in Sep.

WOOD DUCK *Aix sponsa*

North America

A ♂ on the Plym Est 25 Mar – 13 Apr, with a pr on 8 Jun and a ♂ at Buckfast on 13 May, but more intriguing was a ♀ with 5-7 yg, about a week old, near Haytor on 23 Jun (JHB). The latter appears to be the first Devon breeding record.

CHILOE WIGEON *Anas sibilatrix*

Southern South America

Two at Bowling Green Marsh on 27 Dec.

SPECKLED or CHILEAN TEAL *Anas flavirostris*

South America

One at Bowling Green Marsh on 25 Nov.

CHESTNUT TEAL *Anas castanea*

Australia

One at Totnes on 1 Feb and 6 Nov, and probably the same bird in the Clennon Valley on 12 Nov.

WHITE-CHEEKED PINTAIL *Anas bahamensis*

South America and Caribbean

One on the sea at Broadsands on 22 Oct.

PACIFIC BLACK DUCK *Anas superciliosa*

Australia

One at Kenwith NR 11-12 Jun (captured on video).

BUDGERIGAR *Melopsittacus undulatus*

Australia

One in a Derriford (Plymouth) garden on 28 Jul.

COCKATIEL *Nymphicus hollandicus*

Australia

One at Mt Gould Hospital, Plymouth on 5 Jan and 14 Nov.

RED-WHISKERED BULBUL *Pycnonotus jocosus*

India and SE Asia

Four summer records of singles, the first three in gardens, and possibly involving the same bird: Brixham on 14-15 Jun; Ivybridge on 19 Jun; Shaugh Prior on 15 Aug; and on rocks at Prawle Pt, an imm, on 28 Aug.

ISLAND CANARY *Serinus canaria*

North Atlantic Islands

One present, and filmed on video, at a garden seed-feeder at Marldon, Torbay, in Jan, Mar, Sep and Nov.

YELLOW-FRONTED CANARY *Serinus mozambicus*

Southern Africa

One on 4 Aug, 20 Sep and 20 Oct at Dawlish Warren.

CHINESE GROSBEAK *Eophona migratoria*

Eastern Palaearctic

One in a Tavistock garden in Dec.

GE-CHEEKED WAXBILL *Estrilda malpoda*

Africa

One, Jun-Dec, in an East Allington garden.

ZEBRA FINCH *Poephila guttata*

Australia

Five records of singles from: Heathfield on 12 Jan; Okehampton on 10 May; Clennon Valley on 8-10 Jun; Derriford, Plymouth on 12 Jul, a patio door collision; and Manaton on 26 Jul, an ad ♂ brought in by cat.

RED-BILLED FIREFINCH *Lagonosticta senegala*

Africa

A ♂ in a Sidmouth garden, Jun-Oct.

PENDING, INCOMPLETE & DELETED RECORDS

PENDING RECORDS

- FEA'S PETREL. One off Berry Hd & Hope's Nose on 17 Jul. Provisionally included in the Systematic List as a Fea's/Zino's Petrel, but submitted to BBRC as a Fea's.
- 'CONTINENTAL' CORMORANT *P. c. sinensis*. All records.
- LAUGHING GULL. One on the Plym Est on 4 Aug.
- RED-RUMPED SWALLOW. One on Lundy on 26 Oct.
- 'SIBERIAN' CHIFFCHAFF *P. c. tristis*. All records.

RECORDS REQUIRING MORE INFORMATION

The following records for 2001 are as yet unsubstantiated and require further documentation, ideally from the finder. They concern BBRC rarities or DBWPS Listed Species. Please send supporting details to the recorder as soon as possible, so that accepted records can appear as additional records in the Systematic List of DBR 2002.

- PURPLE HERON. One on the Teign Est on 6 Jun.
- MARSH HARRIER. One at Seaton Marshes LNR 17-23 Dec.
- QUAIL. One on Lundy on 7 Jul.
- 'BALTIC GULL' *L. f. fuscus*. One at Budleigh Salterton on 4 May.
- TAWNY PIPIT. One at Dawlish Warren on 8 May.
- YELLOW-BROWED WARBLER. One at Dawlish Warren on 27 Oct.
- DUSKY WARBLER. One at Otterton Ledge on 15 Oct and at White's Bridge on 20 Oct.

DELETIONS OF PREVIOUSLY ACCEPTED RECORDS

- LITTLE CRAKE. One on Lundy 12-14 Sep 1952. BBRC decision based on erroneous statement that Baillon's Crake excluded by lack of white markings on upperparts, and no real statement of colour of underparts but for lack of white barring.
- LEAST SANDPIPER. One on Lundy 24-26 Sep 1957. BBRC decision based on absence of proper plumage description which prevented even age-determination.
- WOODCHAT SHRIKE. One at Exford on 21 Apr 1989. On the grounds that Exford is in Somerset!

EARLY AND LATE DATES FOR MIGRANTS

The following table shows the earliest and latest dates on which certain migrants were recorded in 2001, as well as the extreme earliest and latest Devon records for comparison (based on data collated by Ellicott & Stone (1989) and amended with records from subsequent *Devon Bird Reports*). As always, the problem with several species is distinguishing between wintering/summering birds and true migrants; probable wintering/summering birds have been asterisked, but known ones have been omitted. In cases where the date for 2001 tops the previous extreme (in only one case this year – a late Little Ringed Plover), the former is given in bold and the latter in parentheses. Some species have been added to the list for 2001, and thanks are due to PMM for both suggesting these and providing their extreme dates.

Table 121: Early & Late Date Migrants

Species	Earliest 2001	Extreme earliest	Latest 2001	Extreme latest
Garganey	5 Apr	1 Jan 1958*	26 Sep	23 Nov 1958*
Honey Buzzard	15 May	24 Mar 1977	15 Jul	20 Oct 1985
Montagu's Harrier	12 May	27 Mar 1977	No records	30 Oct 1988
Osprey	12 April	12 Feb 1978	20 Nov	6 Dec 1998
Hobby	5 Apr	13 Mar 1985	22 Oct	9 Dec 1982
Kentish Plover**	22 Mar	17 Mar 1991	23 Oct	16 Nov 1991
Dotterel	11 May	4 Apr 1991	10 Nov	30 Dec 2000
Little Ringed Plover	10 Apr	7 Mar 2000	**12 Oct**	(24 Sep 1976)
Wood Sandpiper	4 May	19 Mar 1990	10 Sep	21 Dec 1996*
Sandwich Tern	12 Feb	27 Jan 1998*	20 Nov	23 Dec 2000*
Roseate Tern	1 May	18 Apr 1970	19 Oct	6 Nov 1982
Common Tern	13 Apr	22 Mar 1970	19 Oct	27 Dec 1963
Arctic Tern	22 Apr	20 Mar 1999	15 Oct	25 Nov 1979
Little Tern	24 Apr	2 Apr 1959	24 Aug	11 Nov 1973
Black Tern	24 Apr	13 Apr 1979, 1993	23 Oct	10 Dec 2000
Turtle Dove	28 Apr	30 Mar 1981	29 Sep	9 Dec 1999*
Cuckoo	5 Apr	4 Mar 1984	17 Aug	10 Dec 1956
Nightjar	27 Apr	2 Apr 1977	6 Sep	18 Oct 1954
Swift	17 Apr	14 Mar 1952	29 Aug	16 Dec 1955
Hoopoe**	19 Mar	25 Feb 1978	No records	10 Dec 1992
Wryneck	10 May	21 Mar 1978	16 Oct	4 Nov 1984

Sand Martin	8 Mar	22 Feb 1988	23 Oct	23 Nov 1957
Swallow	13 Feb	31 Jan 1974*	13 Dec	27 Dec 1966*
House Martin	18 Mar	1 Jan 1987*	27 Oct	31 Dec 1953*
Tree Pipit	28 Mar	9 Mar 1980	21 Oct	10 Nov 1982
Yellow Wagtail	14 Apr	27 Jan 1952*	27 Oct	1 Dec 1957
Nightingale	26 Apr	6 Apr 1985	30 Jun	31 Oct 1979
Redstart	29 Mar	17 Mar 1998	21 Oct	13 Nov 1957,1977
Whinchat	24 Apr	1 Jan 1955*	2 Nov	30 Dec 1973*
Wheatear	14 Mar	early Jan 1978*	4 Dec	6 Dec 1994
Ring Ouzel	19 Apr	17 Jan 1954*	7 Nov	27 Dec 1988*
Grasshopper Warbler	13 Apr	2 Apr 1977	15 Oct	4 Nov 1963
Sedge Warbler	11 Apr	12 Mar 1977	13 Oct	17 Dec 1978
Reed Warbler	21 Apr	29 Mar 1997	5 Nov	17 Nov 1984
Lesser Whitethroat**	3 Apr	3 Apr 1998	18 Oct	23 Dec 1994*
Whitethroat	7 Apr	14 Mar 1968	28 Oct	26 Dec 1984*
Garden Warbler	20 Apr	1 Feb 1973	1 Nov	8 Dec 1973
Wood Warbler	24 Apr	31 Mar 1999	28 Aug	19 Sep 1964
Willow Warbler	22 Mar	4 Mar 1975	12 Nov	6 Dec 1999
Spotted Flycatcher	2 May	9 Feb 1975*	19 Oct	10 Nov 1982
Pied Flycatcher	26 Mar	24 Mar 1977	14 Oct	17 Nov 1985
Fieldfare	28 Sep	29 Jul 1972*	7 Apr	30 May 1960
Redwing	29 Sep	1 Sep 1973	13 May	24 Jun 1987*
Brambling	13 Oct	21 Aug 1989	2 Jun	8 Jun 1972
Snow Bunting	30 Sep	27 Aug 1998	29 Feb	9 May 1968

* = probable wintering/summering birds (as opposed to migrants).
** = not including known wintering birds.

REFERENCES AND SOURCES OF INFORMATION

ANON (2000). Taxonomic Changes. *British Birds* **93**: 464.

ANON (2002a). *Birds of Conservation Concern in the U.K.* RSPB, Sandy.

ANON (2002b). New from BOU…. *British Birds* **95**: 597.

ANON (1994). *Biodiversity: the UK Action Plan.* HMSO, London.

BOU (1992). *Checklist of Birds of Britain and Ireland.* 6th edition. British Ornithologists' Union, Tring.

BOU (1998). *The British List: the official list of the birds of Great Britain, with lists for Northern Ireland and the Isle of Man.* British Ornithologists' Union, Tring.

BWP. See under CRAMP S. *et al.*

BUTCHER, R. (2002). Exmoor Bird Report 2001. *Exmoor Naturalist* **28**: 15-30. Exmoor Natural History Society, Minehead.

CORDREY, L. (ed.) (1997). *Action for Biodiversity in the South-West. South-West Regional Biodiversity Action Plan.* The South-West Biodiversity Partnership, Exeter.

CRAMP S., SIMMONDS, K. E. L. & PERRINS, C. M. (eds.) (1977–1994). *The Birds of the Western Palearctic.* 9 Vols. Oxford University Press, Oxford. [referred to as '*BWP*' in the text].

DARTMOOR BIRD REPORT, THE. See under SMALDON, R.

DAWLISH WARREN, THE BIRDS OF. See under LAKIN, I & RYLANDS, K.

DNPA (2001). *Dartmoor Moorland Breeding Bird Survey 2000.* DNPA, Bovey Tracey.

DUDLEY (1993) *BTO Bird Recording Handbook*, BTO.

ELLICOTT, P. & STONE, C. W. (1989). Early and late dates of migrants, 1988. *Devon Bird Rep. 1988*, **61**: 82.

GEARY, S (2001). *Dartmoor Moorland Breeding Bird Survey 2000.* Report to DNPA.

GIBBONS, D. W., REID, J. B. & CHAPMAN, R. A. (1993). *The New Atlas of Breeding Birds in Britain and Ireland: 1988–1991.* Poyser, London.

GREGORY, R. D. *et al.* (2002). The population status of birds in the United Kingdom, Channel Islands and Isle of Man: an analysis of conservation concern 2002-2007. *British Birds* **95**: 410-450.

HOLMES, J S, & STROUD, D A. (1995). Naturalised birds: feral, exotic, introduced or alien? *British Birds* **88**: 602–603.

KNOX, A G, COLLINSON, M, HELBIG, A J, PARKIN, D T & SANGSTER, G (2002). Taxonomic recommendations for British birds. *Ibis* **144**: 707-710.

LAKIN, I. & RYLANDS K. (2000). *The Birds of Dawlish Warren.* Privately published.

LUNDY BIRD REPORT. See under TAYLOR, A.

MADGE, S. (2002). Caradon Birds in 2001. in *Caradon Wildlife* **18:** 36-77. Caradon Field and Natural History Club

MOORE, R. (1969). *The Birds of Devon.* David & Charles, Newton Abbot.

MUSGROVE, A, POLLITT, M, HALL, C, HEARN, R, HOLLOWAY, S, MARSHALL, P, **ROBINSON**,

J & CRANSWICK, P (2001). *The Wetland Bird Survey 1999-2000 Wildfowl and Wader Counts*. BTO/WWT/RSPB/JNCC, Slimbridge.

OGILVIE, M & the Rare Breeding Birds Panel (2002a). Rare breeding birds in the United Kingdom in 2000. *British Birds* **95**: 542-582.

OGILVIE, M & the Rare Breeding Birds Panel (2002b). Non-native birds breeding in the United Kingdom in 2000. *British Birds* **95**: 631-635.

ROGERS, M & the Rarities Committee (2002). Report on rare birds in Great Britain in 2001. *British Birds* **95**: 476-528.

ROSIER, A. (1995). *A Checklist of the Birds of Devon.* DBWPS.

RSPB (1997). *Seabirds of South Western Waters.* RSPB & EN, Exeter.

SITTERS, H. P. (1988). *Tetrad Atlas of the Breeding Birds of Devon.* DBWPS, Yelverton. [referred to as '*Tetrad Atlas*' in the text].

SMALDON, R (2002). *The Dartmoor Bird Report 2001*. Dartmoor Study Group.

TAYLOR, A. M. (in press). Birds on Lundy 2001. *Lundy Field Society 51st Annual Report*. Lundy Field Society.

TETRAD ATLAS. See under SITTERS, H P.

TUCKER, G. M. & HEATH, M. H. (1994). *Birds in Europe: their conservation status.* BirdLife International, Cambridge.

TUCKER, V. (1995). *Birds of Plymouth.* DBWPS.

VOOUS, K. H. (1977). *List of Recent Holarctic Bird Species.* BOU, London.

LIST OF OBSERVERS & CONTRIBUTORS

With apologies for any ommisions.

A

P Abbot (PA)
R Adams (RAda)
J&R Aley (JRA)
P H Aley (PHA)
M F Allen (MFA)
S Allen (SA)
D Amas (DA)
J Armsworth (JArm)
J Arthur (JAr)
J D Avon (JDA)
S J Ayres (SJA)
H Ayling (HA&DB)
H Ayshford (HA)

B

J B Bailey (JBB)
M R A Bailey (MRAB)
R E Bailey (REB)
C & S Balch (C&SB)
N Baldock
D K Ballance (DKB)
A S C Barker (ASCB)
J Barker (JBa)
G Barlow (GBa)
R W Barrow (RWBa)
C R Bath (CRB)
Paul Baxter (PBx)
R Behenna (RBe)
J H Bennett (JHB)
B Bewsher (BB)
S Biggs (SBg)
G Blackman (GB)
Mr Blackman (MrB)
R W Bone (RWBo)
P Boulden (PBo)
K G Bowcock (KGB)
P Bowden (PBd)
R Bowers (RBo)
L Brian (LBr)
N Briden (NBri)
Sister Bridget (SBr)
C Brooks (CBr)
E Brookes (EJB)
M Brooking (MBk)
V Bubear (VBu)
P & L Buck (P&LB)
P & H Burn (P&HB)
R W Burn (RWBu)
R Burridge (RBu)
A R Burton (ARB)

C

M J Cammack (MJC)
D E D Campbell (DEDC)
P Carah (PC)
G E Carr (GEC)
G Cavanagh (GCa)
C Chadwick (CCh)
R Champion (RCh)
D Churchill (DC)
R J Cleevely (RJCl)
M H Clements (MHC)
L Collins (LC)
C Coombes (CC)
S L Cooper (SLC)
D A Cope (DAC)
D J Cotton (DJCo)
N W Cottle (NWCo)
D R Cox (DRCx)
V Cozens (VC)

D

J Daniels (JDs)
M Darlaston (MD)
A M J Davey (AMJD)
J B Davidson (JD)
P Davies (PDv)
M D Daw (MDD)
W J Deakins (WJD)
J Denning (JDe)
J R Diamond (JRD)
R Doble (RD)
P Dodd (PDd)
Mrs M K Doyle (MKD)

E

Alun Edwards
P W Ellicott (PWE)
P Elliott (PEl)
B R Ellis (BRE)
D Elphick (DEk)
P Emony (PEy)
P Erskine (PEr)
M Evans (ME)
P Exley (PE)

F

I Farrell (IF)
J Fell (JFe)
T Folland (TF)
Miss A Ford (AF)
J E Fortey (JEF)
F Freshney (FF)
D Friend (DFr)

G

J Gale (JGa)
J Gardiner (JGar)
S Geary (SGe)
K German (KGe)
D J Glaves (DJG)
M Glover (MG)
S Goetsch (SGt)
C Good (CG)
P F Goodfellow (PFG)
M R Goss (MRG)
D A Gray (DAG)
R Greenwood (RGd)
D Gregory (DGr)
C Grundy (CGy)

H

G Hackston (GH)
R M Halliwell (RMHa)
A Hancock (AHa)
K Handford (KHf)
D Hargrave (DHa)
C Hassall (CH)
S Hatch (StH)
D Hawes (DHaw)
B B Heasman (BBH)
D W Helliar (DWH)
R Hibbert (RHi)
P J Hopkin (PJH)
S Hopper (SHo)
R J Hubble (RJH)
H & J Huggins (H&JH)
D L Humphries (DLH)
Dr L H Hurrell (LHH)
R H Hurrell (RHH)

I

M Ingram (MI)

J

D Jackson (DJa)
K Jago (KJ)
D J Jarvis (DJJ)
G & B Jasper (G&BJ)
A M Jewels (AMJ)
D Jones (DJ)
R A J Jones (RAJJ)
P Jones (PJs)
R J Jones (RJJ)
D I Julian (DIJ)
R Jutsum (RJ)

K

R N Kelsh (RNK)
F H C Kendall (FHCK)
I Kendall (IK)
A Kennard (AK)
R Khan (RK)
M Kipling (MK)
M Knott (MKn)

L

I Lakin (IL)
J Lambert (JLb)
M Langman (ML)
per M Langman (pML)
J R Lansdell (JRL)
M I Lawrence (MIL)
P Leigh (PLe)
C Lever (CL)
R J Lillicrap (RJL)
S Lister (SL)
A J Livett (AJL)
K Lopez (KLo)
E J Lovesey (EJL)

M

B MacDonald (BMc)
P Madgett (PM)
P Mallett (PMa)
P & R Mallett (P&RM)
T Marchese (TM)
B A Marsh (BAM)

Mrs E M Marsh (EM)
Mrs H Marshall (HM)
R G Marshall (RGM)
P M Mayer (PMM)
R McCarthy (RMcC)
G McKittie (GMc)
J Millen (JMi)
J Miller (JMr)
C Mills (CMi)
M G Mitchell (MGM)
I Moore (IM)
D H W Morgan (DHWM)
A & P A Morley (A&PAM)
B C Morris (BCM)
K E Mortimer (KEM)
K & J W Mortimore (K&JWM)

N

Brother Nicholas (BrN)
J C Nicholls (JCN)
S Nickols (SNk)
R M Normand (RMN)
P Nunn (PN)

O

R J Olliver (RJO)

P

P Page (PP)
A J & S A Park (A&SP)
Mrs M A Phillips (MAP)
P M Phillips (PMPh)
A Pierce (APc)
C Pincher (CP)
N Pinhorn (NPi)
D J Price (DJP)
Miss B O B Primmer (BOBP)
Miss B P R Primmer (BPRP)
C J Proctor (CJP)
N & R Pryor (N&RP)

R

C Ralph (CR)
J M Randall (JMR)
N D Rawlings (NDR)
C H Read (CHR)
J & H Read (J&HR)
P J Reay (PJR)
D Reeves (DRe)
T Reid (TRd)
A Rennells (AR)
M Reynoulds (MRe)
C J Rhodes (CJR)
A M Ritchie (AMR)
N P Roberts (NPR)
A Roetemeiger (ARoe)
D Rogers (DR)
A Rosier (ARo)
B N Rossiter (BNR)
K Rylands (KRy)

S

A J Salter (AJSa)
M L H Sampson (MLHS)
P Sanders (PSa)
A K Searle (AKS)
M S Shakespeare (MSS)
B Sharkey (BSh)
D Shaw (DSw)
C Shelley (CSh)
A R Simkins (ARS)
S Skinner (SSk)
M A Slade (MAS)
J & T Sleep (J&TS)
R Smaldon (RS)
J G Smale (JGS)
P Smale (PSm)
P J Smale (PJS)
D Smallshire (DS)
J R Smart (JRS)
Alec Smith (ASm)
T H Smith (THS)
P Spittle (PSpt)
J A C St Leger
A Stagg OBE (ArSt)
G Stead (GS)
P A Stidwill (PAS)
B H Stone (BHS)
C Stubbs (CSt)
P L Stubbs (PLS)
D J Suckley (DJSu)
G P Sutton (GPS)
A Swash (ARHS)
A J Symons (AJS)

T

D H W Taylor (DHWT)
I Taylor (IT)
B Thorne (BT)
R C Thornett (RCT)
J Tidball (JTi)
M Todd (MTd)
E Townsend (ET)
R K Treeby (RKT)
G G Trenerry (GGT)
N S Trout (NST)
S R Tucker (SRT)
V R Tucker (VRT)
W H Tucker (WHT)
M & C Turner (M&CT)
M W Tyler (MWT)

V

F Vagges (FV)
G A Vaughan (GAV)
W E Vaughan (WEV)
A J Vickery (AJV)
G Vernall (GV)
J W Vaughan (JWV)

W

J Wainwright (JWn)
I J Waite (IJW)
R Wakely (RWk)
Col Walford (ColW)
J M Walters (JMW)
Miss H Walton (HWa)
N C Ward (NCW)
S Warman (SWa)
G E C Waterhouse (GECW)
A F G Weedon (AFGW)
R Wheatland (RWhe)
T W Whiley (TWW)
B D White (BDW)
R J White (RJW)
W J White (WJW)
B Whitehall (BW)
J E Wicks (JEW)
A K Williams (AKW)
D Williams (DWi)
M S Wolinski (MSW))
A Wolstenholme (AWo)
H A Woodland (HAWo)
J & J Woodland (J&JW
D Wood (DWo)
S Wordsworth (SWor)
A Worth (AW)
A Wright (AWri)

Y

S M R Young (SMRY)

CONTRIBUTOR ORGANISATIONS

Berry Head National Nature Reserve (BHNNR)
Bird Guides (BG)
Dartmoor Study Group (per RS) (DSGp)
Dawlish Warren NNR Log (per IL)
Exmoor Natural History Society (ENHS)
Hawk and Owl Trust (per N Dixon) (HOT)
Lundy Field Society (per A Taylor) (LFS)
Prawle Log (per PMM) (PL)
Royal Society for the Protection of Birds (per AJB) (RSPB)
Slapton Log (per DEk) (SLog)
South Devon Seabird Trust (per J Bradford) (SDST)
Wildfowl and Wetlands Trust (per P Stidwill) (WWT)
Yarner Wood NNR (per P Page) (YWNNR)

GAZETTEER OF DEVON SITES

The following is an alphabetical list of all the sites with records that are in the Devon database for 2001. The grid references usually refer to sites marked on the 1:50,000 or 1:25,000 OS maps, although some sites are not included on these maps. In some cases the old area description (pre-1990) used in DBRs before that date have been included in the site name.

Although most of the 3,000 sites listed have been checked for accuracy as part of the on-going review of all local and parent site names and grid references, the Data Manager would be pleased to receive any corrections, updates and/or missing grid references.

Abbey Sands, Torquay, SX905632
Abbeyford (R Okement), SX5997
Abbeyford Woods, SX5897
Abbotsbury, Newton Abbot, SX854715
Abbotsham, SS424264
Abbotsham Cliffs, SS410277
Abbotskerswell, SX855690
Abbrook Pond, Kingsteignton, SX863744
Airey Point, Braunton Burrows, SS449330
Aish Tor, SX704715
Aish, South Brent, SX692607
Aish, Stoke Gabriel, SX842589
Alfington, nr Ottery St Mary, SY114981
Aller, SX888688
Aller Brook, Newton Abbot, SX871707
Aller Down, Sandford, SS825045
Aller Park, Newton Abbot, SX875696
Aller Vale, Newton Abbot, SX878693
Aller Wood, Cullompton, ST0406
Allerton, Loddiswell, SX712498
Alley,
Alphington, SX917898
Alston Cross, Ashburton, SX776714
Alverdiscott, SS520252
Amicombe Brook, SX575837
Amicombe Hill, SX571862
Anchor,
Anchor Bank, Barnstaple, SS545329
Anchor Woods, Barnstaple, SS5734
Andrew's Wood, Loddiswell, SX707516
Annery, SS456228
Anstey Common,
Anstey Gate, SS837298
Anstey's Cove, SX935647
Anvil Corner, Holsworthy, SS373042
Appledore, SS463307
Archbrook, nr Combeinteignhead, SX9172
Archerton, nr Postbridge, SX637793
Arlington Beccot, SS619415
Arlington Court Estate, SS608405
Arlington Court Lake, SS608405
Armarda Park, nr Torquay, SX892648
Arms Tor, SX539864
Ash Mill, nr Bishop's Nympton, SS784234
Ashburton, SX755699
Ashclyst Forest, SY003995
Ashcombe, sx913795
Ashculme, nr Hemyock, ST147149
Ashford, SS5335
Ashilford, nr Cadeleigh, SS925091
Ashill, ST088114
Ashill NR, nr H Metcombe, SY066922
Ashmill, nr Roadford, SX398957
Ashprington, SX819571
Ashreigney, SS630136
Ashton-Chudleigh Bridge, SX844843
Ashwater, nr Holsworthy, SX385954
Assycombe Brook Head, SX657817
Assycombe Hill, SX664821
Atherington, SS591231
Ausewell Wood, SX730715
Aveton Gifford, SX692470
Avon Dam, SX679652
Avon Estuary, SX675450
Avon Head, SX650695
Avonwick, SX725583
Awliscombe, ST135017
Axe Cliffs, SY260897
Axe Estuary, SY254905
Axe Farm (heronry), SY257912
Axe Marshes, SY252910
Axe Valley,
Axminster, SY295985
Axmouth, SY258912
Axmouth (Golf Course), SY259899
Axmouth Marsh, SY236913
Axmouth undercliff, SY260897
Aylesbeare, SY039921
Aylesbeare Common, SY058902
Aylesbeare/Harpford Commons, SY058902
Ayrmer Cove, SX638455
Aysh Green, SX667897
Aysh Green, Throwleigh, SX667896
Ayshford, ST048152
Babbacombe, SX931655
Babbacombe Bay, SX9469
Babbacombe-Petit Tor,
Bachelor's Hall, Princetown, SX601737
Bachelors Hall Bridge, SX601737
Backs Wood, Bickleigh (R Exe), SS951092
Badger's Holt, nr Dartmeet, SX673736
Badgworthy Lees, SS787448
Badgworthy Water, SS791450
Badgworthy Wood, SS795455
Bagga Tor, SX548806
Baggy Point, SS419406
Bagtor Down, nr Haytor, SX762758
Bagtor Mill, SX769756
Ball Gate, S Brent, SX671614
Bampton, SS955222
Ball Gate, S Brent, SX671614
Bampton, SS955222
Bampton Down, SS995211
Bantham, SX670437
Barbrook, SS715477
Barcombe Down, SS759318
Barnstaple, SS558330
Barnstaple (Rock Park), SS562325
Barnstaple Bay, SS4134
Barton, SX905671
Barton Pines, SX846612
Batson Creek, Salcombe, SX743394
Batson, nr Salcombe, SX733397
Batson, Salcombe, SX734397
Battery Gardens, nr Brixham, SX192569
Battisborough, SX603468
Battisborough Cross, SX599483
Batton, nr Hallsands, SX806397
Batworthy, SX661866
Batworthy Corner, SX715853
Baxworthy Cross, nr Hartland, SS282224
Beacon Hill, Newton Ferrers, SX574467
Beacon Hill, Sidbury, SY111913
Beacon Hill, Torbay, SX859618
Beacon Plain, SX664595
Beacon Point, Erme mouth, SX617458
Beacon Point, nr Outer Hope, SX674406
Beaford Bridge, SS542142
Beaford Moor, SS5714
Beam, nr Gt Torrington, SS472207
Beara Common, nr South Brent, SX703617
Beara Down, SS523395
Beardown Farm, Two Bridges, SX604755
Beardown Tors, SX605775
Beardown Woods, Two Bridges, SX602759
Beatland Corner, SX548624
Beaworthy, SX461995
Becka Falls, SX761801
Beckabrook, Manaton, SX765800
Beckaford Bridge, Manaton, SX756799
Beechwood, nr Sparkwell, SX580582
Beer, SY230892
Beer Cliffs, SY227879
Beer Head, SY227879
Beesands, SX819410
Beesands Ley, SX819410
Beesands-Start Point, SX820408
Beeson, SX812407
Bell Rock, Maidencombe, SX928678
Bell Tor, SX7377
Bellehill, Kingsbridge, SX7445
Bellever, SX655774
Bellever Plantation, SX6576
Bellever Tor, SX645765
Bellevue Plantation, Gittisham, SY133975
Belstone, SX619935
Belstone Cleave, SX629938
Belstone Tor, SX614921
Bench Tor, nr Venford, SX692716
Bengie Tor (=Bench Tor), SX692716
Bennett's Cross, SX680817
Bere Alston, SX445666
Bere Ferrers, SX459635
Berra Tor, nr Buck. Monachorum, SX478691
Berry Barton, Branscombe, SY185887
Berry Barton, nr Branscombe, SY185887
Berry Castle, Black Dog, SS803093
Berry Head, SX945565
Berry Pomeroy, SX828612
Berrynarbor, SS561467
Bewley Down, nr.Chardstock, ST281061
Bickerton Fish Ponds, SX813389
Bickham Bridge, Diptford, SX726553
Bickham, nr Roborough, SX495654

Bickington, nr Ashburton, SX795725
Bickington, nr Barnstaple, SS534325
Bickleigh Bridge, R Plym, SX527618
Bickleigh Mill, R Exe, SS937076
Bickleigh, nr Plymouth, SX521622
Bickleigh, nr Tiverton, SS942072
Bicton, SY070862
Bicton Common, SY038861
Bicton Lake, SY070862
Bideford, SS450265
Bideford (The Cleve), SS456285
Bideford Bay, SS425315
Bideford Quay, SS455267
Bigbury, SX667466
Bigbury Bay, SX667466
Bigbury-on-Sea, SX651444
Bindon, nr Axmouth, SY270905
Birch Tor, nr Warren House, SX686814
Birchettes Wood, Hembury, SX718682
Bish Mill, nr South Molton, SS741253
Bishop's Tawton, SS566304
Bishopsteignton, SX910738
Bittaford, SX666570
Bittleford Down, SX7075
Black Dog, nr Kennerleigh, SS806097
Black Down Common, ST115163
Black Dog, nr Kennerleigh, SS806097
Black Down Common, ST115163
Black Down, Exmoor, SS730361
Black Down, nr Mary Tavy, SX5082
Black Dunghill, SX582775
Black Head, SY085826
Black Head, Otterton, SY086826
Black Hill, SX604847
Black Hill, nr Manaton, SX763789
Black Ridge, SX595854
Black Tor, nr Avon Dam, SX681636
Black Tor, nr Burrator, SX573718
Black Tor, nr Meldon, SX568895
Black Torrington, SS465055
Black-a-tor Copse, SX567890
Blackabrook, R Plym, SX568642
Blackaller Quarry, SX729913
Blackaller, nr Bovey Tracey, SX736838
Blackaton Brook, SX6978
Blackaton Cross, sx698778
Blackaton Down, SX703786
Blackawton, SX806510
Blackawton (Woodlands Pk), SX805509
Blackborough, ST096090
Blackbrook Head, SX581777
Blackbrook River, Princetown, SX595743
Blackbury Castle, SY188924
Blackcat, R Exe, SS934217
Blackdown Common, ST115165
Blackdown Rings, SX7242
Blackingstone Rock, SX786857
Blacklane Brook, SX6267
Blackpool, SX853479
Blackslade Down/Mire, SX735757
Blackslade Mire - Top Tor, SX735760
Blackstone Point (N Coast), SS603487
Blackstone Pt, SX536461
Blagdon, nr Hartland, SS234268
Blagdon, nr Paignton, SX857610
Blanchdown Wood, Gunnislake, SX422730
Blanksmill Bridge, SX726410
Blanksmill Creek, SX733411
Blaxton, SX466633
Blaxton Creek, SX466633
Blaxton Marsh, Tavy Est, SX465629
Blaxton Point, SX466633
Blaxton Quay, SX466633
Blaxton Wood, SX469635
Bleak House, SX559865
Blegberry, SS226259
Bolberry (village), SX691393
Bolberry Down, SX685387
Boldford Bridge, R Carey, SX364883
Bolham Weir, nr Tiverton, SS948152
Bolt Head, SX726360
Bolt Tail, SX666397
Bondleigh, nr North Tawton, SS658040
Bonehill area, SX725775
Bonhill Water,
Bonny Cross, ST016228
Boohay, nr Kingswear, SX9052
Boreston, nr Halwell, SX770537
Boringdon, SX539578
Borough Valley, Lee Bay, SS485455
Borrow Pit, Axmouth, SY250907
Boulders Tor, SX527782
Bovey Heath Great Plantation, SX821755
Bovey Heath SSSI, SX824764
Bovey Tracey, SX815785
Bovey Valley Woods NNR, SX781799
Boveycombe Head, SX6982
Bovisand, SX492496
Bow, SS724017
Bow Creek, nr Ashprington, SX825564
Bow Cross, Staverton, SX810651
Bowcombe Creek, SX747435
Bowd, SY117900
Bowden, nr Drewsteignton, SX767720
Bowden, nr Kingsbridge, SX7644
Bowden, nr Stoke Fleming, SX843490
Bowermans Nose, SX743804
Bowling Green Marsh, SX973875
Boyton Mill, R Tamar, SX331921
Bradford, nr Holsworthy, SS421073
Bradiford Reserve, SS5434
Bradiford Water, Barnstaple, SS548342
Bradley Wood, Ogwell, SX845709
Bradninch, SS999040
Bradridge (R Tamar), SX333935
Bradworthy, SS324140
Bradworthy Common, SS325148
Brakehill Plantation, SX530475
Brampford Speke, SX926985
Brandis Corner, nr Holsworthy, SS411040
Brandy Head, SY089832
Branscombe, SY197887
Branscombe Mouth, SY207880
Branscombe Sewage Works, SY200885
Branscombe to Weston Mouth, SY190878
Branscombe Sewage Works, SY200885
Branscombe to Weston Mouth, SY190878
Brat Tor - Lyd Valley, SX545855
Bratton Clovelly, SX463919
Bratton Fleming, SS645377
Braunton, SS488365
Braunton Burrows, SS450355
Braunton Great Field, SS480357
Braunton Marsh, SS472345
Bray Valley, SS695426
Brayford, SS687348
Breakneck Hole, SS715410
Bremridge, nr Ashburton, SX786701
Brenamoor Common, Belstone, SX623939
Brendon Common, SS770445
Brent Fore Hill, SX6661
Brent Hill, SX700617
Brent Moor, SX674639
Brent Tor, SX471804
Brentor, SX482814
Bridestowe, SX514895
Bridford (village), SX815865
Bridford Woods, SX802881
Bridfordmills, SX833872
Bridge Reeve, Chulmleigh, SS663133
Bridgend, Noss Mayo, SX555482
Bridgerule, SS273028
Brim Brook, SX587870
Brimford Bridge, SS285173
Brimpts Farm, Dartmeet, sx668739
Brimstone Down, SX666874
Brisworthy Burrows, SX563645
Brisworthy plantation, SX556657
Brisworthy Ponds, SX552652
Brixham, SX924568
Brixton, SX555522
Brixton (Set-aside land), SX558527
Brixton Tor, SX540521
Broad Down, nr Farway, SY172939
Broad Down, nr Postbridge, SX615804
Broad Hole, SX592786
Broad Moor, SX684598
Broad Sands, nr Berrynarbor, SS5647
Broadaford, SX692763
Broadaford Farm, SX693763
Broadall Gulf, SX605640
Broadall Head, SX606638
Broadall Lake, Dendles Waste, SX612625
Broadbury (Mansditch Farm), SX497947
Broadbury area, SX483958
Broadclyst (old Rly Stn), SX992953
Broadclyst (village), SX984973
Broadhembury, ST103047
Broadhempston, SX808663
Broadmeadow, Teignmouth, SX925728
Broadmoor Farm, SX527785
Broadnymett (pond), SS703009
Broadpath, Braunton, SS483364
Broadsands, SX896575
Broadsands-Elberry Cove, SX896580
Broadwoodwidger, SX411892
Broomhill, Harford, SX637585
Browns House, nr Postbridge, SX616798
Brownston, SX699527
Brownstone, nr Coleton, SX904507
Buck's Cross, SS348230
Buck's Mills, SS358234
Buckfast, SX742674
Buckfastleigh, SX737662
Buckland Abbey, SX488689
Buckland Beacon, SX735731
Buckland Brewer, SS419209
Buckland Bridge, SX719719
Buckland Common, SX734739
Buckland Filleigh, SS465091
Buckland Ford, SX661661
Buckland in the Moor, SX720731
Buckland Monachorum, SX490684
Buckland, nr Thurlestone, SX678438
Buckland-tout-Saints, SX758461
Bude Aqueduct, SS301095
Budlake, Broadclyst, SS984001
Budleigh Salterton, SY070819
Budshead Creek, SX461602
Budshead Wood, SX462598
Bulkamore, nr Rattery, SX745628
Bull Point, SS462469
Bullpark, SX6073
Bullycleaves Quarry, nr Ashbtn,
Bulmore Cross, nr Axminster, SY293945
Burford, nr Clovelly, SS301224

Burgh Island, SX647438
Burlescombe, ST076167
Burley Wood, SX496875
Burn Valley, nr Butterleigh, SX970077
Burrator Res, SX555687
Burrator Wood, SX553674
Burridge Moor, Chawleigh, SS7412
Burrington, SS638167
Burrington (Grain mill), SS638167
Burrow Cross,
Burrow Farm, nr Stoke Canon, SX939992
Bursdon Moor, Hartland, SS274202
Burston, SS713024
Bus Lawn Water, SS790467
Bush Down, SX6072
Butter Brook (Harford), SX642592
Butter Cove, Thurlestone, SX662432
Butter Hill, SS707441
Butterdon Down, SX751884
Butterleigh, SS973081
Buttern Hill, SX658894
Buttshill Copse, nr Torquay, SX855615
Bye Cross, nr North Buckland, SS473402
Bymore Wood, R Walkham, SX485780
Bystock NR, SY032844
Cadbury, Thorverton, SS9105
Cadeleigh, SS915080
Cadhay House,
Cadover Bridge, SX555647
Cadover clay pools, SX553650
Cadworthy Wood, SX544642
Caen Estuary, SS483340
California Cross, SX704530
Calveslake Tor, SX6067
Camp Gate, Okehampton, SX588928
Cann Quarry, SX524596
Cann Woods, Plympton, SX531595
Canonteign Falls, SX835825
Capstone Point, Ilfracombe, SS519481
Capton, SX827532
Carey river nr Virginstow, SX037092
Carey Valley (part of), SX040096
Cargreen, SX436626
Caroline Farm, SX668812
Caroline Marsh, SX666811
Carswell, nr Mothecombe, SX590477
Castle Cross, nr Beaworthy, SX460975
Castle Drogo, SX722900
Castle Drogo (Hunters Path), SX722900
Castle Quay, SS554333
Cater's Beam, SX6368
Caton Cross, Bickington, SX783717
Caton, nr Ashburton, SX781719
Cator, SX683770
Cator Common, SX673778
Cator Gate, SX6876
Catstor Down, SX5465
Cawsand Bay, SX450505
Cawsand Beacon, SX636915
Central Dartmoor, SX6083
Chaddlewood, nr Plympton, SX556562
Chagford, SX700877
Chagford Bridge, SX700877
Chagford Common, SX675828
Chain Bridge, nr Bampton, SS939209
Chalk Ford, SX685681
Challaborough, SX6545
Challacombe (Exmoor), SS694410
Challacombe Down (Dartmoor), SX690798
Challenger Farm (Membury), ST282045
Challon's Combe, A'ton Gifford, SX677485
Channings Wood,

Chantry, nr Aveton Gifford, SX703491
Chapel Wood, Spreacombe, SS482415
Chapelton, SS579264
Chapman Barrows, SS695435
Chapmans Well, nr Holsworthy, SX360938
Chapple Moor, Sampfd Courtenay, SS625003
Chardstock, ST300044
Charford, SX724585
Charleton Bay, SX749413
Charton Bay, SY3090
Chawleigh, SS712126
Cheldon (on Little Dart R), SS734134
Chelston, Torquay, SX902635
Chenson, SS702095
Chercombe Bridge, Ogwell, SX833710
Cheriton Bishop, SX773930
Cheriton Coombe, SX646909
Cheriton Fitzpaine, SS868062
Cheriton Ridge, SS7446
Cherry Brook Hotel,Two Bridges, SX622760
Cherrybrook Head, SX619802
Cheston, nr S Brent, SX681586
Chevithorne, nr Tiverton, SS976155
Childe's Tomb, SX6270
Chillaton, SX433819
Chillaton, Tavistock, SX432819
Chillington, SX793429
Chilsworthy, nr Holsworthy, SS328063
Chinkwell Tor, SX730782
Chiselbury Bay, SY093845
Chitterley, nr Silverton, SS943045
Chittlehamholt, SS649208
Chittlehampton, SS638255
Chivelstone, SX783388
Chivenor Airfield, SS494343
Chivelstone, SX783388
Chivenor Airfield, SS494343
Cholake, SX618785
Chorland Farm, SS994099
Christow, SX832850
Chudleigh, SX870795
Chudleigh Knighton, SX836775
Chudleigh Knighton Heath, SX836775
Chulmleigh, SS687143
Churchstow, SX712459
Churston, SX895565
Churston Ferrers, SX903559
Clam Bridge, R Bovey,
Clannaborough Fm, Throwleigh, SX662913
Clannaborough, Bow, SS745025
Clapham, SX895871
Clapham, Kennford, SX8985
Clapper Bridge, SX628772
Clawton nr Holsworthy, SX354991
Clayhidon, ST163156
Clearbrook, nr Yelverton, SX523657
Cleave Wood (Ponsworthy), SX707737
Clennon Valley, SX885592
Clifford, SS306209
Clifford Bridge, Dunsford, SX781988
Clovelly, SS318248
Clovelly Cross, Clovelly, SS313233
Clovelly Ponds, SS315225
Clyng Mill, SX628490
Clyst Estuary, SX974874
Clyst Honiton, SX988937
Clyst Hydon, ST035015
Clyst Marshes, SX974882
Clyst St George, SX982887
Clyst St Mary, SX973909
Coaxdon, ST311009
Cockington, SX895638

Cockleford Bridge,
Cockleridge, nr Bigbury, SX667442
Cocks Hill, SX5679
Cockwood, SX975806
Cockwood (harbour & estuary), SX975806
Cockwood Marsh, SX972806
Coffin Wood, SX5481
Coffinswell, SX891685
Cofflete Creek, SX544500
Cofflete Creek Quarry, SX545503
Cofton, nr Cockwood, SX968804
Colaton Raleigh, SY075875
Colaton Raleigh Common, SY050875
Cold East Cross, SX741743
Coldridge, SS697077
Colebrooke, SS770001
Coleton area, SX910508
Coleton Fishacre, SX910508
Collapit Creek, SX734417
Collard Tor, SX557621
Collaton Raleigh Common, SY0487
Collaton St Mary, Paignton, SX867601
Collaton, Salcombe, SX719394
Colleton Mills Cross, SS666157
Colleybrook, Peter Tavy, SX532773
Colleytown, nr Lopwell, SX465652
Collipriest Weir, Tiverton, SS953119
Collyton Farm, SX565674
Columbjohn, Killerton, SX960998
Coly Valley, SY230945
Colyford, SY252925
Colyford Marshes, SY2592
Colyton, SY245941
Colyton (sewage works), SY255931
Combe Martin, SS586464
Combe, nr Brixton, SX547520
Combe, nr Scorriton, SX703681
Combe, nr Soar, SX718384
Combeinteignhead, SX903715
Combestone Island, SX675723
Combestone, nr Dartmeet, SX670725
Combpyne, SY290925
Combshead Tor, SX5868
Combshead Tor, nr Burrator, SX587688
Compton, SX869647
Conies Down Tor, SX589791
Cookworthy Moor, SS410014
Cookworthy Plantation, SS4100
Cool Stone (Taw Estuary), ss472318
Coombe Cellars, SX902723
Coombe Farm, nr Exmouth, SY008851
Coombe Fishacre, SX8464
Coombe Fm, Barnstaple, SS594334
Coombe Valley,
Coombe, nr Castle Drogo, SX720897
Copplestone, SS026768
Corbyn's Head, Torquay, SX907633
Cornborough, SS420283
Cornbury Hill, SS743066
Corndon Down, SX692852
Cornwood, SX606597
Cornworthy Moor, SX825557
Corringdon Ball, SX672609
Cornworthy Moor, SX825557
Corringdon Ball, SX672609
Coryton's Cove, Dawlish, SX963760
Coryton, nr Lewtrenchard, SX456836
Cosdon Beacon, SX636915
Cosdon Stone Rows, SX644918
Cotleigh, nr Honiton, ST205023
Couchill Farm, Seaton, SY232907
Countess Wear, SX947890

Countess Wear Sewage Works, SX947890
Countisbury, SS747497
Cousins Cross, nr South Pool, SX787397
Cove, SS950197
Cove Bridge, nr Huntsham,
Cove Cleeve, SS959204
Cow and Calf, Woody Bay, SS665497
Cowley Bridge, Exeter, SX907955
Cowsic Head, SX594804
Cowsic Intake, SX595767
Cox Tor, SX532762
Coxe's Cliff, nr Branscombe, SY177880
Crab Ledge, SY092839
Crabrock Point, Man Sands, SX925532
Craddock, nr Uffculme, ST085125
Cramber Down, SX593713
Cramber Tor, SX584713
Cranbrook Down,
Crane Hill, SX622691
Cranmere Pool, SX603858
Crapstone, nr Yelverton, SX503678
Crazywell Pool, SX582705
Crealy Park, SY002905
Crediton, SS830005
Crediton area, SS830005
Cripdon Down, SX734803
Criptor (Newtake), SX5572
Criptor Ford, Sampford Spiney, sx556727
Criptor Ford/Ingra Tor, SX5572
Crockern Tor, SX615758
Crofts Plantation, SX600697
Crook Plantation, Yettington, SY055865
Cross Furzes, Buckfastleigh, SX699668
Crow Beach, SS4632
Crow Neck, Taw Estuary, SS464321
Crow Point, Taw Estuary, SS466319
Crow Tor, SX605788
Crowdon Copse, nr Boreston, SX768541
Crownhill Down, SX570602
Croyde, SS446392
Croyde Bay (Downend), SS432388
Cruwys Morchard, nr Tiverton, SS875125
Cuckoo Rock, SX584687
Cudlipptown, SX521790
Cullever Steps, nr Belstone, SX608922
Cullompton, ST020072
Culm Davy, nr Culmstock, ST124151
Culm Grassland Sites, NW Devon,
Culm Valley (Collmptn-Uffclme),
Culm Valley (Hele-Cullompton), ST0104
Culm Valley (Rewe-Bradninch), SS965010
Culm/Exe confluence, SX933970
Culmbridge, Hemyock, ST139139
Culmstock, ST102138
Culmstock Beacon, ST110151
Culverhole Slip, Seaton, SY275895
Curtery Clitters, SX5990
Cussacombe Common, SS8030
Cut Combe, SX588838
Cut Hill, SX598828
Cylinder Bridge, R Yealm,
Daccombe, SX902681
Daddyhole Cove, SX926628
Dainton, Ipplepen, SX855667
Dalditch, Budleigh Saltn, SY047836
Dalwood, ST2400
Dalwood (sewage works), ST251998
Dalwood Hill, ST2399
Dalwood Hill, Axminster, SY239995
Dalwood, Axminster, ST247004
Damage Hue Rock, SS469468
Dane's Brook, SS825316
Dane's Wood, nr Killerton, SX967991
Danes Hill, nr Dalwood, ST257008
Danger Point, SY081820
Darlick Moor,
Dart Estuary, SX8756
Dart Gorge, SX725724
Dart Valley, SX716718
Dartington, SX798627
Dartmeet, SX672732
Dartmoor area,
Dartmouth (offshore),
Dartmouth (town), SX879515
Daveytown, SX549733
Dawlish (Langdon Hospital), SX961790
Dawlish (town), SX963766
Dawlish Bay, SX963766
Dawlish Warren, SX987795
Dead Lake Head, SX565784
Deal's Brake, Ugborough Beacon, SX6758
Dead Lake Head, SX565784
Deal's Brake, Ugborough Beacon, SX6758
Dean Burn, Dean Wood, SX715646
Dean Prior, SX730635
Deancombe Farm, SX579688
Deancombe, nr Buckfastleigh, SX722644
Deancombe, nr Burrator, SX577688
Deckler's Cliff, SX755366
Decoy Country Park, SX865703
Decoy Lake, Newton Abbot, SX865703
Deep Swincombe, SX643718
Deeper Marsh, SX7171
Denbury, SX824688
Dendles Wood, SX616619
Denham Bridge (R Tavy), SX477678
Denham Bridge Woods (R Tavy), SX476678
Dennings Down Fm, nr Axminster, ST2902
Derriton, nr Holsworthy, SS335031
Devils Elbow, Princetown, SX581729
Devon,
Devon Exmoor,
Devonport Leat,
Dewerstone Rock, Shaugh Prior, SX537638
Dewerstone Wood, Shaugh Prior, SX536638
Dick's Well, SX552860
Didworthy, nr South Brent, SX684622
Diggers Cross, nr Combe Martin, SS575445
Dinger Tor, SX586881
Dippertown, nr Lifton, SS429848
Diptford, SX728567
Ditsworthy Warren, SX584663
Dittisham, nr Dartmouth, SX8655
Dittisham, nr Walkhampton, SX535707
Dockwell Hole, SX695645
Dockwell Ridge, SX692642
Doddiscombsleigh, SX858866
Doe Tor, nr Lydford, SX543848
Dogmarsh Bridge, Chagford, SX713894
Dolton, SS575123
Dolton Beacon, SS590136
Dolton DWT Reserve, Beaford, SS552130
Doone Valley=Badgworthy Water, SS795455
Dotton, nr Newton Poppleford, SY085885
Double Bluff,
Double Dart Gorge, SX695719
Double Locks Wetland,
Double Waters, nr Tavistock, SX476699
Dousland, SX538685
Down End, nr Kingswear, SX894504
Down Thomas, nr Wembury, SX504500
Down Tor, SX580694
Downlands Plantation, ST095075
Drakeland Corner, nr Hemerdon, SX5759
Drewsteignton, SX736909
Drewston Wood, SX7490
Dripping Well, Lee Bay, SS486463
Drizzle Combe, SX589670
Duck's Pool (Plym Head), SX625680
Ducks' Pool, SX6267
Dulford, nr Cullompton, ST065060
Dumpdon Hill, Honiton, ST175043
Dunchideock, SX880872
Dunkeswell, ST142078
Dunkeswell Turbury, ST130058
Dunnabridge, Two Bridges, SX645744
Dunscombe Cliff, Sidmouth, SY155877
Dunsdon Farm NNR, SS295080
Dunsford, SX813892
Dunsford area, SX813892
Dunsford Woods, SX804884
Dunsland Cross, Holsworthy, SX405039
Dunstone, SX594515
Duntz Valley, SS435205
Durl Head, nr Berry Head, SX940557
Duty Point, nr Lee Bay, SS695497
Easdon Forest, North Bovey, SX741828
Easdon Tor, SX729823
East Allington, SX770485
East Anstey, SS867265
East Anstey Common, SS867287
East Ash, SX680911
East Bovey Head, SX693822
East Budleigh, SY070846
East Budleigh Common, SY040845
East Burne, nr Goodstone, SX799708
East Charleton, SX763426
East Cornworthy, nr Dittisham, SX847553
East Dart Head, SX609855
East Devon,
East Devon Commons, SY0487
East Down, SS601419
East Emlett, nr Kennerleigh, SS810083
East Glaze Brook, SX668610
East Hill, SX5993
East Hill Strips,Ottery StMary, SY123940
East Leigh, SS698053
East Lyn River, SS7548
East Mill Tor, SX599899
East Ogwell, SX832702
East Okement Farm, SX605913
East Ogwell, SX832702
East Okement Farm, SX605913
East Okement Head, SX605881
East Okement River, SX607930
East Okement Valley,
East Panson, nr Virginstow, SS361923
East Peeke Plantation, SX681589
East Portlemouth, SX748385
East Prawle (new ponds), SX782368
East Prawle (village), SX781365
East Prawle area, SX781365
East Prawle Wood, SX785363
East Putford, SS368164
East Rook (village), Cornwood, SX606608
East Rook Gate, Cornwood, SX606615
East Soar, SX721371
East Tordown, SX6991
East Week, SX666919
East Worlington, SS775136
East-the-Water, SS458265
Eastdon, nr Dawlish Warren, SX977798
Eastern Beacon, SX668588
Ebford, SX982878
Eddystone,
Edginswell, SX883667

Edginswell Marsh, SX883667
Efford House, Holbeton, SX620495
Efford Pond, Holbeton, SX619491
Eggesford, SS681115
Elberry Cove, SX904571
Elfordleigh Ponds, Plympton, SX546582
Ellerhayes, Silverton, SS975020
Elmscott, SS232218
Elsford, nr Bovey Tracey,
Elwill Bay, SS6449
Emmets Post, SX633567
Emsworthy (=Hemsworthy), SX742761
Erme Estuary, SX615475
Erme Head, SX622668
Erme Pound, SX639657
Erme Valley (Dartmoor), SX6463
Erme Valley (S Devon), SX6453
Ermington, SX638531
Ernesettle, SX442595
Escot Pond, nr Talaton, SY083979
Escot, nr Talaton, SY082982
Exbourne, SS604021
Exe Estuary, SX9884
Exe Estuary (Cockle Sand), SX9982
Exe Estuary (Cockwood), SX980886
Exe Estuary (Exmouth), SX998812
Exe Estuary (Exton), SX978865
Exe Estuary (Lympstone), SX988843
Exe Estuary (Powderham - Turf), SX971852
Exe Estuary (Powderham), SX976840
Exe Estuary (Riversmeet), SX9787
Exe Estuary (Starcross), SX979808
Exe Estuary (Starcross-DWarren, SX9888
Exe Estuary (Topsham), SX9987
Exe Estuary (Turf), SX965862
Exe Estuary (Turf-Topsham), SX965865
Exe Estuary (West Mud), SX9686
Exebridge, SS932247
Exeter (Belle Isle Park), SX926914
Exeter (Belmont Park), SX930932
Exeter (city), SX920928
Exeter (Duryard), SX9194
Exeter (Exwick), SX906936
Exeter (Heavitree), SX944925
Exeter (Ludwell Valley), SX9491
Exeter (Marsh Barton), SX930895
Exeter (Mount Wear), SX947893
Exeter (Newtown), SX932931
Exeter (Northbrook Park), SX937905
Exeter (Pinhoe), SX965945
Exeter (Pynes Hill), SX948911
Exeter (Riverside CP pools),
Exeter (Riverside CP), SX920928
Exeter (Sowton Quarry), SX964914
Exeter (Sowton), SX9691
Exeter (St Davids), SX910933
Exeter (St James), SX927937
Exeter (St Leonards), SX9392
Exeter (St Loyes), SX942914
Exeter (St Thomas), SX9191
Exeter (Stoke Hill), SX9394
Exeter (University), SX915942
Exeter (Whipton), SX952938
Exeter (Wonford), SX943919
Exeter Airport (Clyst Honiton), SY002937
Exeter Canal (C'Wear to Turf), SX947890
Exeter Canal WBS,
Exmansworthy Cliff, SS276271
Exminster (Luccombes Ponds), SX934866
Exminster (village), SX945876
Exminster Marshes, SX958875
Exmoor, SS761361
Exmouth (Cocklesand Bay), SX9982
Exmouth (Maer Rocks), SY015797
Exmouth (Mudbank Lane), SX999821
Exmouth (Maer Rocks), SY015797
Exmouth (Mudbank Lane), SX999821
Exmouth (Orcombe Cliffs), SY025797
Exmouth (Phear Park), SY007815
Exmouth (seafront), SY015797
Exmouth (town), SY004802
Exton, SX981864
Exwick, SX9093
Exwick Barton Weir, Exeter, SX906951
Exwick Flood Relief, SX909935
Eylesbarrow, nr Ditsworthy, SX600686
Factory Bridge, nr Ivybridge, SX633556
Fairmile, Ottery St Mary, SY087972
Fallapit, nr East Allington, SX765491
Farley Water, SS743460
Farringdon, SY018912
Farway Country Park, SY186943
Fatherford, nr Okehampton, SX604954
Feather Tor, SX535742
Feather Tor Ford, SX535743
Feniton, SY097993
Fernworthy Res, SX665841
Filham, nr Ivybridge, SX649558
Filleigh, SS663279
Fingle Bridge, SX744900
Finlake, Chudleigh, SX846790
Fire Beacon Hill, Dittisham, SX862539
Fire Stone Cross, South Zeal, SX666930
Fish-in-the-Well, Prawle, SX776355
Fishcombe Point, SX915572
Fisher's Mill, SX971882
Fishlake Mire, SX6468
Five Barrows Cross,
Five Cross Ways, SS880285
Five Fords, R Culm, ST082135
Flat Tor, SX604807
Fleet Mill, nr Totnes, SX826589
Flete Estate, Holbeton, SX628515
Fluxton, nr Ottery St Mary, SY087929
Foggintor area, SX566786
Foggintor Quarries, SX566786
Folly, SS725445
Ford & Fairy Cross, SS405244
Ford Brook, nr Cornwood, SX608621
Ford, nr South Pool, SX788405
Forder, SX672897
Forder Brook, Throwleigh Com., SX6689
Fordsland Ledge, SX5788
Foreland Point, SS754513
Fosterville Tip, Heathfield, SX863762
Fosterville, Kingsteignton, SX863762
Four Firs Cross, SY031865
Fox Tor, SX626698
Foxtor Mires, SX619707
Foxworthy, Lutton, SX694742
Fremington, SS513333
Fremington Pill, SS515328
Fremington Quay, SS515334
Frenchbeer Rock, SX6785
Frenchbeer, Chagford, SX675857
Frogmore, SX775427
Frogmore Creek, SX760417
Frogwell Farm, nr Diptford, SX741556
Froward Point, SX905496
Fur Tor, SX589831
Fursdon, nr Staverton, SX782647
Furzebeam Hill, SS480203
Furzedown Farm, Malborough, SX708384
Furzeleigh, Buckfastleigh, SX745677
Gaddon Down,
Gallaven Brook, SX639880
Gallaven Mire, SX632885
Gallows Cross, Kingsteignton, SX861735
Galmpton Creek, SX880559
Galmpton, nr Brixham, SX889561
Galmpton, nr Hope Cove, SX690403
Gammaton Moor, SS4924
Gammon Head, SX765357
Gara Bridge, SX729535
Gara Point, SX523469
Gara Rock, nr Portlemouth, SX751370
Gartaven Ford, SX635886
Gawton, SX453686
George Nympton, SS7023
Georgham, SS464397
Ger Tor, SX547831
Germansweek, SX440943
Gerston Point, SX740415
Gibbet Hill, SX502811
Gidleigh, SX673883
Gidleigh Common, SX650875
Gidleigh Woods, SX674877
Gittisham, SY134984
Glasscombe Ball, SX659603
Glaze Valley, nr S Brent, SX678599
Glazebrook, nr S Brent, SX689591
Glendenings Quarry, nr Ashbtn,
Gnaton, nr Yealmpton, SX579498
Godborough NR, SS438273
Godsworthy, nr Peter Tavy, SX527775
Godborough NR, SS438273
Godsworthy, nr Peter Tavy, SX527775
Golden Park, Welcombe, SS233201
Goodameavy, SX532647
Goodleigh, nr Barnstaple, SS598342
Goodrington, SX897585
Goodshelter, SX764388
Goodshelter Creek, South Pool, SX764391
Goodstone, Bickington, SX785719
Gooseford, Whiddon Down, SX676918
Goosemoor NR,
Goosewell, nr Ilfracombe, SS555472
Goren Farm, nr Chardstock, ST3202
Gorlofen, nr Brixton, SX567527
Gorvin, Hartland Forest, SS291196
Gotleigh Moor, Smeatharpe, ST188108
Goutsford Bridge, nr Ermington, SX637515
Goveton, nr Kingsbridge, SX754464
Grand Western Canal,
Grand Western Canal (WBS), ST027137
Grand Western Canal(sect 01), SS953124
Grand Western Canal(sect 02), SS974123
Grand Western Canal(sect 03), SS987121
Grand Western Canal(sect 04), SS998127
Grand Western Canal(sect 05), ST005133
Grand Western Canal(sect 06), ST010132
Grand Western Canal(sect 07), ST019130
Grand Western Canal(sect 08), ST027137
Grand Western Canal(sect 09), ST035145
Grand Western Canal(sect 10), ST043149
Grand Western Canal(sect 11), ST054155
Grand Western Canal(sect 12), ST058161
Grand Western Canal(sect 13), ST069169
Grand Western Canal(sect 14), ST070175
Grand Western Canal(sect 15), ST071188
Grand Western Canal(sect 16), ST072194
Grand Western Canal, Halberton, SS998138
Grand Western Canal, Tiverton, SS978122
Gratton Bridge, SX529670
Great Black Hill, SS788458
Great Burland Rocks, SS664495

Great Fossend, Westleigh, ST071177
Great Gnats' Head, SX617679
Great Haldon, SX897839
Great Kneeset, SX589859
Great Links Tor, SX552867
Great Mew Stone, Wembury, SX501474
Great Mis Tor, SX562769
Great Nodden, nr Lydford, SX538874
Great North Wood,
Great Odam Moor Plantation, SS7418
Great Torr, nr Kingston, SX638484
Great Torrington, SS490189
Great Tree, nr Chagford, SX705902
Great Varracombe, SX629842
Greator Rocks, SX747787
Green Combe, SX695835
Green Hill, SX637678
Green Lane End, SX776503
Green Tor, SX562864
Greena Ball, SX564778
Greenswood Fm, Blackawton, SX831503
Greenwell Girt, SX540657
Gren Tor, SX551879
Grendon Farm (nr Soussons), SX685781
Grenofen, SX495715
Grenofen Bridge, SX4971
Grenofen Woods, SX491709
Grey Wethers, SX639832
Grimspound, sx701809
Gullet Farm, South Pool, SX396766
Gullet Plantation, South Pool, SX763393
Gulliford, nr Dawlish, SX947798
Gunnislake Woods, SX438718
Gutter Mire, Ringmoor, SX577673
Gutter Tor, Ringmoor, SX577668
Gypsy Hill, nr Exeter, SX965935
Haccombe, nr Newton Abbot, SX894704
Hackney Marshes, SX873723
Hackpen Hill, Culmstock, ST117120
Hakeford, nr Barnstaple, SS614355
Halberton (Hartnoll Fm), SS993131
Halberton, nr Tiverton, ST006130
Haldon (Ashcombe Cross), SX902795
Haldon (Beggars Bush), SX893795
Haldon (BOP View Point), SX880854
Haldon (Buller's Hill), SX883847
Haldon (Kenton Hill), SX910815
Haldon (Kiddens Plant'n), SX875845
Haldon (North Wood), SX863888
Haldon (nr N Kenwood), SX915825
Haldon (Obelisk), SX925807
Haldon (Racecourse), SX900835
Haldon (Underdown), SX888857
Haldon Forest area, SX890837
Hale Bridge, R Bovey,
Half Moon, nr Crediton, SX893976
Halford, nr Bovey Tracey, SX812745
Hall Farm, nr Harford, SX631597
Halls Fm, nr Higher Metcombe, SY075920
Hall Farm, nr Harford, SX631597
Halls Fm, nr Higher Metcombe, SY075920
Hallsands, SX818385
Halsdon DWT Reserve, SS554129
Halshanger & Bickington, SX758734
Halshanger Common, SX758734
Halshanger, nr Rippon Tor, SX758734
Halshanger-Mountsland area, SX758734
Halsinger Bog, nr Braunton, SS514388
Halspill (Torridge), SS470235
Halstock Wood, SX607935
Halton Quay, SX414655
Halwell, SX777532
Halwell Wood, nr South Pool, SX755408
Halwill Junction, SS444001
Halwill Plantations, SS420015
Halwill, nr Beaworthy, SX428995
Ham, ST238014
Ham Point, SX757415
Hamel Down, SX710793
Hameldown Tor, SX703807
Hanger Down, SX624589
Hanger Mill Valley, Salcombe, SX727387
Hangershell Rock, nr Harford, SX655595
Hangingstone Hill, SX616861
Hannaford Pond, R Dart, SX709707
Hannafords Quay,
Hannington's Cove,
Hapstead, Buckfastleigh, SX718668
Harberton, SX777585
Harbertonford, SX783562
Harbourne Head, SX693651
Harbourneford, SX716623
Harcombe, nr Chudleigh, SX889818
Hardsworthy, nr Bradworthy, SS288165
Hardwick Woods, Plympton, SX530555
Hare Tor, SX551842
Hares Down, SS848214
Hareston, nr Brixton, SX566538
Harford Bridge, nr Peter Tavy, SX505767
Harford CBC, SX631597
Harford Moor, SX647625
Harford Moor Gate, SX643596
Harford Moor Gate-Higher Piles,
Harford, nr Ivybridge, SX638595
Harpford, SY092903
Harpford Common, SY065902
Harraton, nr Modbury, SX675506
Harter Hill, SX603918
Harter Tor, SX602678
Hartland (village), SS250245
Hartland For (Welsford Moor), SS2820
Hartland Forest, SS2720
Hartland Point, SS230278
Hartland Quay, SS223247
Hartland Tor, SX641800
Hartley, Plymouth, SX485575
Hartor Tors, SX604676
Hartridge, nr Upottery, ST179059
Hartyland, SX643795
Hatch Bridge, nr.Loddiswell, SX715473
Hatch Mill (R Tavy), SX470683
Hatchwell, nr Widecombe, SX702775
Hatherleigh, SS542043
Hatherleigh Moor, SS550037
Haven Banks, Exeter, SX920917
Haven Cliff - Dowlands Cliff, SY265895
Hawkchurch, ST340006
Hawkerland Valley, SY054895
Hawkerland village, SY057887
Hawkesdown Hill, nr Axmouth, SY263914
Hawkmoor Farm, SY3799
Hawks Tor, SX554625
Hawkshead Valley, SY053892
Hawridge Walk, nr East Anstey, SS860307
Haydon Common, SX933835
Hayes Wood, E Budleigh, SY050845
Haytor (Rocks & Down), SX757770
Haytown (R Torridge), SS385145
Head Wood, Kingsnympton, SS663179
Headland Warren, SX687812
Headwear Ford, SX622709
Heanton Court, SS514350
Heanton Punchardon, SS501355
Heath Cross, Tedburn St. Mary, SX849944
Heathercombe nr Manaton, SX719812
Heathfield, SX8376
Heathstock, Stockland, ST246029
Heatree Cross, Dartmoor, SX7280
Heddon Down, SX591602
Heddon Mill, Braunton, SS0000
Heddon Valley, SS655475
Heddon's Mouth, SS655497
Hedge Barton, nr Widecombe, SX739793
Hedgemoor Farm, nr Bridford, SX807858
Hele, SS995025
Hele Bay, nr Ilfracombe, SS535481
Hele Bridge, SS547063
Hele Cross, Cornwood, SX612610
Heltor Rock, nr Dunsford, SX8087
Hele Cross, Cornwood, SX612610
Heltor Rock, nr Dunsford, SX8087
Hembury Wood, SX730685
Hemerdon, SX563574
Hemsworthy Gate, SX742761
Hemyock, ST137134
Hen Tor, SX594653
Hendom Cross, ST012214
Hendom Farm, nr Huntsham, ST012212
Hendon Moor, nr Bursdon, SS264187
Henford, SX364949
Henlake Down, SX633573
Hennock (village), SX830809
Hennock area, SX829809
Hennock Res, SX810834
Henscott, SX5183
Herner, SS585267
Herons Wood,
Herring Cove, Maidencombe, SX931694
Hew Down, SX635861
Hexton Woods, Hooe, SX504528
Hexworthy, SX655726
Hexworthy Mine, SX658706
Heybrook Bay, SX497491
High Bickington, SS600205
High Down, SX527850
High House Point, SX742428
High Peak, nr Otterton, SY105859
High Willhays, SX580894
High-house Waste, SX610627
Highampton, SS483043
Higher Ashton, SX856847
Higher Barn Farm, nr Ashburton, SX772702
Higher Cherrybrook Bridge, SX6367
Higher Godsworthy, Peter Tavy, SX530772
Higher Hartor Tor, SX6067
Higher Huntingdon Corner, SX6668
Higher Knowle, Lustleigh, SX794807
Higher Metcombe, SY066922
Higher Piles (enclosure), SX6461
Higher Piles Wood,
Higher Sandygate, SX869756
Higher Soar, SX710379
Higher Tor, Belstone, SX614917
Higher-Lower White Tors, SX619789
Highfield Farm, Topsham, SX965889
Highleigh Bridge, nr Exebridge, SS921228
Highveer Point, SS656499
Highweek, Newton Abbot, SX845720
Hill 563, SX595808
Hillhead nr Brixham, sx905538
Hillhead Quarry Complex, ST0613
Hillsborough, Ilfracombe, SS535478
Hillsford Bridge, SS741479
Hillson's Brake, SX6061
Hilltops,
Hilsea Point, SX540458

Hittisleigh, SX730955
Hoaroak Water, SS739450
Hoccombe Combe, SS780445
Hoccombe Water, SS780434
Hockinston Tor, SX698719
Hockworthy, nr Huntsham, ST038195
Hoist Point, Bigbury, SX632458
Holbeton, SX615503
Holbeton Wood, SX627503
Holcombe, SX955748
Holcombe Burrows, SS757441
Holcombe Down, SX940752
Holcombe Rogus, ST057190
Holdstone Down, SS618480
Holdstone Down, nr Cmbe Martin, SS618472
Hole's Hole, nr Weir Quay, SX430654
Holelake, nr Huntsham, ST022207
Holewell Farm, Walkham valley, sx541712
Hollicombe, Torbay, SX897623
Hollicombe-Torquay harbour, SX9163
Hollocombe Moor, SS6012
Hollow Moor, nr Northlew, SS480015
Hollow Tor, SX731762
Holming Beam, SX591769
Holne (R Dart), SX720715
Holne Bridge, SX730706
Holne Lee, SX685695
Holne Moor, SX675705
Holne Ridge, SX670698
Holne Woods, SX7070
Holsworthy, SS342037
Holsworthy Beacon, SS358083
Holsworthy Woods, SS354015
Holwell Down, SX738775
Holwell Mire, SX746775
Holwell Tor, SX751777
Holy Brook, Buckfast, SX725679
Holyford Wood, SY234922
Home Farm, Kingswear, SX897507
Homer Common,
Homerton Hill, SX563904
Honeybag Tor/Jay's Grave, SX730792
Honeychurch,Sampford Courtenay, SS629028
Honeybag Tor/Jay's Grave, SX730792
Honeychurch,Sampford Courtenay, SS629028
Honiton, ST163005
Hoo Meavy, SX527657
Hood Barton,
Hooe (Tamar Estuary), SX421655
Hooe, nr Plymstock, SX504525
Hook Ebb, SY155875
Hooken, nr Branscombe, SY214886
Hookney Tor, SX698815
Hookway, nr Crediton, SX851987
Hoops, nr Buck's Mills, SS374232
Hope Cove, SX673398
Hope's Nose, SX949637
Hore Wood, Upper Teign Valley, SX790895
Horndon, nr Mary Tavy, SX521801
Horrabridge, SX515700
Horridge, nr Rippon Tor, SX762742
Horse Cove, nr Dawlish, SX961756
Horse Hole, SX601802
Horsebridge, SX405749
Horsey Island, SS475335
Horsey Pond, SS473335
Horseyeatt, nr Walkhampton, sx545703
Horsham Cleave, nr Manaton, SX760814
Horsham Marsh (Tavy), SX471626
Horsham Pond, SX758812
Horsham, Tamerton Foliot, SX471626
Horsley Cove, Prawle, SX786359

Horton Farm, SS300178
Horwood, SS503277
Hound Tor, Manaton, SX743790
Hound Tor, N Moor, SX629890
Houndtor Wood, SX7780
Huccaby, SX659728
Hucken Tor, SX550738
Huckham, SX805403
Huckham, nr Beesands, SX808404
Huckworthy, nr Walkhampton, sx532706
Huddisford Cross, Woolsery, SS314207
Huddisford farms, Woolsery, SS305195
Hughslade Farm, nr Meldon, SX5693
Humber, nr Ideford, SX900753
Hunter's Inn, SS655482
Hunter's Inn (cliffs & woods), SS6548
Hunter's Inn-Woody Bay Hotel, SS655482
Hunter's Tor, SX7289
Hunters Inn-Woody Bay,
Huntingdon Warren, SX658671
Huntsham (village), ST003205
Huntsham area, ST002204
Huntsham Lake, ST008208
Huntshaw Woods, SS495225
Hurdwick, Tavistock, SX472579
Huxham Brake, nr Exeter, SX945968
Huxham, nr Exeter, SX947978
Huxton Cross, SX704443
Ice Works, SX546900
Ide, SX898904
Ideford, SX895773
Ideford Common, SX900785
Ideford Ford, SX895772
Idestone Cross, Dunchideock, SX882886
Ilfracombe, SS515475
Ilkerton Ridge, SS718450
Ilsington, SX785762
Ilton, nr Malborough, SX722402
Incledon, SS478386
Ingra Tor, nr Princetown, SX555722
Ingsdon, nr Seale Hayne, SX815729
Inner Hope, SX676398
Instow, SS473306
Instow Barton Marsh, SS475319
Inwardleigh, SX560994
Ipplepen, SX835666
Irishman's Wall, SX615918
Iron Bridge, nr Stoodleigh, SS943178
Isley Marsh, SS490328
Ivy Cove, nr Coleton Fishacre, SX919507
Ivybridge, SX635563
Ivybridge Wood,
Ivyhouse Cross, SX405938
Ivyhouse School, SX405937
Jacobstowe, SS587016
Jay's Grave, SX731799
Jennetts Res, Bideford, SS442247
Jennycliff, SX490524
Jetty Marsh, Newton Abbot, SX863720
Jordan, nr Ponsworthy, SX7075
Jurston, SX696844
Kelly, Lustleigh, SX796818
Kenn, SX921855
Kennerleigh, SS820075
Kennford, SX915862
Kennick Res, SX803844
Kennon Hill, SX643894
Kentisbeare, ST068082
Kenton, SX958833
Kenwith Castle, Abbotsham, SS435272
Kenwith NR, SS448273
Kenwith Castle, Abbotsham, SS435272

Kenwith NR, SS448273
Kenwith Valley, SS435272
Kersbrook, Budleigh Salteron, SY067831
Kerse, nr Thurlestone, SX689434
Kerswell Cross, ST018207
Kerswell Down, SX872676
Kerswell Farm, nr Huntsham, ST013210
kerswell, nr Cullompton, ST0806
Kerwell, ST0806
Kestor, SX665864
Kettle Plantation,
Killerton, SS974002
Kilmington, SY2798
Kilmington (sand pits),
Kilmington (sewage works), SY278975
Kilminster,
King's Nympton, SS684194
King's Tor, SX556738
Kings Oven, SX674812
Kings Wood, Buckfastleigh, SX720663
Kingsbridge, SX735445
Kingsbridge Estuary, SX743412
Kingsett Down, SX517816
Kingsford Gate, SS740366
Kingshead Farm, Widecombe, SX712777
Kingskerswell, SX878678
Kingskerswell Down, SX887683
Kingsmoor Cross, SS4404
Kingsnympton Park, SS672196
Kingsteignton, SX785735
Kingston, SX636477
Kingswear (town), SX885511
Kingswear Castle, SX885511
Kit Rocks (East Dart R), SX613827
Kitley Pond, SX555513
Kitterford Cross, nr Ugborough, SX690563
Kitty Tor, SX567875
Klondyke, SX593940
Knap Head, SS213187
Knap Mill, R Avon, SX708473
Knattabarrow Pool, sx656645
Knighton Heath, SX836775
Knighton, Wembury, sx528498
Knightshayes CBC, SS963152
Knightshayes Court, Tiverton, SS960153
Knowle, SY052824
Knowle Down, Walkhampton, SX529700
Knowle Hill, B Salterton, SY045825
Knowstone (village), SS826230
Knowstone Moor, SS837217
Labbett's Cross, nr Lapford, SS725117
Labrador Bay, nr Shaldon, SX935708
Labrador to Watcombe (3 kms), SX928690
Ladies Wood -choose right one!,
Ladies Wood NR (N Devon),
Ladies Wood NR, nr S Brent, SX686591
Ladram Bay, SY097850
Lamb's Down, SX695660
Lambhill, nr Willand, ST070145
Lambside, nr Newton Ferrers, SX583472
Lamerton, SX4577
Lana Heath, SS300074
Landcross (R Torridge), SS461237
Landcross (R Yeo), SS458238
Landkey, SS595313
Landscove, SX771666
Langcombe Brook, SX6066
Langcombe Hill, SX618656
Langerstone Point, SX783353
Langford, SX901978
Langston, nr Kingston, SX644490
Langstone Moor, nr Peter Tavy, SX554781

Langstone Rock, Dawlish Warren, SX9878
Langtree (village), SS450156
Langtree Moor, SS4615
Lank Combe, SS780455
Lannacombe, SX802371
Lapford, SS732084
Laployd Plantation, SX805850
Laughter Hole, sx662758
Laughter Tor, nr Bellever, SX654757
Leat Farm, nr Witheridge, SS827167
Leather Bridge, nr Burrator, SX569699
Leather Tor, SX563701
Ledstone, nr Kingsbridge, SX747465
Lee Abbey, SS698494
Lee Bay, nr Ilfracombe, SS480467
Lee Bay, nr Lynton, SS693495
Lee Bay-Ifracombe (coast), SS500472
Lee Bay-Ilfracombe (via resr), SS505455
Lee Bay-Ilfracombe over Tors,
Lee Bay-Mortehoe (via Morte P), SS465468
Lee Mill, nr Ivybridge, SX599557
Lee Moor, SX590645
Lee Moor (settlement), SX572617
Lee Moor Clay Pits, SX565613
Leedon Tor, SX563719
Left Lake, SX646634
Legis Lake, SX567665
Left Lake, SX646634
Legis Lake, SX567665
Legis Tor, SX571655
Leigh Barton, SX720467
Leigh Tor, SX710717
Leighon, SX755792
Lettaford, SX702841
Leusdon, nr Ponsworthy, SX707732
Lewdon, nr Witheridge, SX776106
Lewtrenchard, SX457861
Ley Green, R Teign, SX848753
Ley Wood, Diptford, SX716546
Lich Way, SX579790
Lifton, SX395851
Lifton Park (heronry), SX378843
Liftondown, SX370853
Limebury Farm, Parkham, SS363221
Limebury Pt, nr Salcombe, SX736375
Limers Cross, Combe Raleigh, ST145045
Limsboro Cairn, SX565805
Lincombe, SX738404
Lints Tor, SX580875
Liphill Marsh (Tamar Est), SX441635
Littaford Tors, SX616772
Little Bradley Pond, SX830778
Little Dartmouth, SX875492
Little Goat, Soar Mill Cove, SX708365
Little Haldon, SX918760
Little Hangman, SS585481
Little Hempston, SX812626
Little Hound Tor, SX631900
Little Meshaw Moor, SS7617
Little Sherberton, SX638735
Little Silver, Bickleigh, SS918095
Little Stannon, Postbridge, SX654803
Little Torrington, SS491168
Littlecombe Shoot, SY183878
Littleham Cove, SY040803
Littleham, nr Bideford, SS440234
Littleham, nr Exmouth, SY029813
Littlehempston, SX820627
Littlejoy, Newton Abbot, SX831717
Livaton Pit, SX679935
Livermead, Torquay, SX904628
Liverton, nr Bovey Tracey, SX806752

Loddiswell, SX719485
Loddiswell station, SX731483
Loddiswell-Aveton Gifford, SX710472
Long Ash Moor, SX549743
Long Plantation, Two Bridges, SX592760
Long Quarry Point, SX938651
Long Ray, W of Stoke Point, SX548458
Long Stone Combe,
Longaford, nr Two Bridges, SX616778
Longash Brook, SX5574
Longdown, nr Exeter, sx865912
Longland Barn, SX758832
Longsford, Whitchurch Common, SX5174
Longstone Hill, SX567912
Longstone Manor (Burrator), SX556684
Lopwell Dam/Lake, SX475651
Lovaton, nr Meavy, SX545663
Lower Ashton, SX845844
Lower Bradley Pond, SX827777
Lower Cator, SX688763
Lower Cherry Brook, SX6374
Lower Cherrybrook Br (B3357), SX632748
Lower Greenway Wood, Galmpton, SX875551
Lower Hartor Tor, SX6067
Lower Lea Farm, nr Dalwood, SY2499
Lower Leigh (or Leigh), SX721467
Lower Lowery, Burrator, SX555693
Lower Metcombe,
Lower Netherton, SX890718
Lower Otter Valley,
Lower Otter Valley (habitat 1), SY075825
Lower Otter Valley (habitat 2), SY075825
Lower Otter Valley (habitat 3), SY075825
Lower Otter Valley (habitat 4), SY075825
Lower Piles, Harford, SX643608
Lower Tamar Lake, SS295110
Lower Towsington, Exminster, SX976873
Lower Varracombe, SX632838
Lower White Tor, SX719793
Loxhore, SS617387
Luckcroft Fen, nr Northlew, SS485005
Luckdon, Manadon, SX6284
Lud Gate, Buckfastleigh Moor, SX685674
Ludbrook, nr Ermington, SX659544
Lundy crossing, SS135455
Lundy Island, SS135455
Luppitt, ST169067
Luscombe, Dawlish, SX943768
Luscombe, Harbertonford, SX796572
Luson, nr Holbeton, SX602501
Lustleigh area, SX785812
Lustleigh Cleave, SX762815
Luton, SX902769
Lyd Head, SX557884
Lydford, SX513851
Lyd Head, SX557884
Lydford, SX513851
Lydford Forest, SX490844
Lydford Tor, SX5978
Lydia Bridge, South Brent, SX695608
Lyme Bay (seabird survey), SX9970
Lyme Bay /Torbay,
Lyme Regis undercliff, SY330914
Lympstone, SX988843
Lympstone Common, SY025848
Lyn Down, SS725472
Lynch Common, SX554664
Lynch Tor (nr Mary Tavy), SX566806
Lynmouth, SS725495
Lynton, SS723493
Maceley Cove, SX7635
Mackham, nr Dunkeswell, ST151099

Madford, nr Hemyock, ST145111
Maelcombe, nr E Prawle, SX791364
Maer Farm,
Magpie Bridge, nr Grenofen, SX504703
Maiden Down, ST087160
Maiden Hill, SX587798
Maidencombe, SX926684
Main Head, nr Ugborough Beacon, SX660591
Malborough, SX708398
Malmsmead, SS792477
Malmsmead Hill, SS785467
Mamhead, SX930813
Man Sands, SX923534
Manadon, Plymouth, SX480582
Manaton, SX750813
Manga Brook Head, SX627850
Manga Hill/Rock, SX637858
Manley, nr Tiverton, SS990118
Mannamead, Plymouth, SX485565
Manor Common, SY055921
Marden Down, SX769876
Mardle Head, SX666692
Mardon Down, SX770878
Maristow, nr Lopwell, SX473648
Marjery Cross, nr Ivybridge, SX623548
Marldon, SX868633
Marley Head, SX725602
Marsh, ST255107
Mary Tavy, SX5079
Mary Tavy Common, SX533795
Matford, Exeter, SX928892
Mattiscombe, SX808422
Maypool, SX876545
Mead, nr Ashburton, SX779709
Meadfoot Beach, SX932630
Meavy, Yelverton, SX540673
Meddon, SS277179
Meddrick Rocks, Bigbury Bay, SX622456
Meeth, SS548083
Mel Tor, nr Poundsgate, SX695725
Melbury Plantation, SS380198
Melbury Res, SS385200
Meldon, SX560923
Meldon Hill, Chagford, SX696862
Meldon Quarry, SX570927
Meldon Res, SX5691
Meldon Woods, SX565925
Meltor, SX693725
Meltor Wood, SX693723
Membury (sewage works), ST274021
Membury Court, ST2603
Membury, nr Stockland, ST277032
Merafield, Plympton, SX525559
Merripit, nr Postbridge, SX657803
Merrivale, SX546753
Merrivale Quarry, SX770712
Merrymeet, SX696928
Merton Moors, SS510125
Merton, nr Torrington, SS527123
Meshaw Moor NR, SS7617
Metcombe, SY083919
Metherall, SX673839
Michelcombe, nr Holne, SX696690
Mid Soar, SX713375
Mid-Otter WBS, SY092898
Middle Rocombe, SX910693
Middle Staple Tor, SX545756
Middle Stoke, Holne, SX696705
Middle Tor, SX669859
Middleham, E Lyn River,
Middlesworth Plant'n, Burrator, SX574689
Milber, nr.Newton Abbot, SX875702

Milford, SX403863
Milford, nr Hartland, SS232225
Milford, nr Lifton (heronry), SX410865
Milfordleigh,
Mill Bay Cove, Kingswear, SX894502
Mill Bay, East Portlemouth, SX741382
Mill End (R Teign), SX706884
Millburn Valley, A'ton Gifford, SX681471
Millhayes, Stockland, ST235036
Milton Abbot, nr Tavistock, SX407794
Milton Damerel, Holsworthy, SS385107
Milton Abbot, nr Tavistock, SX407794
Milton Damerel, Holsworthy, SS385107
Mis Tor, SX563770
Modbury, SX657517
Molland, SS807284
Molland Common, SS830305
Molland Cross, nr Brayford, SS712333
Monckswood, Wembury, SX539492
Monkleigh, SS455208
Monkokehampton, SS583055
Monks Withecombe, SX696893
Monkton, ST188031
Moor Farm, nr Peter Tavy,
Moorgate, nr Ugborough Beacon, SX677593
Moorgate, Okehampton, SS591932
Moorhaven Hospital, Bittaford, SX667575
Moorhouse Ridge, SS825310
Moorlands Farm, W Dart, SX624736
Moorshop, SX5174
Moortown, nr Gidleigh, SX665891
Moortown, nr Tavistock, SX527738
Morchard Bishop, SS770077
Morchard Bishop (Watcombe), SS784064
Morchard Bishop (Weeke), SS761064
Morchard Rd Stn, Down St Mary, SS750051
Morchard Road, SS571049
Morebath, SS955250
Moreleigh, SX767528
Moreton, nr Holsworthy, SS285085
Moretonhampstead, SX753861
Morganhayes Covert, nr Seaton, SY220925
Morte Point, SS440456
Mortehoe, SS460455
Mortehoe-Lee(via inland route),
Morwellham, SX445697
Mothecombe, SX609478
Mountsland, nr Rippon Tor, SX759740
Mouth Mill, nr Clovelly, SS298267
Muddiford, SS563383
Muddilake, nr Two Bridges, SX623753
Mullacott Cross, Ilfracombe, SS512444
Murchington, nr Chagford, SX694881
Musbury, SY274946
Mutter's Moor, SY104879
Mutterton, ST034052
Myrtleberry Cleave, SS735488
N Devon Link Rd (A361),
Naker's Hill, SX640689
Narrator Plantation, SX575685
Nat Tor, SX572673
Nat Tor Down, SX6482
Nat Tor, nr Willsworthy, SX545824
Natsworthy Manor, SX7280
Nattor Down, nr Willsworthy, SX542828
Nattor, nr Sheepstor, SX571673
Neadon Cleave, Manaton, SX755818
Nether Bridge, nr Liftondown, SX348876
Nether Exe, SS937001
Netherexe Pond, Kingsteignton, SX862742
Netherton, SX892713
Nethway, nr Kingswear, SX904522
Netton Down, nr Stoke Point, SX555462
Netton Island, Newton Ferrers, SX554456
New Barn Farm (heronry), SX451619
New Bridge (Black-a-ven Brook), SX596904
New Bridge (R Creedy,NStCyres), SX881985
New Bridge (R Dart, nr Holne), SX711709
New Bridge (R Taw,nr BpTawton), SS570282
New Bridge, R Teign, SX849765
New Cross Pond, Kingsteignton, SX863737
New Waste, Cornwood, SX628612
Newbridge (R Dart, nr Holne), SX711709
Newcombe, nr Blackborough, ST102038
Newhouse Farm (trout lakes), SX738535
Newleycombe Lake, SX582698
Newport, Barnstaple, SS566322
Newton Abbot, SX713860
Newton Abbot Cemetery, SX853702
Newton Ferrers, SX545480
Newton Poppleford, SY086898
Newton St Cyres, SX8898
Newton St Petrock, SS412123
Nodden Gate, nr Lydford, SX530864
Nodden Hill, SX538870
Norsworthy Bridge, SX567695
North Bovey, SX7484
North Devon Coast, SS0000
North Devon Link Road, SS5633
North Hessary Tor, SX578742
North Hill, ST098064
North Huish, nr Diptford, SX713565
North Molton (village), SS737298
North Molton Ridge, SS778325
North Plantation, SX597733
North Pool, SX775412
North Sands, Salcombe, SX728384
North Tawton, SS664017
North Teign River valley, SX645868
North Wilborough, SX870664
North Wood, Dartington, SX785634
North Wilborough, SX870664
North Wood, Dartington, SX785634
North Wood, Shaugh Prior, SX546640
North Wyke, SX660985
Northam, SS448291
Northam Burrows, SS445304
Northcote, Honiton, ST175015
Northleigh, SY196959
Northlew, SS505992
Noss Creek, SX880531
Noss Mayo, SX548474
Noss Mayo-Stoke Pt coast path, SX5445
Nuns, SX606699
Nuns Cross, SX606699
Nurston Farm, nr Buckfastleigh, SX718639
Nymet Rowland, SS712082
O Brook, SX660711
Oakery Br, Princetown, SX595742
Oakford, SS910214
Oakford Cross, SS9121
Oakfordbridge, SS919219
Oakhay Barton, Stoke Canon, SX937982
Occombe, SX874631
Ockerton Court, SX6086
Oddicombe, SX926658
Off Cove, nr Bolt Head, SX715365
Offwell, SY195995
Ogwell Cross, Newton Abbot, SX713860
Oke Tor, SX613899
Okehampton, SX589950
Okehampton (Abbeyford Woods), SX589975
Okehampton area, SX589950
Okehampton Camp, SX588928
Okehampton Common, SX580908
Okehampton Woods, SX5894
Okement Hill, SX604875
Okenbury Lake, Kingston, SX641471
Old Hill, SX664629
Old Mill Creek, SX865523
Oldaport Wood, SX635495
Oldborough, nr Morchard Bishop, SS773063
Older Bridge, Nuns Cross, SX5970
Ollsbrim Cross, Corndon Down, SX687734
Orcheton Pond, SX630494
Orcheton Weir, SX628492
Orcheton Wood, SX624492
Orcombe Fields, Exmouth, SY022798
Orcombe Heights, Exmouth, SY030798
Orcombe Point, Exmouth, SY020795
Ore Stone/Thatcher Rock, SX950629
Orley Common, SX825665
Otter Est (White Bridge),
Otter Estuary, SY075824
Otter Estuary to Otterton WBS, SY076840
Otter Fields, SY073824
Otter Head,
Otter Valley,
Otterton, SY080855
Otterton (South Farm), SY079828
Otterton Cliffs, SY083823
Otterton Ledge, SY077817
Otterton Weir, SY080855
Ottery St Mary, SY099956
Outer Hope, SX676403
Overbecks, nr Salcombe, SX729374
Owley (village), nr S Brent, SX676597
Owley Farm, nr S Brent, SX677597
Owley Gate, nr Harford, SX670598
Paignton, SX895610
Pamflete Pond, Mothecombe, SX618481
Pamflete Wood, Mothecombe, SX620485
Pancrasweek, SS297058
Pancrasweek, Holsworthy, SX297058
Panson Mill Farm, SX368924
Paradise Copse, ST0101
Park Farm, SX734427
Parke, Bovey Tracey, SX808785
Parkham, SS388213
Parracombe, SS6844
Parson & Clerk, Dawlish, SX960747
Parson's Cottage, Two Bridges, SX615755
Passage House, Teign Estuary, SX880724
Pathfinder Village, SX835935
Pattesons Cross, SY096977
Payhembury, ST090017
Paynes Bridge, nr Throwleigh, SX6591
Peak Hill CP, Sidmouth, SY108870
Peak Plantation, Dartington, SX785612
Peaked Tor Cove, Torquay, SX9263
Peamore, nr Exeter, SX919880
Peartree Cross, Ashburton, SX750693
Peartree Point, nr Start, SX820366
Peat Cot, SX605715
Peat Tie Bog, Thornworthy, SX0000
Pedley Wood (heronry), SS777130
Peek Hill, nr Burrator, SX558699
Peekhill (habitation), SX545698
Penhill, SS521332
Penhill Marshes, SS530335
Penhill Point, SS517345
Penhill Marshes, SS530335
Penhill Point, SS517345
Penn Beacon, SX598629
Penn Moor, SX608637
Pennycomequick, nr Tavistock, SX515740

Penquit, Ivybridge, SX650543
Peppercombe, nr Buck's Mills, SS380244
Pepperdon Down, SX777855
Peter Tavy, SX513776
Peter Tavy Great Common, SX548771
Peterhayes Farm, nr Stockland, ST2406
Peters Marland, nr Torrington, SS478135
Petit Tor Cliffs, Babbacombe, SX927662
Petit Tor-Shaldon,
Petre's Cross, SX654655
Petrockstowe, SS513095
Pew Tor, SX532735
Piddledown Common, SX725899
Pig's Nose Valley, SX765363
Pil Tor, SX735759
Pilchard Cove, nr Strete, SX840463
Piles Copse, SX644620
Piles Hill, Ugborough Moor, SX653608
Piles Valley, Harford, SX641610
Pilton, nr Barnstaple, SS556337
Pinhay Bay, SY320907
Pitland Corner, nr Tavistock, SX472777
Pitley Hill, nr Ashburton, SX770773
Pizwell, nr Postbridge, SX668785
Place, nr Okehampton, SX568940
Plaster Down, SX515724
Plym Est (Saltram tidal pool), SX519562
Plym Estuary, SX506554
Plym Ford, SX611684
Plym Head, SX622683
Plym Steps, SX603672
Plym Valley, SX585660
Plym Valley (Colwill Wood), SX5260
Plym Woods, SX530595
Plymbridge, SX524587
Plymouth (at sea), SX4447
Plymouth (Beacon Park), SX470575
Plymouth (Budshead Wood LNR), SX462598
Plymouth (Cattedown), SX494535
Plymouth (Cattewater), SX494533
Plymouth (Central Park), SX472560
Plymouth (Chelson Meadows), SX509549
Plymouth (City Centre), SX477547
Plymouth (city), SX4656
Plymouth (Crownhill), SX490585
Plymouth (Derriford), SX460534
Plymouth (Devil's Point), SX459534
Plymouth (Devonport), SX447560
Plymouth (Dockyard),
Plymouth (Drake's Island), SX468528
Plymouth (Efford Marsh LNR), SX512569
Plymouth (Efford), SX5056
Plymouth (Elburton), SX526528
Plymouth (Estover), SX510591
Plymouth (Ford Park Cemetery), SX477560
Plymouth (Glenholt), SX505607
Plymouth (Greenbank), SX484552
Plymouth (Ham Woods), SX463577
Plymouth (Honicknowle), SX466585
Plymouth (Hooe Lake), SX500528
Plymouth (Laira), sx500557
Plymouth (Leigham),
Plymouth (Lipson), SX500557
Plymouth (Mannamead), SX4856
Plymouth (Marsh Mills), SX518562
Plymouth (Mayflower Marina), SX459538
Plymouth (Milehouse), SX467562
Plymouth (Millbay Docks), SX469537
Plymouth (Mount Batten), SX485533
Plymouth (Mount Gould), SX493550
Plymouth (Mount Wise), SX457542
Plymouth (Mutley), SX483557
Plymouth (North Prospect), SX463568
Plymouth (Oreston), SX500534
Plymouth (Pennycomequick), SX470557
Plymouth (Peverell), SX475569
Plymouth (Pomphlett), SX510538
Plymouth (Southway), SX4860
Plymouth (St Budeaux), SX4458
Plymouth (Staddon Heights), SX492516
Plymouth (Stoke), SX465556
Plymouth (Stonehouse), SX466543
Plymouth (Sutton Harbour), SX485543
Plymouth (Torpoint Ferry), SX445551
Plymouth (Turnchapel), SX495532
Plymouth (West Hoe), SX473536
Plymouth (Weston Mill), SX442571
Plymouth (Whitleigh Wood), SX476602
Plymouth (Widewell), SX499619
Plymouth (Woodland Wood LNR), SX467595
Plymouth area, SX4460
Plymouth Breakwater, SX470504
Plymouth Hoe, SX481537
Plymouth Sound, SX475520
Plympton (Triumphal Arch), SX530579
Plymouth Sound, SX475520
Plympton (Triumphal Arch), SX530579
Plympton Farm, nr Malborough, SX714386
Plympton, Plymouth, SX540565
Plymstock, Plymouth, SX517528
Plymtree, ST052028
Poltimore, SX965969
Pondsbury, sS135455
Ponsford, ST002074
Ponsworthy, SX702740
Poole Farm,
Popple's Bridge (R Yealm), SX598543
Pork Hill, SX518749
Portledge Pond, nr Bideford, SS391245
Portlemouth Down, SX745372
Portworthy, SX554602
Posbury, nr Crediton, SX814976
Postbridge, SX647789
Postbridge area, SX648789
Pottery Pond, Bovey Tracey, SX812773
Poughill, SS857085
Poundsgate, SX705722
Powder Mills, SX628769
Powderham, SX967842
Powderham (R Kenn), SX973833
Powderham Corner, SX974844
Powderham Park, SX968836
Powderham-Turf, SX963860
Powler's Piece, nr Melbury, SS372185
Prawle Area, SX773350
Prawle Point, SX773350
Prawle Point to Pig's Nose, SX762362
Prawle Pt to Start Pt, SX772349
Prawle to Lannacombe, SX785355
Preston, nr Teigngrace, SX855747
Preston, Torbay, SX896616
Prestonbury Common, SX7490
Prince Hall, SX625742
Princetown, SX587737
Princetown Sewage Works, SX587737
Pudsham Down, SX730747
Pullabrook Wood, Bovey Valley, SX792796
Pulley Ridge, SS449324
Pullhayes Fm, nr East Budleigh, SY071841
Pupers Hill, SX673675
Puslinch, nr Yealmpton, SX568510
Putford (East & West), SS364160
Putsborough Sands, SS445407
Putsborough, nr Croyde, SS447402
Pynes Weir, nr Exeter, SX918962
Pyworthy, SS313029
Quickbeam Hill, SX653648
Quintin's Man, SX621838
Quither Common, nr Lifton, SX499803
Quoditchmoor Plantation, SX4099
R Avon (water meadows), SX700473
R Axe,
R Caen, SS4834
R Clyst,
R Culme,
R Dart,
R Dart (Buckflgh-tidal limit),
R Dart Country Park, SX7370
R Erme,
R Exe,
R Exe WBS, SS953129
R Kenn, SX975833
R Lyd,
R Mardle, Buckfastleigh, SX7366
R Meavy (Burrator), SX5567
R Meavy (Burrator-Goodameavy), SX529647
R Meavy WBS, SX527680
R Okement, Okehampton, SX5895
R Plym WBS, SX533637
R Sid, SY1389
R Swincombe, SX635724
R Tamar,
R Taw,
R Teign WBS, SX781898
R Torridge,
R Torridge, Torrington, SS485187
R Walkham,
R Yarty,
R Yarty WBS, ST273000
R Yeo, SS564340
Rackenford Moor, SS858208
Rackerhayes, Kingsteignton, SX865725
Raddick Hill, SX575713
Raddick House, SX570701
Raddick Plantation, Burrator, SX5770
Radford Lake, SX505528
Radford Park, SX506529
Radford Woods, SX505528
Radford/Hooe area, SX505528
Radworthy Combe, SS693430
Ramshorn Down, SX795739
Ramsley Common, SX653928
Rasherhayes (clay pits), SX8672
Rattery, SX741615
Rattle Brook, nr Bleak House, SX561852
Rattery, SX741615
Rattle Brook, nr Bleak House, SX561852
Rattlebrook Hill, SX555855
Raven's Cove, nr Start Point, SX822367
Raybarrow Pool, SX639901
Raymond's Hill, SY325965
Rectory Wood, Drewsteignton, SX7390
Rectory Wood, Littleham, SS446231
Red Brook, nr Brent Moor, SX665626
Red Lake, SX6466
Red-a-ven Brook, SX575913
Red-a-ven Valley, SX575913
Redhill Quarry, ST0818
Redlake Brook, SX640664
Redlap, nr Stoke Fleming, SX873487
Renney Lentney, Heybrook bay, SX493488
Rewe, SX945993
Rickham, SX750377
Riddiford, SS612101
Riddlecombe, SS612139
Riddon Ridge, SX665765

Rifton Wood, nr Stoodleigh, SS900185
Rillage Point, nr Ilfracombe, SS542487
Ringleshutte Mine, SX675702
Ringmoor Down, SX570664
Ringmoor, nr Sheepstor, SX570664
Ringmore, nr Bigbury, SX652458
Ringmore, nr Shaldon, SX925722
Ringsdon Clump, Kenton, SX949835
Rippon Tor, SX747755
Rival Tor=Rippator, SX643882
Riverford Bridge (R Dart), SX773637
Rixdale Farm, Dawlish Water, sx943777
Roadford Res, SX428912
Roborough Down, SX5165
Roborough, nr Gt Torrington, SS577171
Roborough, nr Plymouth, SX503621
Rockford, SS756478
Rockham Bay, SS455460
Rodney Bay, Orcombe, SY019794
Rolle Quay (R Yeo), SS554334
Rolster Bridge, Harbertonford, SX771561
Roncombe Valley, Sidbury, SY150925
Rook Tor, SX603617
Rook Wood, nr Cornwood, SX599613
Rook, nr Cornwood, SX604608
Roos Tor, SX547767
Rosemoor, nr Great Torrington, SS5018
Rough Grey Bottom, Dunkeswell, ST154078
Rough Tor, SX606798
Round Pound, nr Kestor, SX664868
Roundball Hill, Honiton, SY156991
Roundham Head, Torbay, SX897600
Rousdon, Seaton, SY295913
Routen,
Routrundle, SX555717
Rowden Cross, nr Widecombe, SX701763
Rowtor, SX593916
Royal Hill, nr Princetown, SX616727
Ruddycleave Water, SX730739
Ruelake Pit, SX639883
Ruffwell Cross, Thorverton, SS945017
Rumleigh, SX448683
Rundlestone, SX575750
Runnage (nr Soussons), SX668792
Rushford Bridge, Chagford, SX705883
Rushford Wood, nr Chagford, SX702898
Rushlade Common, SX742732
Ryders Hill, SX6669
Saddle Bridge (Hexworthy), SX667719
Saddle Tor, SX751764
Salcombe, SX735385
Salcombe Bar, SX734371
Salcombe Estuary, SX747391
Salcombe Hill, Sidmouth, SY142880
Salcombe Mouth, SY147877
Salcombe Regis, SY148889
Salmon Pool, R. Exe, SX930907
Salt Duck Pond, SS505331
Saltern Cove, Torbay, SX896585
Salterns (East-the-Water), SS462268
Saltram Estate, SX515555
Saltram Marsh,
Saltram Park, SX515555
Saltram Wood, Saltram Estate, SX5155
Sampford Courtenay, SS632013
Sampford Peverell, ST035145
Sampford Spiney, SX535725
Sandford (Ashridge Farm), SS825063
Sandford (Beacon Cross), SS798033
Sandford area, SS828025
Sandford, nr Crediton, SS828025
Sands Copse, nr Kingsteignton, SX868758
Sandy Bay, nr Exmouth, SY035797
Sandy Cove, Watermouth, SS5548
Sandy Hole, SX614819
Sandy Hole Pass, E Dart, SX621816
Sandy Park, SX712895
Sandygate, nr Exeter, SX967915
Sandygate, nr Kingsteignton, SX868749
Sandygate, nr Exeter, SX967915
Sandygate, nr Kingsteignton, SX868749
Sandyway, SS794334
Saunton, SS455376
Saunton Beach, SS443370
Saunton Down, SS438383
Saunton Sands, SS443370
Sawmills Pond, Kingsteignton, SX861746
Scabbacombe, SX919519
Scad Brook, SX667599
Scanniclift Copse, SX842858
Scob Hill, nr Brendon, SS752467
Scobbiscombe, Kingston, SX630469
Scoble, nr South Pool, SX759398
Scorhill Circle, SX655875
Scorhill Down, SX654873
Scorriton Down, SX687684
Scorriton, nr Holne, SX704685
Scouts Hut Wood, nr Gutter Tor, SX581673
Seale Hayne, Newton Abbot, SX827731
Season Point, Wembury, SX527478
Seaton, SY252901
Seaton (tramline), SY252940
Seaton Hole, SY235896
Seaton Marshes, SY252905
Sequers Bridge (Erme), SX632519
Seven Lords Lands, nr Haytor, SX742763
Seven Stones Cross, nr Bigbury, SX658490
Shaden Moor, Shaugh Prior, SX548636
Shaldon, SX935720
Shaldon Ness, SX9472
Shallowford, SS713450
Shallowford & Thornworthy Coms, SS7144
Shapley, SX698825
Shapwick Quarry, SY313918
Sharkham Point, Brixham, SX937545
Sharp Tor, nr Castle Drogo, SX728898
Sharp Tor, nr Dartmeet, SX685729
Sharp Tor, nr Harford, SX649617
Sharp Tor, nr Lydford, SX550848
Sharper's Head, SX786357
Sharpham House, nr Totnes, SX825578
Sharpham Marsh, nr Totnes, SX818585
Sharpham Wood (heronry), SX833572
Sharpitor, nr Bolt Head, SX730365
Sharpitor, nr Burrator, SX557702
Shaugh Beacon, SX559632
Shaugh Br-Bickley Br (3km), SX533637
Shaugh Br-Legis Tor (4.4km), SX555645
Shaugh Bridge, Shaugh Prior, SX533637
Shaugh Moor, SX5663
Shaugh Prior, SX540633
Shavercombe Brook, SX598658
Shavercombe Head, SX605653
Shebbear, SS440093
Sheeps Tor, SX566682
Sheepstor (village), SX560677
Sheepwash, SS486063
Sheepwash, nr Hatherleigh, SS487063
Sheldon, ST120086
Shell Jetty, SS480327
Shell Top, SX598638
Shelly NR, nr South Zeal, SX655939
Shelstone Tor, SX557898
Sherberton Common, SX6973
Sherberton Farm, SX646735
Sherford, SX778443
Sherford, nr Frogmore, SX778443
Sherpa Marsh, SS487350
Sherpa Pond, SS487350
Shiddicombe Wood,
Shilley Pool, SX651912
Shillingford St George, SX904879
Shilstone, SX703907
Shilstone Moor, Drewsteignton, SX699906
Shilstone Tor, Throwleigh, SX659902
Shipley Bridge, nr S Brent, SX681629
Shipload Bay (Hartland), SS247275
Shipload Bay, Hartland, SS245275
Shircombe Brake, SS843305
Shoalstone Point/Beach, SX936569
Shobrooke Lake, SS855015
Shobrooke Park, SS855015
Shortacombe (Exmoor), SS766340
Shortwood Common, E Budleigh, SY054839
Shovel Down, SX657857
Shute, SY253975
Shuttamoor, SX8282
Shutterton, SX969789
Shuttleton Common, ST127114
Sidborough NR, Morchard Bishop, SS774061
Sidbury, SY140916
Sidford, SY135900
Sidmouth, SY125875
Sillery Sands, Lynton, SS736497
Sillick Moor, SS340007
Silverton, SS957028
Site 2174,
Sittaford Tor, nr.Fernworthy, SX634831
Skaigh Valley, R Taw, SX6293
Sittaford Tor, nr.Fernworthy, SX634831
Skaigh Valley, R Taw, SX6293
Skaigh Warren, SX635935
Skern, SS455310
Skerraton, nr Harbourne Head, SX705647
Skir Gut, SX648707
Slapton (Dere Bridge), SX820441
Slapton (Lower Ley),
Slapton (Marsh Lane), SX816445
Slapton (sea),
Slapton (Upper Ley),
Slapton (village), SX8245
Slapton Ley, SX824431
Slapton Sands, SX830444
Slocombeslade, SS775470
Small Brook, SX625905
Smallacombe (Exmoor), SS815290
Smallacombe Bridge, SS815290
Smallacombe Rocks, SX754783
Smallbrook Head, SX685647
Smallbrook Plains (Avon Dam), SX687650
Smallhanger Waste, SX580592
Smeardon Down, SX523780
Smeatharpe, ST195105
Smith Hill, SX632759
Smugglers Cove, SX936710
Snapes Point, SX746393
Snider Park Plantation, SX663737
Snowdon, SX669684
Soar area, SX713373
Soar Mill Cove, SX697375
Soar Mill valley, SX697378
Sorley, nr Kingsbridge, SX732469
Sourton, SX534903
Sourton Down, SX544919
Sourton Quarry, SX524897
Sourton Tors, nr.Meldon, SX544899

Soussons Down, SX677796
Soussons Marsh, SX688788
Soussons Plantation, SX677795
South Allington, SX793387
South Brent, SX798602
South Devon area,
South Hams area,
South Hessary Tor, SX597723
South Huish Marsh, SX680415
South Langston, nr Kingston, SX648484
South Milton (village), SX699428
South Milton Ley, SX685422
South Milton sands, SX676415
South Milton/South Huish area, SX685422
South Molton, SS712259
South Moor, Northlew, SX530973
South Pool, SX776404
South Pool - Ford, SX765397
South Pool Creek, SX760395
South Tavy Head, SX595820
South Tawton, SX653945
South Tawton Common, SX645914
South Wonford, nr Holsworthy, SS380088
South Zeal, SX651935
Southcombe, nr Widecombe, SX714763
Southcott, nr East-the-Water, SS469275
Southdown valley, Soar, SX697378
Southdown, nr.Brixham, SX923538
Southdown, nr.Malborough, SX700385
Southleigh, nr.Seaton, SY205935
Southway Barton, Culmstock, ST090134
Southwood Farm, Teignmouth, SX934758
Sparkwell, SX582579
Speke's Mill, Hartland, SS225237
Spinster's Rock, SX703908
Spitchwick, Poundsgate, SX708725
Splatford, nr Kennford, SX9085
Spreacombe, nr Braunton, SS485414
Spreyton, SX699968
Spriddlestone, nr Brixton, SX540518
Squabmoor, SY0484
St Anchorite's Rock, sx591473
St Anns Chapel, SX664472
St Cyres Hill, ST146025
St Davids, Exeter, SX010933
St Giles in the Wood, SS535190
St Giles on the Heath, SX360902
St Mary's Bay, Brixham, SX932553
St Saviours Bridge,
Staddiscombe, SX513513
Staddon Point, SX486506
Stafferton,
Stafford Moor, SS595112
Stafford Moor Fishery, Dolton, SS515112
Staffords Bridge,nr Upton Pyne, SX923963
Stag's Head, nr S Molton, SS675278
Stall Moor, SX625640
Stalldown, SX635623
Standon Hill, SX557816
Stanlake Brook, SX570706
Stanlake Plantation, SX569706
Stannon Tor, SX645811
Staple Tors, Merrivale, SX541756
Stannon Tor, SX645811
Staple Tors, Merrivale, SX541756
Starcross (Exe Estuary), SX977819
Starcross (Golfing Range), SX975812
Starcross (village), SX977819
Starehole Bottom to W Bolberry, SX697375
Starehole Valley, SX725366
Start area, SX822372
Start Bay, SX8343
Start Farm, SX818375
Start Point, SX8337
Statts Bridge, SX667805
Statts House Hill, SX622825
Statts Marsh, SX6680
Staverton, SX794641
Staverton Bridge, SX784637
Stedcombe, nr Axmouth, SY264919
Steeperton Tor, SX619888
Steeple Cove, nr Bolt Head, SX704367
Steer Point, Yealm Estuary, SX545498
Stenga Tor, SX567881
Steps Bridge, SX804884
Steps Bridge Woods, Dunsford, SX804884
Sterridge Valley, SS555448
Stewdon Moor, SS487018
Stibb Cross, SS430148
Sticklepath, Barnstaple, SS552327
Sticklepath, Okehampton, SX640941
Stiddicombe, SX677450
Stingers Hill, SX632862
Stockers Clay Quarry, Hemerdon, SX572587
Stockland, ST245046
Stockleigh Pomeroy, SS879036
Stoke Beach, SX564461
Stoke Canon, SX937978
Stoke Canon-Rewe (R Culm), SX945985
Stoke Fleming, SX863485
Stoke Gabriel, SX847575
Stoke Gabriel (mill pond), SX847575
Stoke Gregory,
Stoke Point, SX562458
Stoke Pt to Gara Pt, SX535462
Stoke Rivers, SS6335
Stoke Woods, nr Exeter, SX930962
Stokeinteignhead, SX916705
Stokeley Barton Pond, SX817427
Stokeley Pond, SX818427
Stokeley Pond, Torcross, SX816427
Stokenham, SX808428
Stone Tor, SX648858
Stoneyford, Colaton Raleigh, SY064887
Stony Cross, Alverdiscott, SS514256
Stoodleigh, SS923188
Stover CP, SX835752
Stowford, Colaton Raleigh, SY060871
Stowford, Ivybridge, SX641570
Stowford, Sidmouth, SY114898
Straight Point, SY038795
Stretchford, Buckfastleigh, SX763643
Stretchford, nr Staverton, SX761644
Strete Gate, nr Slapton, SX835455
Strete Ralegh, nr Whimple, SY047954
Strete, nr Slapton, SX8447
Stumpy Post Cross, SX743474
Summerwell Plantn, Hartland, SS280207
Sutcombe, SS347117
Swale Cove, SX558460
Swanton GC,
Swell Tor, SX560733
Swelltor Quarries, SX560733
Swimbridge, SS6230
Swincombe (Exmoor), SS695415
Swincombe Farm, SX642725
Swincombe Intake Works, SX632718
Swincombe Valley, SX635723
Swincombe, nr Hexworthy, SX640726
Swinesbridge Weir, Tiverton, SS949139
Sydenham (nr Lifton), SX428839
Sydenham Damerel, SX409760
Syon Abbey, nr South Brent, SX725612
Talaton, SY068996
Tamar Bridge (Plymouth), SX435588
Tamar Estuary, SX435635
Tamar Lakes, SS290115
Tamar/Tavy Estuary, SX4461
Tamerton Creek, SX452605
Tamerton Foliot, SX471613
Tamerton Lake, SX460603
Tavistock, SX482744
Tavistock (kelly College), SX488752
Tavy Cleave, SX554830
Tavy Estuary, SX461630
Tavy Head, SX5982
Tavy Hole, SX582817
Tavy Rly Bridge, SX450615
Taw Estuary, SS468318
Taw Estuary (Ashford), SS525348
Taw Estuary (Broad Sands), SS568476
Taw Estuary (Chivenor), SS507346
Taw Estuary (Crow Point), SS466319
Taw Estuary (Chivenor), SS507346
Taw Estuary (Crow Point), SS466319
Taw Estuary (Foxhole), SS519349
Taw Estuary (Heanton Court), SS513349
Taw Estuary (Isley), SS490328
Taw Estuary (Pottington), SS550333
Taw Estuary (White House),
Taw Head, SX609859
Taw Marsh, SX620905
Taw/Torridge Estuary, SS468318
Tawny Wood NR, nr Dalwood, ST2403
Tawstock, SS555298
Tawton Gate, SX680849
Tedburn St Mary, SX815941
Teign Est (Flow Point), SX912727
Teign Estuary, SX920725
Teign-e-ver Bridge, SX654871
Teigncombe, nr.Chagford, SX672872
Teigngrace, SX854743
Teignhead, SX615839
Teignhead clapper bridge, SX639845
Teignhead Farm, SX635844
Teignmouth, SX9473
Teignmouth Golf Course, SX918758
Templeton, SS887140
Ter Hill, SX642707
Territory 5K SE,
Territory 5K W,
Territory G6, Bigbury Bay,
Tetcott, SX335965
Thatcher Rock/Ore Stone, SX944629
The Henroost Mine, SX652711
The Mounts, nr Kingsbridge, SX757488
The Old Quay,
The Sheepfold, SX644808
The Warren, Kingswear, SX897499
The Warren, Noss Mayo, SX529469
The Warren, Soar, SX708368
Thorn, SX683902
Thorn Hill, SS727437
Thornbury, SS400084
Thornbury, Holsworthy, SS400085
Thorndon Cross, nr Okehampton, sx531940
Thorne Moor, Stibb Cross, SS412162
Thornehillhead, Stibb Cross, SS414165
Thornworthy (Dartmoor), SX671849
Thornworthy (Exmoor), SS710458
Thornworthy Tor (Dartmoor), SX664852
Thorverton, SS924020
Three Barrows, SX653626
Three Beaches,
Three Boys (standing stone), SX6585
Throwleigh, SX668908

Throwleigh Common, SX655905
Thurlestone, SX676424
Thurlestone Bay, SX668418
Thurlestone Marsh, SX677425
Tidwell Mount, E Budleigh, SY063836
Tipton St John, SY090917
Titchberry, nr Hartland Point, SS244270
Titcombe Wood, nr Loddiswell, SX734501
Tiverton, SS960125
Tiverton (Amory Park), SS9512
Tiverton (Exe Salmon Ponds), SS948135
Tiverton (People's Park), SS9512
Tiverton (R Exe), SS954125
Tiverton area, SX9612
Tiverton Jnctn (old rly stn), ST033114
Tiverton Parkway (Rly Stn), ST045139
Tolcis Farm, nr Axminster, ST277008
Tongue End, SX624948
Top Tor, SX736763
Topsham, SX967871
Topsham Bridge, nr Loddiswell, SX733511
Topsham to Turf, SX964870
Tor Rocks Quarry, SX641592
Tor Royal, SX603733
Tor Wood, Erme Estuary, SX6248
Tor Wood, Sourton, SX535892
Torbay, SX9162
Torbryan, nr Ipplepen, SS820668
Torcross, SX823420
Torgate Woods, Princetown, SX599731
Torgate, Princetown, SX599731
Torquay (harbour), SX917634
Torquay (Ilsham Valley), SX937635
Torquay (town), SX915645
Torquay harbour to Daddyhole,
Torr, SX617606
Torr Wood , nr Kingston, SX625487
Torre Abbey & Livermead,
Torre Abbey & Torquay Harbour,
Torre Abbey - Broadsands,
Torre Abbey Sands, SX910636
Torridge (Beam weir), SS475206
Torridge Bridge,
Torridge Est (Cleave), SS457285
Torridge Est (Westleigh), SS4628
Torridge Est (Zeta Berth), SS468293
Torridge Estuary, SS468282
Torridge Est (Zeta Berth), SS468293
Torridge Estuary, SS468282
Torridge/Hartland Forest, SS2820
Torrington, SS4919
Torrington Commons, SS485195
Torrs, nr Ilfracombe, SS507477
Totnes (Brutus Br - Weir), SX806610
Totnes (Castle), SX799606
Totnes (Snipe Island), SX805612
Totnes (town), SX803603
Totnes Cross, Halwell, SX780542
Totnes Weir, SX800612
Tottiford Res, SX810834
Tower Hill, SX370904
Tracey, nr Honiton, ST151016
Trago Mills, nr Bovey Tracey, SX825745
Trenchford Res, SX805825
Trendlebere Down, SX773795
Trendlebere Res, SX773795
Trentishoe, SS647487
Trentishoe Down, SS633478
Trentworthy, SS289150
Trew's Weir, Exeter, SX924915
Trinity Hill, nr Axminster, SY307957
Triss Combe, SS815295
Trowlesworthy Tor, SX578645
Trowlesworthy Warren, SX575647
Trusham, SX855822
Tuckermarsh, nr Bere Alston, SX446676
Tunhill Rocks, SX7375
Turf Hotel, SX965862
Turners Wood Garden NR, SS438272
Twitchen, SS790305
Two Bridges, SX608750
Two Pots, SS532446
Tytherleigh, ST318038
Uffculme, ST069128
Ugborough (village), SX677557
Ugborough Beacon, SX667592
Ugborough Moor, SX655620
Ugbrooke, nr Chudleigh, SX875780
Umberleigh, SS608237
Umborne, SY231972
Uncle Abs House, SX656639
Undercliffs NNR, nr Rousdon, SY297901
Underdown, nr Dunchideock, SX888857
Up Exe, nr Thorverton, SS942024
Uphams Plantation, SY045865
Uplowman, nr Tiverton, ST013155
Uplyme, SY325935
Upottery, ST203077
Upper Tamar Lake, SS285120
Upton Hellions, nr Crediton, SS843033
Upton Pyne, SX911977
Urgles, Wigford Down, SX535649
Valley of Rocks (Lynton), SS705495
Valley of Rocks - Hunter's Inn, SS673495
Valley of the Rocks (Meldon), SX0000
Velator, SS485357
Vellake, SX555905
Venford Moor, SS862295
Venford Res, SX685710
Venn Ottery, SY081913
Venn Ottery Common, SY066916
Venn, nr Morchard Bishop, SS776057
Vennbridge Farm, SX961815
Venngreen, nr Holsworthy, SS377112
Venny Tedburn, nr Crediton, SX822975
Vention, nr Croyde, SS454414
Venton, Whiddon Down, SX695910
Vergyland Combe, SX596870
Virginstow, SX377927
Vitifer, SX671803
Vixen Tor, SX542742
Vogwell, nr Manaton, SX725817
Volehouse Moor Farm, SS340162
Wadbrook Farm, nr Axminster, ST3201
Wadbrook, nr Axminster, ST325018
Wadden Brakes, Haldon, SX895800
Waddeton, nr Churston, SX874568
Wagg's Plot, Axminster, ST315013
Walkham Bridge,
Walkham Head, SX5881
Walkham Valley, S of Merrivale, SX545740
Walkhampton (village), sx533697
Walkhampton Common, SX573723
Wall Park (heronry), SX755408
Walla Brook, SX675760
Wallaton Cross, SX793483
Walls Hill, Torquay, SX935651
Warborough Plantation, SX960823
Ware, nr Lyme Regis, SY326917
Warleigh Marsh, SX465625
Warleigh Point, SX444610
Warleigh Woods, SX448610
Warmore House Bridge,Exebridge, SS935260
Warren Copse, nr Dawlish, SX971777
Warren Hill, nr Huntsham, ST978222
Warren House area, SX675810
Warren House/Bennett's Cross, SX675810
Warren House area, SX675810
Warren House/Bennett's Cross, SX675810
Warren Point, Ernesettle, SX443606
Warren Point, Thurlestone, SX668420
Warren Point, Yealm Estuary, SX554507
Washbourne, SX798547
Washfield Weir, nr Tiverton, SS944160
Washford Pyne, SS812118
Wat Combe, SS787474
Wat Combe Water, SS7847
Watcombe, nr Torquay, SX925675
Water Cleave, SX7681
Water Hill, SX671813
Waterhead, nr South Pool, SX768387
Waterleat, nr Ashburton, SX751722
Watermouth, nr Ilfracombe, SS555483
Watern Combe, SX624867
Watern Oke, SX566836
Watern Tor, Upper N.Teign, SX629868
Watersmeet, nr Lynmouth, SS743486
Watervale, Lydford, SX518837
Way Down, nr Chagford, SX685892
Waytown, Broadhempston, SX793673
Weare Giffard, SS475222
Weatherdon Hill, SX651589
Webbery Wood, SS5026
Wedlake, nr.Peter Tavy, SX537774
Week Ford, Hexworthy, SX662724
Weir Quay, SX433650
Welcombe, SS228184
Welcombe Mouth, SS213180
Wellsfoot Island, SX7070
Wellswood, Torquay, SX930641
Welltown, nr Walkhampton, SX541701
Welsford Moor, SS275207
Welstor Common, SX738731
Wembury, SX517485
Wembury (HMS Cambridge), SX502481
Wembury (sea), SX5147
Wembury Beach, SX515484
Wembury Cliff, SX525481
Wembury coast; Plym to Yealm, SX502481
Wembury Point, SX502481
Wembury to Heybrook, SX505485
Wembury Wood, SX540503
Wembworthy, SS663099
West Alvington, SX723438
West Anstey Common, SS850293
West Buckland, nr Barnstaple, SS658314
West Charleton, SX754426
West Charleton Bay, SX748415
West Charleton Marsh, SX749417
West Cleave, Belstone, SX608941
West Dart (Intake), SX608780
West Dart Head, SX601816
West Dartmoor area,
West Down, SS508420
West Down, Budleigh Salterton, SY045814
West Down, Tavistock, SX480705
West Glaze Brook, SX663608
West Hill, SY065935
West Lyn Valley, SS720485
West Mill Tor, SX588910
West Ogwell, SX822700
West Okement Valley, SX565890
West Paignton (BBS transect), SX8760
West Prawle, SX768377
West Putford, SS359158
West Raddon, nr Shobrooke, SS893023

West Rook, SX604614
West Webburn Valley, SX688775
West Withecombe, SX698901
West Woods, nr Puslinch, SX555503
West Yelland, SS475322
Westacott, nr Barnstaple, SX5832
Westcombe Beach, SX633458
Westcombe Valley, SX637460
Westcott, nr Cullompton, ST020043
Westcott, nr Moretonhampstead, SX789869
Westerland Valley, SX864615
Western Beacon, Bittaford, SX655576
Western Common, SS732365
Westhay Farm, nr Hawkchurch, ST3500
Westlake, SX623537
Westleigh (Torridge Est), SS468294
Westleigh, nr.Tiverton, ST062172
Weston Cliff, nr Branscombe, SY172882
Weston Combe, nr Branscombe, SY162884
Weston Ebb, SY175877
Weston Mouth, SY164879
Weston Mouth-Branscombe, SY180882
Westward Ho!, SS431292
Westweekmoor, SX408935
Weycroft, SY9930
Whacka Tor, Hickley Plain, SX663621
Wheal Jewell Res, Mary Tavy, SX503796
Wheathill CP, E Bud Common, SY040847
Wheatley, SX894916
Whetcombe, SX846816
Whetcombe Barton Quarry, SX846816
Whetcombe, SX846816
Whetcombe Barton Quarry, SX846816
Whidborne Copse, Torquay, SX937635
Whiddon Down, SX691925
Whiddon Park, Drewsteignton, SX722893
Whiddon scrubs, Ashburton, SX754718
Whiddon Valley, Barnstaple, SS579316
Whiddon, nr Beaworthy, SX475996
Whimple, SY270905
Whimple, nr Ottery St Mary, SY0497
Whipcott, nr Westleigh, ST072185
Whitchurch Common, Merrivale, SX535747
Whitchurch Down, Tavistock, SX502735
Whitcombe, Kenn, SX928843
White Barrows, SX665652
White Down Copse (NT), ST005015
White Horse Hill, SX6185
White House, Taw Estuary, SS469331
White Moor, SS815300
White Ridge, SX648822
White Tor, SX542787
Whitechurch Cove (Bolt Tail), SX672391
Whitecleaves Quarry (Buckfstl),
Whitefield Down, Muddiford, SS5639
Whitehill, nr Stoke Gabriel, SX858587
Whitehorse Hill, SX616854
Whiteleigh Meadow, SS417028
Whitemoor Marsh, SX651891
Whiterocks Down, SS875296
Whitestone (Dartmoor), SX508785
Whitestone, nr Tedburn St Mary, SX868936
Whiteworks, SX612710
Whitford, nr Axminster, SY260957
Wickeridge, nr Ashburton, SX787697
Widecombe in the Moor, SX718768
Widewell, nr Torcross, SX815421
Wigford Down, SX545648
Wiggaton, Ottery St Mary, SY102937
Wild Goose, Poundsgate, SX7072
Wild Tor, SX624877
Wildbanks Hill, SX618812
Wildtor Well, SX627876
Willand, ST035105
Willey, nr South Tawton, SX644955
Willingcott, SS485432
Willingstone Rock, SX756888
Willsworthy, SX535818
Willsworthy Cross, Kenn, SX937850
Wilmington, SY212998
Wilton, nr South Pool, SX775393
Wind Tor, SX709757
Winkleigh, SS633080
Winney's Down, SX621826
Winneys Down Brook, SX629817
Winslade, SS385190
Wiscombe Park, ST188928
Wishford Fen, SX9995
Wistlandpound Res, SS645417
Wistman's Valley, SX6177
Wistman's Wood, SX613773
Withecombe, nr Chagford, SX699893
Witherdon Wood, nr Roadford Re, SX430959
Witheridge, SS805144
Witheridge Moor, SS859150
Withleigh, nr Tiverton, SS915127
Withycombe Raleigh Common, sy035830
Wollake, SX629669
Wollaton Plantation, SX561529
Wolleigh Bridge, SX799797
Wonwell Beach, Erme Est, SX617472
Woodbarrow Hangings, SS712428
Woodbridge, Farway, SY189950
Woodbury, SY0187
Woodbury Castle, SY034874
Woodbury Common, SY035868
Woodbury Salterton, SY013890
Woodcock Hill, SX5587
Woodcombe Farm, E Prawle, SX787373
Woodcombe Valley, E Prawle, SX787373
Woodhead, nr Branscombe, SY204901
Woodland, SX791688
Woodleigh, SX738488
Woodrow, SX928970
Woodrow Barton, SX920968
Woody Bay, SS679493
Woolacombe, SS455438
Woolacombe Down, SS464420
Woolacombe Sand, SS4542
Woolfardisworthy, nr Crediton, SS828087
Woolfardisworthy, nr Hartland, SS332210
Woolsery area, SS3322
Woolston, nr Kingsbridge, SX717417
Wooston, nr Mardon Down, SX764890
Worlington, nr Instow, SS305480
Worlington, nr Witheridge, SS723135
Worston,
Wotter, SX558618
Wrafton, nr Braunton, SS494356
Wrangaton, nr Bittaford, SX675580
Wrescombe, nr Puslinch, SX565502
Wrangaton, nr Bittaford, SX675580
Wrescombe, nr Puslinch, SX565502
Wrigwell, nr Denbury, SX812715
Wringapeak (Woody Bay), SS673495
Wringcliff Bay, SS701498
Yadsworthy, nr Cornwood, SX632608
Yalberton, SX864586
Yalberton Tor, SX868592
Yar Tor, SX678740
Yarcombe, ST245082
Yardworthy, nr Chagford, SX679851
Yarner Wood NNR, SX780787
Yartor Down, SX677732
Yarty Valley, nr Membury, ST265015
Yealm Estuary, SX545495
Yealm Falls (Steps), SX616639
Yealm Head, SX615647
Yealmbridge, SX5952
Yealmpton, SX579516
Yealmpton (Furzehill), SX583536
Yealmpton (Stoneycross), SX580526
Yealscombe, SS793463
Yelland, SS495320
Yelland (Home Fm Marsh), SS497330
Yelland (powerstation), SS482325
Yelland Pond, SS495320
Yellowford, nr Thorverton, SS925008
Yellowmead Down, SX570683
Yelverton, SX525678
Yennadon Down, SX548687
Yeo Bridge, SS428231
Yeo Valley, nr Barnstaple, SS6035
Yeoford, SX785990
Yes Tor, SX582902
Yettington, SY054857
Youlditch, nr Meldon, SX553921
Youlstone Plantation, SS273164
Zeal Gulley, nr Avon Dam, SX684648
Zeal Hill, nr Shipley Bridge, SX687635
Zeal Monachorum, SS7204

GUIDANCE ON RECORD SUBMISSIONS

All records for the previous year should be submitted by 31 January at the latest. However, the Recorder encourages records to be sent in **at any time of the year, ideally on a regular basis**. This not only enables interesting records to be extracted for the 'Bird News' in *The Harrier*, but also helps to avoid the end-of-year bottleneck in transferring records to the database, which in turn can delay the production of the final report.

Basic guidance on how to submit records, and on the sort of records required for each species, is given in *A Checklist of the Birds of Devon* (Rosier, 1995), and more general information is available in the *BTO Bird Recording Handbook* (Dudley, 1993).

If submitting records on paper, please try to use the new, revised, **standard recording forms** (or photocopies), which are available from the Recorder. Records can also be accepted in **electronic form** (in a Microsoft **Excel** spreadsheet on diskette or via email) and this is now the preferred method; please contact either the Recorder or Data Manager for details on acceptable formatting.

The format for record submissions is flexible in that records can be accepted in any sequence as the computer will sort them out, but please don't forget to give your details (name, address and telephone number) and provide the following information for each record:

Date: it is important to accurately date **all** sightings (use Start and End Date and Start and End Time if appropriate).

Species: use standard English names from *DBR* (can be abbreviated using standard BTO codes, but the full English name must be included with the first use of the code to avoid any confusion/error). The new DBWPS database is now sorted using Euring code numbers and these can be used as well if known.

Location: use standard names from the Site Gazetteer published in the latest *DBR* wherever possible, but **an accurate OS grid reference is now essential** and should include the 2 letter grid code, followed by 2, 3 or 4 figure eastings and then 2, 3 or 4 figure northings.

Number: be as accurate as possible, and exercise caution by giving *minimum* numbers, or approximate *limits* such as c.1000 or 1000+ and make it clear if the number is an estimate rather than a count.

Observations: age/sex, flight direction, habitat, unusual behaviour, weather etc; in some cases it will also be useful to record the *time* to help sort out the real number of birds in situations with multiple sightings at different locations.

DBWPS is currently redesigning all its databases and data capture and reporting forms for both paper and electronic recording. Please contact the Data Manager or the county recorder for the most up to date forms and recommended formats.

Rare and scarce species, and/or those which are difficult to identify, require adequate supporting evidence in order to be accepted. Full descriptions of all **national rarities** should be forwarded to the Recorder, who will pass them on to the British Birds Rarities Committee. Ideally these rarities should be submitted on special forms obtainable from the Recorder or the BBRC Secretary. For **county rarities (lists A and B)**, supporting details must be submitted on a separate sheet and forwarded to the County Recorder. For a complete list of all Devon Birds, with their Devon status, please refer to the full Devon County List at the end of this report.

Supporting details may also be required for uncommon subspecies, out of season migrants and species outside their normal range, e.g. on Lundy.

WETLAND BIRD SURVEY REPORT FOR DEVON 2001

Phil Stidwill

WeBS is a monitoring scheme for non-breeding waterbirds in the UK and aims to provide the principal data for the conservation of their populations and wetland habitats. The main objectives are to assess the size of non-breeding waterbird populations in the UK, to assess trends in their numbers and distribution and to assess the importance of individual sites for waterbirds. Those results also form the basis for informed decision-making by conservation bodies, planners and developers and contribute to the sustainable and wise use and management of wetlands and their dependent waterbirds.

WeBS is a partnership scheme of the British Trust for Ornithology (BTO), the Wildfowl & Wetlands Trust (WWT), Royal Society for the Protection of Birds (RSPB) and Joint Nature Conservation Committee (JNCC). Synchronised counts are conducted once per month primarily from September to March, but if possible, all waterbirds are counted every month throughout the year so that the survey can build up a population index for breeding and migratory birds as well as winter populations. The national results are summarised in annual reports, the latest available being *The Wetland Bird Survey 1999-2000, Wildfowl and Wader Counts* by Andy Musgrove *et al*., published by the Wildfowl & Wetlands Trust in 2001. Data for 2001 will not be available until the report due in 2002 and will then only cover January to March as the WeBS year runs from April to March.

In Devon, the Exe estuary remains our only site of *international importance* as it regularly exceeds the qualifying level of 20,000 wildfowl and waders. The rank order of the other estuaries is given in Table 122 of *DBR* 2000. The sites counted in 2001 are listed in Table 122, and the account that follows is a summary of Devon WeBS data based on 2001 counts.

Table 122. Sites counted in 2001 (observers' initials in brackets)

Estuaries		**Reservoirs & Lakes**			
Avon	(RWBo)	Arlington	(DP)	Beesands	(GB)
Axe	(DEDC)	Bicton Lake	(JWy)	Burrator Res	(JBp)
Dart	(SO'D,BAP)	Clennon Valley Lake	(SO'D)	Decoy Lakes	(NIH)
Erme	(H&JH)	Fernworthy Res	(MGM)	Elfordleigh Ponds	(RHW)
Exe	(DJP *et al*)	Hennock Res	(DJP)	Huntsham	(RGd)
Kingsbridge	(GECW *et al*)	Kenwith NR	(DC)	Kitley Pond	(AJP)
Otter	(GV)	Lopwell Dam	(PFG)	Meldon Res	(GAV)
Plym	(PAS)	Portworthy Dam	(SM)	Radford Lake	(PFG)
Tamar [1] & Tavy	(GlG *et al*)	Roadford Res	(AWGJ *et al*)	Shobrooke Lakes	(CL)
Taw/Torridge	(AJV *et al*)	Slapton Ley	(RiA)	Stover CP	(JDA)
Teign	(NIH)	Tamar Lakes	(MSS)	Tiverton Canal	(PC)
Yealm	(AJP)	South Huish area [2]	(PJR)	Venford Res	(RCh)
		Wistlandpound Res	(CPr)		

[1] *Only counts for Tamar Estuary above the Tamar Bridge (sections 1-5 including parts of both Devon and Cornwall) are included in this report*

[2] *South Huish area = Thurlestone Marsh, South Milton Ley and South Huish Marsh*

Wildfowl (including divers, grebes, cormorants, herons and rails)

September saw the start of the winter influx coinciding with the relaxation of the access restrictions in some areas. The greatest number was present in December with just over 18,500 wildfowl counted on Devon wetlands.

Despite the restrictions, **Little Egret**, which peaked at 405 in September, was recorded in all but May and outnumbered **Grey Heron** in all but May and June. The increase in numbers of Mute Swans was sustained with four counts in excess of 200 birds. Nationally, **European White-fronted Goose** numbers have been steadily dwindling during the 1990s and a low count of nine birds in Devon is not unexpected. Naturalised **Canada Geese** are, in contrast, faring much better, with a peak of almost 3,000 birds in September again up on the previous year. **Dark-bellied Brent Geese**, whose numbers nationally have been affected by a continued run of poor breeding success, peaked at 1314 birds, down on last year. **Shelduck** counts in Devon continue to fall, but **Pintail** numbers increased slightly. **Mallard** numbers in Devon remained fairly constant.

A total of 36 species of ducks, swans and geese and 15 other species were recorded on the WeBS counts during 2001. The monthly totals of the main species are given in Table 123.

Species occurring on WeBS counts in small numbers (and not listed in Table 123) were: **Black-throated and Great Northern Divers**, **Black-necked Grebe**, **Black Swan**, **Whooper Swan**, **Bar-headed Goose**, **Snow Goose**, **Pink-footed Goose**, **Egyptian Goose**, **American Black Duck**, **Ruddy Shelduck**, **Marbled Duck**, **Ringed-necked Duck**, **Red-crested Pochard**, **Scaup and Smew**. (See Systematic List for further details).

Table 123. Summary of WeBS counts for the main 'wildfowl' species in Devon in 2001 (including divers, grebes, cormorant, herons and rails)

Species	Jan	Feb	Mar	Apr	May	Jun	Jul	Aug	Sep	Oct	Nov	Dec
Little Grebe	107	100	4	-	-	-	2	12	29	61	76	103
Great Crested Grebe	33	37	2	2	2	4	4	5	10	63	55	33
Slavonian Grebe	6	2	-	-	-	-	-	-	-	-	-	-
Cormorant	332	278	4	6	1	7	11	224	193	432	460	329
Shag	3	12	-	-	-	-	-	3	60	-	3	46
Little Egret	125	167	1	5	-	4	21	268	405	254	268	136
Grey Heron	90	87	-	2	2	10	16	44	154	135	111	93
Spoonbill	1	8	-	-	-	-	-	-	-	3	6	2
Mute Swan	156	161	4	6	4	36	31	167	210	242	255	240
White-fronted Goose	1	-	-	-	-	-	-	-	-	-	-	9
Greylag Goose	4	4	-	-	-	-	-	1	1	3	2	5
Canada Goose	2499	1065	17	23	45	72	175	1142	2985	2272	2541	2304
Barnacle Goose	7	3	-	-	-	-	-	2	-	1	-	1
Dark-bellied Brent	978	1173	-	-	-	-	-	1	53	1102	1314	871
Light-bellied Brent	13	4	-	-	-	-	-	-	8	-	-	1
Shelduck	1134	1188	-	32	33	18	54	51	126	123	478	879
Mandarin	1	2	2	-	-	-	4	22	3	-	-	-
Wigeon	4303	3547	-	2	-	-	-	141	1533	4551	4230	5929
Gadwall	27	8	-	2	-	-	-	-	20	74	87	121
Teal	1708	1053	-	6	-	-	-	126	628	1125	1147	2534
Mallard	2269	1447	55	69	97	193	225	1058	2210	2879	2628	2427
Pintail	100	23	-	-	-	-	-	-	3	30	113	95
Shoveler	116	83	-	-	-	-	-	9	17	36	59	123
Pochard	114	34	5	1	1	1	1	1	2	50	100	199
Tufted Duck	224	213	4	1	4	4	3	6	16	124	170	276
Eider	20	36	-	-	-	4	-	2	2	3	1	9
Long-tailed Duck	3	-	-	-	-	-	-	-	-	-	-	-
Common Scoter	3	-	-	-	-	1	-	1	-	-	7	-

Goldeneye	67	56	-	-	-	-	-	-	-	-	2	51
Red-breasted Merganser	165	193	-	11	-	-	-	-	4	29	131	127
Goosander	44	38	-	-	-	-	-	-	16	-	21	37
Ruddy Duck	1	-	-	-	-	-	-	-	-	8	13	14
Water Rail	8	8	-	1	-	-	-	1	2	3	7	5
Moorhen	166	175	15	9	14	31	24	71	179	196	189	236
Coot	494	505	6	5	3	4	7	65	239	1508	1517	1413

Waders

A total of 31 species of wader was recorded, two more than in 2000. As always most waders are passage migrants or winter visitors, and because of access restrictions both spring and autumn passage were under-recorded in 2000. Among the winter visitors, **Avocet** peaked at 780 in December (a big increase on the 2000 peak of 552), and **Dunlin** continued to increase in 2001, if only slightly.

Species recorded on WeBS 2001 counts in small numbers and not included in Table 124 were: **Little Ringed Plover**, **Kentish Plover**, **Baird's Sandpiper**, **Purple Sandpiper**, **Jack Snipe**, **Lesser Yellowlegs**, **Wood Sandpiper and Grey Phalarope**. (See Systematic List for further details). The monthly totals for all other species are given in Table 124.

Table 124. Summary of WeBS counts for the main wader species for Devon in 2001

Species	Jan	Feb	Mar	Apr	May	Jun	Jul	Aug	Sep	Oct	Nov	Dec
Oystercatcher	3555	2886	-	4	11	21	14	3863	4385	2398	2858	3589
Avocet	555	434	-	-	-	-	-	-	8	1	589	780
Ringed Plover	117	88	-	-	-	-	-	53	184	145	216	157
Golden Plover	1534	2070	-	-	-	-	-	-	1	93	1786	1380
Grey Plover	495	422	-	-	-	-	-	6	10	60	301	418
Lapwing	5447	1685	-	-	-	-	24	36	48	197	1018	2350
Knot	268	57	-	-	-	-	-	35	33	-	7	94
Sanderling	81	30	-	-	-	-	-	1	114	25	160	72
Little Stint	1	-	-	-	-	-	-	2	16	12	1	-
Curlew Sandpiper	-	-	-	-	-	-	-	1	12	2	4	-
Dunlin	5735	5333	-	-	-	-	2	332	1067	956	6048	9481
Ruff	1	-	-	-	-	-	-	1	4	4	1	-
Snipe	307	195	1	1	-	-	3	8	6	38	123	187
Black-tailed Godwit	691	764	-	-	-	-	-	404	537	799	785	755
Bar-tailed Godwit	470	406	-	-	-	-	-	25	181	260	411	536
Whimbrel	1	1	-	-	-	-	4	36	16	1	-	1
Curlew	2419	2379	-	-	-	-	75	1374	3093	2823	2370	1772
Spotted Redshank	8	6	-	-	-	-	-	-	-	1	-	2
Redshank	1297	1265	-	-	-	-	19	509	909	981	1256	936
Greenshank	54	70	-	1	-	1	-	70	138	85	60	56
Green Sandpiper	9	6	-	-	-	-	7	14	8	11	13	8
Common Sandpiper	6	6	-	3	-	3	20	30	25	19	13	13
Turnstone	163	322	-	-	-	-	-	-	2	77	106	121

Gulls, terns and Kingfisher

As gull counts are not carried out on all sites, and some of the figures are estimates, it is difficult to assess the 'real' populations but the data in Table 125 clearly show again how important our estuaries are for wintering gulls.

Gulls and terns recorded on WeBS 2001 counts in small numbers and not included in Table 125 were: **Little Gull, Ring-billed Gull, Yellow-legged Gull, Iceland Gull, Kittiwake and Little, Common, and Black Terns**. (See Systematic List for further details).

Table 125. Summary of WeBS counts for the main gull and tern species, and Kingfisher, for Devon in 2001

Species	Jan	Feb	Mar	Apr	May	Jun	Jul	Aug	Sep	Oct	Nov	Dec
Mediterranean Gull	2	3	1					1	3	2	4	1
Black-headed Gull	8694	5051	51	5		11	400	2359	7331	7443	8445	6945
Common Gull	388	167						17	24	58	187	206
Lesser Black-backed Gull	196	197	98	7		20	160	919	1075	908	2168	184
Herring Gull	3100	2273	17	176	202	409	1652	2556	7953	4560	3662	2951
Great Black-backed Gull	182	285		5	5	12	22	339	636	523	278	291
Sandwich Tern				5	6			162	85	2		
Kingfisher	8	7	1	1		1	3	3	18	14	19	19

Reporting sightings of colour-marked wildfowl

WWT co-ordinates the colour marking of wildfowl in the UK on behalf of the BTO. Part of this responsibility is to channel sightings of marked birds from members of the public to those responsible for the ringing projects. To do this a certain amount of information about the sighting is required and without this it is not always possible to trace the sighting to the correct ringer or inform the observer of the origins and movements of the bird they have seen. These sightings can be sent to 'Colour-marked Wildfowl' WWT Slimbridge Gloucester GL2 FBT or email colourmarkedwildfowl@wwt.org.uk.

Also a plea from Dr. C.M Eising: if you see a wing-tagged gull could you please send details of sightings to: Dr. C.M Eising,Department of Animal Behaviour, P.O.Box 14, 9750 AA Haren, The Netherlands. (Tel: ++31 50 3632069 Fax: ++ 31 50 3635205 or email: C.M.Eising@biol.rug.nl. As an example of the interesting information arising from reporting tagged birds, a wing-tagged **Black-headed Gull** noted by a DBWPS member at Yelverton had been tagged on a programme in the Netherlands where nest sites were flooded and Drs C M Eising and others had incubated the eggs and later released the offspring into the wild.

Acknowledgements

I would like to thank all counters who have supported the WeBS over the past year. Please continue with the prompt return of survey forms, i.e. after the March count and again after the December count as this will help greatly in producing reports as quickly as possible.

BREEDING BIRD SURVEY 1994-2000 – DEVON UPS AND DOWNS

Peter Reay

Background

The BTO's Breeding Bird Survey (BBS) has been running since 1994 and has now taken over from the Common Birds Census as the method for monitoring the numbers of Britain's common breeding birds. National results are summarised each year in *BTO News* (e.g. Raven & Noble 2001) and *The State of the UK's Birds* (e.g. Gregory *et al.* 2002); the latter also includes longer-term trends from 1970. More detailed analysis and regional results appear in *The BBS Reports* (e.g. Noble *et al.* 2001) and Devon results, together with a brief explanation of BBS, have appeared in *Devon Bird Reports* since 1994.

The virtual absence of BBS data for 2001 is in some ways frustrating because the idea of the survey is to provide continuous monitoring of the bird populations from data collected each year. On the other hand, the lull enforced by the Foot & Mouth access restrictions provides a useful opportunity to reflect on what has been learned so far from the BBS, and to review what bird populations have been up to between 1994 and 2000. It is these aspects that are the subject of this article.

The first step was to find out what data are available. In the *DBR*s there are seven years of data, mostly comprising: % occurrence (the % of the surveyed squares in which each species occurs); mean numbers (the total number of birds recorded in all squares divided by the number of squares); Devon indices (calculated from mean numbers after setting the 1994 value at unity). From the latter it is easy to see whether a species has increased or decreased between 1994 and any other year, because the index will be greater than unity for an increase and less than unity for a decrease. Indices are an important output from BBS and appear in the review publications for the UK, component countries and regions. Unfortunately, **the Devon indices appearing so far have not been calculated in the same way as these other indices, and most importantly cannot be used to assess the statistical significance of any recorded changes**. Fortunately, however, official Devon indices do exist and I am indebted to Mike Raven of the BTO for providing these. They have been calculated using all the field data collected during the survey work, not just the mean numbers, and are given in Table 126 for all species recorded in an average of at least 20 squares during the seven-year period. **For almost all purposes, it is recommended that these indices replace those given in the earlier reports, and that any conclusions on population change deduced from the latter are treated with extreme caution**.

Data are available at the BTO to determine whether any of the year-to-year changes in the indices are statistically significant, but the usual procedure is just to compare the current year with a) the previous year and b) the baseline year of 1994. In fact, there was only one significant change between 1999 and 2000 in Devon, a **Linnet** decrease, but there perhaps needs to be an explanation for the unusually high 1999 index value from which the decrease has occurred. Mike Raven has commented on this as follows:

> *"The relatively small sample sizes obtained from a single county mean that results are prone to be influenced by particularly large counts. Linnets are always difficult to census, as they do not form obvious territories and form flocks even in the spring and summer. Therefore a single large feeding flock could influence the results from a small sample."*

This article, therefore, is solely concerned with looking at changes between 1994 and 2000 and it is important to realise that, whatever conclusions are reached, nothing will be implied about the significance of changes from year to year within that period or the time either before 1994 (no BBS data) or after 2000 (no data and no crystal ball!).

Thus for the 34 species occurring in at least 20 Devon squares, the % increases and decreases from 1994 to 2000 are also shown in Table 126 (column 3). **Those changes that are statistically significant are shown in bold – basically this means that we are very confident that these are real changes and**

not chance effects (95% confident in statistical terminology, meaning that there is only a one in twenty chance that the result could have arisen by chance). **Statistical significance is affected both by the number of squares sampled, the extent of the differences between the two years and the variation in numbers between squares.** For species occurring in fewer than 20 squares, the statistical methodology is such that it is not possible to determine whether any changes in the indices are significant, so no indices are provided. However, although a species may not occur in 20 squares in Devon it may well occur in more than 20 squares in the SW region. In such a case it may be possible to say that a species has shown a significant change in the SW, which might imply that the Devon change (as long as it's in the same direction!) is real – and may be worthy of further targeted fieldwork. The same applies to data for England, and this is why the *significant* % changes for the SW and England have also been included in Table 126, though more weight has been given to the SW changes to support (cautiously) any inconclusive Devon data.

One final point follows on from the last as it concerns differences between areas. It must be accepted that, for example, a species which is increasing in England as a whole, may be decreasing in some areas of England (see Song Thrush data in Raven & Noble (2001) for example), and just because a species may be showing an overall decrease in the SW, it does not follow that it will also be decreasing within Devon. On a smaller scale, the data presented here represents an overview or average picture of the whole county; it is perfectly feasible that in a local area numbers may not follow the county trend.

Results

Increases

- Four species showed significant Devon increases (and mostly significant SW and England increases too): **Pheasant**, **Woodpigeon**, **Swallow** and **House Martin**.
- Five species showed insignificant Devon increases, but significant SW (and mostly England) increases: **Dunnock**, **Blackcap**, **Goldcrest**, **Jackdaw** and **Greenfinch**.

Decreases

- Four species showed significant Devon decreases: **Skylark**, **Carrion Crow**, **House Sparrow** and **Linnet**. These also mostly showed significant SW and England decreases, but not for House Sparrow in SW, and note that for England, Carrion Crow showed a significant *increase*.
- Three species showed insignificant Devon decreases, but significant SW (and except for Chaffinch, England) decreases: **Willow Warbler**, **Starling** and **Chaffinch**.

Probably stable species (no compelling evidence of change in Devon)

- Three species showed insignificant Devon and SW increases (but significant increases in England): **Collared Dove**, **Wren** and **Blackbird**.
- Two species showed insignificant Devon and SW decreases (but significant decreases in England): **Swift** and **Yellowhammer**.
- Thirteen species showed insignificant Devon changes, and either no significant SW change, or a significant change in the *opposite direction*: **Buzzard**, **Herring Gull**, **Pied Wagtail**, **Robin**, **Song Thrush**, **Mistle Thrush**, **Chiffchaff**, **Whitethroat**, **Blue Tit**, **Great Tit**, **Magpie**, **Rook** and **Goldfinch**.

Unknowns

The following terrestrial breeding species comprise those recorded in less than 20 Devon squares (*and there are many other Devon breeding species not picked up at all by BBS*). We do not know what the Devon populations are doing, but they are grouped in order to indicate SW and England changes which *may* also apply to Devon.

- *Possible increases*. Three species showed significant SW increases: **Stock Dove**, **Great Spotted Woodpecker** and **Marsh Tit**. Nine further species showed significant England (but not SW) increases: **Red-legged Partridge**, **Green Woodpecker**, **Sand Martin**, **Grey Wagtail**, **Redstart**, **Stonechat**, **Sedge Warbler**, **Long-tailed Tit** and **Raven**.

- *Possible decreases.* Four species showed significant SW decreases: **Feral Pigeon**, **Cuckoo**, **Lesser Whitethroat** and **Bullfinch**. Six further species showed significant England (but not SW) decreases: **Kestrel**, **Grey Partridge**, **Turtle Dove**, **Wood Warbler**, **Spotted Flycatcher**, and **Willow Tit**.
- *Stable.* Twenty three species showed no significant SW or England change: **Sparrowhawk**, **Little Owl**, **Tawny Owl**, **Kingfisher**, **Woodlark**, **Tree Pipit**, **Meadow Pipit**, **Dipper**, **Whinchat**, **Ring Ouzel**, **Wheatear**, **Cetti's Warbler**, **Grasshopper Warbler**, **Reed Warbler**, **Garden Warbler**, **Pied Flycatcher**, **Coal Tit**, **Nuthatch**, **Treecreeper**, **Jay**, **Siskin**, **Lesser Redpoll**, and **Reed Bunting**.

Discussion

Limited data have been presented. They cover only terrestrial breeding species that have occurred on average in at least one BBS square since 1994. They refer to the whole county rather than to individual sites or habitats. And they represent only two years, 1994 and 2000. Nevertheless, within those constraints they are the best, and in most cases the only, data available.

At a general level, it is interesting to note that for no species were the indices the same in the two years, although **Mistle Thrush** came close at a 1% increase (not surprisingly, an insignificant one!). So changes are the norm, whether they are real or the result of 'errors' in counting (an error here can simply mean not having enough squares). Another general point is that in those 34 species for which Devon indices have been calculated, only eight showed significant changes between 1994 and 2000, with the four increases balancing the four decreases. There is no evidence for significant changes in the majority of species, especially when we add in all those species for which Devon BBS data are inadequate.

Of the four species showing significant Devon increases, two, **Pheasant** and **Woodpigeon**, are game birds and there may be independent data from shoots which could be used to support or refute the population increases suggested by BBS, but Pheasant counts are very prone to extreme results, caused by large scale releases occurring in or near a square, especially with small sample sizes (Mike Raven, pers. comm.). For one of the other two species, **Swallow** and **House Martin**, there is, fortuitously, independent data in the form of the results of the DBWPS House Martin survey organised by Roger Thornett – doubly fortuitous since this survey also began in 1994. Roger has kindly provided data from the 18 observers who have counted nests at the same sites in both 1994 and 2000. The result was a 50% increase from 320 in 1994 to 480 in 2000. The BBS showed a much larger increase of 133% in **House Martin** numbers, but bearing in mind the differences in survey techniques and area coverage, this actually represents good agreement on the changing status of this species and should inspire in doubters at least a little confidence in the BBS approach.

Of the four species showing significant Devon decreases, **Skylark**, **House Sparrow** and **Linnet** have been flagged nationally as problem species (Red-listed in Birds of Conservation Concern (Anon 2002)). The surprise inclusion here is **Carrion Crow** especially as the changes are not matched in either the SW or England indices. Perhaps not many will lament a declining crow, but it would be interesting to have some independent information on this species and whether the perceived decline in this persecuted species is in some way linked to the increase in **Pheasant**.

Taking the results as a whole, there are two extremes of interpretation possible. On the one hand, one could argue that over the seven-year period there has been very little change, stability is the rule and there is little to worry about. On the other hand, since some significant changes have been demonstrated, including decreases, there must be many more changes among the large number of species for which there is not enough data – therefore we must make a huge effort at county level to collect data on these species independently of BBS. As in most cases, a moderate approach, a compromise between these extremes, is likely to be the way forward. But the future, whether in terms of numbers of birds or the surveying behaviour of birdwatchers, certainly looks interesting!

Acknowledgements

Thanks to Mike Raven of the BTO for supplying the Devon indices and for his comments on a draft of this article, and to John Woodland, the BTO RR, for making the contact. Thanks also to Simon Geary for his comments and to Roger Thornett for supplying the results of the House Martin Survey. In addition, John Woodland adds the following comment.

> *"As the BTO RR, I am appreciative of the work and interest that Peter has put into this article. As he mentions, these data are not only the best available, but the only data available for such comparisons. Their existence is entirely due to the support and dedication of the 80 plus Devon birders who in many cases have been covering their BBS sites year by year since the survey began in 1994. "*

References

ANON (2002). *Birds of Conservation Concern in the UK.* RSPB, Sandy.

GREGORY, R D *et al.* (2002). *The State of the UK's Birds 2001.* The RSPB, BTO, WWT and JNCC, Sandy.

NOBLE, D G, RAVEN, M J & BAILLIE, S R (2001). *The Breeding Bird Survey 2000,* Report Number **6**. Published by BTO, JNCC and RSPB.

RAVEN, M J & NOBLE, D G (2001). The Breeding Bird Survey: 1994-2000. *BTO News* **237**: 12-14.

Table 126: BBS data for Devon 1994-2000 showing for each species: mean number of squares occupied; % change in Devon, SW and England indices (significant changes in bold) from 1994 to 2000; and Devon indices 1994-2000. *Only those terrestrial species with a mean occupancy of at least five squares are included.*

mean number of squares		% changes in indices from 1994 to 2000 (significant in bold)			Devon Indices 1994-2000						
Species		**Devon**	**SW**	**England**	**index 1994**	**index 1995**	**index 1996**	**index 1997**	**index 1998**	**index 1999**	**index 2000**
Sparrowhawk	8										
Buzzard	40	-11	**23**	**46**	1.00	0.91	1.25	1.15	0.83	1.14	0.89
Pheasant	40	**37**	**37**	**15**	1.00	1.02	1.56	1.09	1.18	1.31	1.37
Herring Gull	28	-6			1.00	1.65	1.76	0.99	1.01	0.86	0.94
Feral Pigeon	8		**-27**								
Stock Dove	16		**40**								
Woodpigeon	51	**27**	**11**	**5**	1.00	1.03	1.16	1.28	1.56	1.42	1.27
Collared Dove	22	3		**19**	1.00	0.85	1.06	0.98	1.38	0.76	1.03
Cuckoo	14		**-36**	**-31**							
Swift	23	-31		**-19**	1.00	0.86	0.86	0.75	0.56	0.47	0.69
Green Woodpecker	14			**31**							
Great Spotted Woodpecker	16		**47**	**48**							
Skylark	40	**-24**	**-15**	**-19**	1.00	1.08	0.87	0.87	1.00	0.93	0.76
Swallow	45	**36**	**34**	**15**	1.00	1.01	1.22	1.31	0.79	1.43	1.36
House Martin	25	**133**	**34**		1.00	1.40	1.84	1.62	1.17	0.83	2.33
Tree Pipit	5										
Meadow Pipit	12			**-12**							
Grey Wagtail	6			**36**							
Pied Wagtail	27	-2		**28**	1.00	1.12	1.21	1.13	1.35	1.22	0.98
Wren	52	15		**11**	1.00	1.19	0.98	0.82	1.06	1.25	1.15
Dunnock	46	6	**14**	**6**	1.00	1.08	1.12	0.95	1.00	1.23	1.06
Robin	51	-5		**18**	1.00	1.08	1.08	0.87	0.88	0.92	0.95
Stonechat	6			**85**							
Blackbird	53	12		**10**	1.00	1.02	1.09	1.03	1.13	1.21	1.12

continued...... mean number of squares		% changes in indices from 1994 to 2000 (significant in bold)			Devon Indices 1994-2000						
Species		**Devon**	**SW**	**England**	**index 1994**	**index 1995**	**index 1996**	**index 1997**	**index 1998**	**index 1999**	**index 2000**
Song Thrush	44	-5	**14**	**7**	1.00	0.97	1.05	0.82	1.07	0.97	0.95
Mistle Thrush	22	1	**-32**		1.00	0.73	1.42	0.80	1.16	0.68	1.01
Whitethroat	27	-13	**32**	**25**	1.00	1.01	1.34	0.95	1.14	1.01	0.87
Garden Warbler	13										
Blackcap	38	11	**23**	**24**	1.00	1.14	0.93	1.05	1.27	1.42	1.11
Chiffchaff	42	-1			1.00	1.00	0.98	1.07	1.24	0.87	0.99
Willow Warbler	36	-16	**-30**	**-9**	1.00	1.14	1.13	0.86	0.92	1.15	0.84
Goldcrest	20	67	**52**	**65**	1.00	2.64	1.81	0.96	1.24	1.46	1.67
Spotted Flycatcher	8			**-24**							
Long-Tailed Tit	17										
Marsh Tit	5		**91**								
Coal Tit	17										
Blue Tit	50	-4			1.00	1.06	1.28	1.14	0.99	0.96	0.96
Great Tit	46	-4	**18**	**16**	1.00	0.97	0.98	1.05	1.09	1.36	0.96
Nuthatch	15										
Treecreeper	10										
Jay	18			**-17**							
Magpie	47	8			1.00	1.12	1.23	1.16	1.15	0.96	1.08
Jackdaw	38	25	**17**	**27**	1.00	1.07	1.08	0.92	0.91	1.20	1.25
Rook	36	35			1.00	1.36	1.60	0.99	1.71	3.49	1.35
Carrion Crow	53	**-29**		**12**	1.00	0.86	1.45	0.78	0.88	0.82	0.71
Raven	15			**88**							
Starling	30	-6	**-15**	**-21**	1.00	0.98	1.11	1.14	0.49	0.61	0.94
House Sparrow	43	**-29**		**-12**	1.00	0.83	0.82	0.71	0.73	0.94	0.71
Chaffinch	51	-10	**-8**	**6**	1.00	0.99	1.09	0.95	0.83	0.93	0.90
Greenfinch	41	2	**43**	**34**	1.00	1.20	1.01	1.19	1.16	1.26	1.02
Goldfinch	39	-23			1.00	0.89	1.22	1.18	0.61	1.31	0.77
Linnet	29	**-65**	**-21**	**-19**	1.00	1.14	0.94	0.70	0.45	2.18	0.35
Bullfinch	15		**-38**	**-24**							
Yellowhammer	25	-25		**-13**	1.00	1.12	0.99	1.19	1.34	0.54	0.75

N.B. Devon indices for Swallow and House Sparrow are plotted in the Systematic List as Figures 24 & 26, showing the means given in Table 126 and also 95% confidence intervals provided by BTO

SUMMARY OF *Seabird 2000* IN DEVON

Simon Geary

Introduction

Seabird 2000 was the third comprehensive survey of breeding seabirds in the UK and Ireland (UK/IRE). The survey actually ran between 1999 and 2001, and provided the latest national population estimates of our breeding seabirds following those attained by Operation Seafarer 1969–70 and the Seabird Colony Register in 1985–87.

Seabirds in Devon have been censused in all three surveys and, allowing for some differences in survey methods and coverage, this has enabled the overall population trends of each species to be assessed.

This report presents estimates of breeding populations over the three surveys, and attempts to assess the county populations in a regional and national context. At the time of writing regional and national totals for *Seabird 2000* were unavailable (Ian Mitchell (JNCC), pers. comm.). Therefore, to put Devon's seabirds in a wider context, *Seabird 2000* data for Devon are compared with national data from the Seabird Register 1985–87, but as a result the relative differences in populations need to be treated with caution.

The report also addresses seabird breeding distribution within the county and identifies key sites and coastal units for each species.

Collection of data

Seabird 2000 was completed by over 30 volunteers, and overseen by county coordinator, Ken Partridge. A combination of land and sea-based counts was used, the latter along the north coast. Survey methodology was similar to that used for the Seabird Register 1985-87. The coastline was divided into discrete subsections using obvious geographical features and separate counts were conducted of all species per subsection. Subsections were given unique name and code, and six-figure grid references were recorded for start and end points in order to avoid double counting or omission of coastline.

In general both accurate counts of birds and estimates of birds hidden from view were recorded. Individual birds, e.g. Fulmar, were classed as occupying a breeding site if the bird was sitting at a site considered large enough to support an egg. Prescribed count periods (time of day/year) and count units (see below) were used for each species as described by Walsh *et al.*(1995).

Treatment of data

Raw data were kindly supplied by JNCC. Data arrived in various count units, including estimates of pairs, individuals on land, apparently occupied sites/nests/ territories (aos/aon/aot), and estimates of hidden sites/pairs. For analysis purposes, all counts of aos/aon/aot are classed as breeding pairs, and estimates of hidden pairs have been included in the totals. Counts of individual auks have been converted by the appropriate standard factor to estimate pairs. Conversion factors were applied as follows: Guillemot and Razorbill (individuals x 0.67 = pairs); and Puffin (individuals x 0.5 = pairs) (Lloyd *et al.*1991). All population estimates quoted in this report represent breeding pairs.

Species and breeding habitat

Devon has 11 species of breeding seabird – Fulmar, Manx Shearwater, Cormorant, Shag, Lesser Black-backed Gull, Herring Gull, Great Black-backed Gull, Kittiwake, Guillemot, Razorbill, and Puffin. The populations are very small in national terms with only Cormorant and Herring Gull exceeding 1% of the UK/IRE population in 1985–87. Lundy Island holds good numbers of all species except Cormorant, which is absent, and both Manx Shearwater and Puffin are now confined to the Island. The remaining nine species breed in varying concentrations around the 465 km of mainland coastline.

Although Devon has a predominantly rocky coastline, there are relatively few stretches of cliff with suitable structure to support colonial seabirds. Thus, large colonies are few, and most species have dispersed populations. There are also few offshore islands, and this limits the amount of predator free, undisturbed breeding habitat preferred by many seabirds. Extensive beach and dune systems, favoured by terns, are scarce. Although such habitat exists at the mouths of two major estuaries, both sites incur high levels of disturbance because of human activities, and no terns currently breed.

Table 127: Populations of breeding seabirds in Devon 1969–2001

	National Survey		
	1969/70	1985/87	1999/01
Fulmar	380	894	473
Manx Shearwater	100	1200	166
Cormorant	178	292	194
Shag	138	290	260
Lesser Black-backed Gull	101	180	426
Herring Gull	12752	2296	3839
Great Black-backed Gull	106	206	163
Kittiwake	2228	1663	1204
Guillemot	1657	2210	2631
Razorbill	1002	828	762
Puffin	41	20	7
Total pairs	**18683**	**10079**	**10125**

Table 128: Comparison of county, regional and national populations using 1985/87 Seabird Register data

	Devon	SW Region	UK,IRE
Fulmar	473	2818	571000
Manx Shearwater	166	2200	275000
Cormorant	194	728	11700
Shag	260	2200	47300
Lesser Black-backed Gull	426	6257	88500
Herring Gull	3839	8146	206000
Great Black-backed Gull	163	1555	23400
Kittiwake	1204	5768	544000
Guillemot	2631	3619	806100
Razorbill	762	1453	122000
Puffin	7	151	467000
All species	**10125**	**34895**	**3162000**

SW Region = Isles of Scilly; Cornwall; Devon; Dorset; Somerset; Avon.

Table 129: Cities, towns & villages with roof-nesting Herring Gulls

Locale	Pairs	Coastal Unit*
Paignton	231	Torbay
Brixham	225	Torbay
Torquay	208	Torbay
Teignmouth	173	Hope's Nose-Dawlish Warren
Plymouth	119	Plymouth-Prawle Pt
Exeter	108	Hope's Nose-Dawlish Warren
Dartmouth	87	Start Pt (north)-Berry Head
Goodrington	79	Torbay
Exmouth	76	Exmouth-Seaton
Dawlish	69	Hope's Nose-Dawlish Warren
Seaton	56	Exmouth-Seaton
Sidmouth	54	Exmouth-Seaton
Totnes	36	Start Pt (north)-Berry Head
Shaldon	28	Hope's Nose-Dawlish Warren
Budleigh Salterton	26	Exmouth-Seaton
Beer	19	Exmouth-Seaton
Kingswear	16	Start Pt (north)-Berry Head
Illfracombe	15	Braunton Burrows-Foreland Pt
Kingsbridge	15	Plymouth-Prawle Pt
Newton Abbot	11	Hope's Nose-Dawlish Warren
Salcombe	11	Plymouth-Prawle Pt
Stoke Gabriel	6	Start Pt (north)-Berry Head
Bishopsteignton	4	Hope's Nose-Dawlish Warren
Galmpton	4	Torbay
Starcross	8	Hope's Nose-Dawlish Warren
Axmouth	2	Exmouth-Seaton
Bigbury-on-Sea	2	Plymouth-Prawle Pt
Dartington	2	Start Pt (north)-Berry Head
Honiton	2	Exmouth-Seaton
Sidford	1	Exmouth-Seaton
East Portlemouth	1	Plymouth-Prawle Pt
Kenton	1	Hope's Nose-Dawlish Warren
Total pairs	**1695**	

**See Table 131 for comparison*

Table 130: Sites / Supersites holding at least 1% of Devon's breeding seabirds

	Lundy	Berry Head	Woody Bay	Hall-sands	Wring-apeak	Thatcher Rock	Cow & Calf	Ore Stone	Straight Point	Wemb-ury Mew Stone	Total	% Devon
Fulmar	190	6	19	15			22		3		255	58
Manx Shearwater	166										166	100
Cormorant			1			12	2			84	99	51
Shag	55	6	1			30		15		47	154	59
Lesser Black-backed Gull	406										406	96
Herring Gull	777		40	36	18	190	50	120	13		1244	32
Great Black-backed Gull	34					40				25	99	61
Kittiwake	237	27	5	495	22			11	165		962	80
Guillemot	1573	476	137		291		151	3			2631	100
Razorbill	637		95		10		12				754	99
Puffin	7										7	100
Total pairs	**4082**	**750**	**412**	**546**	**489**	**272**	**316**	**151**	**181**	**156**	**6777**	**67**
% Devon seabirds	**40**	**7**	**4**	**5**	**5**	**3**	**3**	**2**	**2**	**2**		

Percentages are rounded to nearest 1%

Table 131: Breeding seabird populations by mainland coastal unit

	Plymouth-Start Pt (south)	Start Pt (north)-Berry Head	Torbay	Hope's Nose-Dawlish Warren	Exmouth-Seaton	Welcombe Mouth-Hartland Pt (west)	Hartland Pt (east) Apple-dore	Braunton Foreland Pt	All units total	% Devon Pop.
Fulmar	12	19	43	36	29	13	37	94	283	60
Cormorant	123		45	14	9			3	194	100
Shag	70	6	109	6	12			2	205	79
Lesser Black-backed Gull	2	1	1	1	15				20	4
Herring Gull	404	174	1257	360	547	21	20	279	3062	80
Great Black-backed Gull	35	3	79	1	3			8	129	79
Kittiwake		267	506		165			29	967	80
Guillemot*		476	3					578	1057	40
Razorbill*								125	125	16
Total pairs	**646**	**946**	**2043**	**418**	**780**	**34**	**57**	**1118**	**6042**	**60**
% Devon seabirds	**6**	**9**	**20**	**4**	**8**	**< 1**	**1**	**11**		
				South Devon total	**4833**	**North Devon total**		**1209**		

Overview of significant population changes
See Tables 127 & 128 for overall population estimates.

Fulmar

The Devon population appears to have declined by *c.*47% since 1987. This contrasts with the national population trend, which has continued to show an increase in many areas since the Seabird Register 1985-87.

The Lundy population has remained stable. However, major declines have occurred along the mainland north coast, where there has been a decline from 400–500 pairs to 144 (-64 to -71%). A particularly dramatic decline appears to have occurred around Lynmouth Bay, where there were 270 pairs in 1985–87 but only 78 in 1999-01. This is against the long-term population trend for this stretch of the north coast, which has been increasing since initial colonisation in 1958. For example, between Lynmouth and Heddon's Mouth there were three pairs in 1958 increasing steadily to *c.*280 pairs in 1987, 323 in 1990 (Manning, 1991), and c.400 in 1995 (Grant, 1995).

Fulmar - *Mike Langman*

Manx Shearwater

Confined to Lundy Island. Estimates of its population since 1960 have ranged widely from 100 to an optimistic 7000 pairs (Price & Booker, 2001). The estimate of 1200 pairs in 1985-86 (Taylor, 1985) is unlikely to be reliable because it may have included nonbreeding birds and birds from other colonies (D Price, pers. comm.). In 2001, a detailed RSPB survey estimated the population to be *c.*166 pairs, with a rather fragmented distribution, concentrated in three main areas. The survey also found evidence of predation around burrows. Rats are thought to be one of the main predators (Taylor, 1985 and 1989).

Lesser Black-backed Gull

A considerable increase in population has occurred on Lundy Island. Here, there were an estimated 40 nesting pairs in 1957, but the cessation of control measures in the 1970s has enabled the population to grow during the past two decades to 186 pairs in 1986, 328 in 1996 and now over 400 pairs (Price & Booker, 2001). There were 17 roof-nesting pairs, 15 in Exeter, and one at Dartmouth and Torquay. A survey of roof-nesting gulls in 1994 found three in Exeter and one at Torquay and Totnes (Woodland, 1996).

Herring Gull

In all contexts this is Devon's most widespread 'seabird.' The population has increased by 39% since 1985–87, but is still well down on the 12,750 pairs in 1969.

The population of the open coast (excluding Lundy's 777 pairs) is *c.*1,350 pairs but no previous estimate is available for comparison.

Roof-nesting gulls

The roof-nesting population now accounts for 44% of the total. Again, there is no comparative data because in 1985–87 coverage of town rooftops was not comprehensive. Nationally, the increase in roof-nesting birds was rapid during the 1970s/80s (Lloyd *et al.*1991). Herring gulls have nested on coastal buildings since at least the 1920s but this is a relatively recent and increasing phenomenon in many coastal towns. During fieldwork for the Devon Atlas (1984-87) roof-nesting was recorded inland at Totnes, Chudleigh, Exeter and Axminster (Stormont, in Sitters, 1988).

Long-term data suggest that roof-nesting gulls in Devon have spread and increased. From 1969, when there were 48 pairs at seven sites, the numbers increased to 141 pairs at seven sites in 1976, 903 pairs at 16 sites in 1994 (Woodland, 1996) and *Seabird 2000* found 1715 pairs at 32 sites. Coverage on the four surveys was variable both in terms of towns surveyed and probably coverage within each town. This has undoubtedly affected totals and data shows variable trends for some towns. However, at Torquay, for example, the numbers have increased each time from one to 44 to 145 and 208, and many towns have shown increases since 1994.

Behavioural adaptations have enabled Herring Gull to take advantage of artificial cliff ledges provided by buildings. Key to this is supplementary (or new) food resources, such as at landfill sites, e.g. Chelson Meadow and Heathfield, but also increasing quantities of discarded human food in towns.

Table 129 displays data for main population centres of urban gulls. Note the significance of Torbay, perhaps a reflection of a large volume of supplementary food scavenged from tourists, the fishing industry and landfill sites.

Kittiwake

Another species in decline. The population declined by 28% between 1969 and 1985–87 (Grant, in Sitters, 1988), mainly due to loss of birds from Hope's Nose and a decline on Lundy. During the same period birds began nesting at Start Point, Soar Mill Cove and Woody Bay, and there were increases at Anstey's Cove and Berry Head. *Seabird 2000* has recorded a further *c.*28% decline since 1985–87. The main losses are from Lundy (-67%), Berry Head (-82%) and Ore Stone (-85%). However, a new colony has been established at Straight Point (*c.*165 pairs; 14% of county total) and the Start Point/Hallsands area now holds 735 pairs (+2196%). Individual mainland colonies are prone to annual fluctuations (Grant, in Sitters, 1988) and, over the years, some colonies have declined, whilst others have become established. The fate of the Kittiwake in Devon is therefore difficult to predict.

Puffin

The other species confined to Lundy Island. This is the most critically endangered species in county terms, with an estimate of 13 birds (7 pairs) present in June 2000. However, it is possible that due to the time of surveying a number of adults may have been incubating eggs and thus were uncounted. Lundy puffins have been in decline for some years in line with other colonies in southern Britain. In 1939, the

Lundy population was estimated at 3500 pairs but this was down to 400 by 1954 (Price, in Sitters, 1988). Since the early 1980s the population has probably been less than 100 pairs but has continued declining (Price & Booker, 2001).

The remaining species have remained relatively stable since 1985-87.

Key Sites/Supersites

Table 130 summarises population counts at key sites holding ≥1% of Devon's seabird population.

Lundy is the most important site holding *c.*40% of the county's seabirds. This includes all of the county's Manx Shearwater and Puffin, 96% of Lesser Black-backed Gull, 60% of Guillemot and 84% of Razorbill.

The remaining key sites support a modest 33% of the county's seabirds, including the remainder of the auk populations and 80% of Kittiwake. Relatively fewer Fulmar and large gulls are covered by the key sites and these species have more dispersed populations. Many Herring Gulls nest on buildings in coastal towns and cities (Table 129), and the Torbay conurbation holds around 44% of the roof-nesting birds.

Besides Lundy there are three key island sites – Thatcher Rock, Ore Stone (in Torbay) and Wembury Mewstone – each holding 2–3% of Devon's seabirds. These are important breeding grounds for Shag and Great Black-backed Gull, and the Wembury Mewstone is the most important site for Cormorant holding 43% of the county's population.

Seabird distribution by mainland Coastal Unit

In an attempt to identify key areas of coastline for different species the mainland coast has been divided arbitrarily into discrete units. It is acknowledged that other permutations are possible. However, using distinctive geographical breaks in the coastline eight units are identified (Table 129). Overall, Torbay holds 20% of mainland seabirds but the data are skewed by the huge Herring Gull population. Excluding Herring Gull, Torbay holds similar overall numbers to the Start Point-Berry Head coastline, and in terms of overall numbers and key species these two units are by far the most important. They support important county populations of Shag, Kittiwake and Guillemot.

The south Devon coast from Plymouth to Prawle Point supports relatively few birds except Cormorants and large gulls, with the majority of Herring Gull nesting around Plymouth. The majority of Devon's Cormorant population is found along the south coast units, and this is considered to be mainly because of more favourable nesting habitat in comparison to the more precipitous north coast (Stevens, in Sitters, 1988). Further east, the Exmouth to Seaton coast supports 14% of Kittiwake, and small numbers of Shag and Cormorant from Straight Point to Otterton Ledge and Brandy Head to Chiselbury Bay, respectively.

In the north, Welcombe Mouth to Hartland Point and Hartland Point to Appledore are the least important units with only Fulmar and Herring Gull breeding. The remaining unit, Braunton to Foreland Point, holds the majority of north coast birds, with the core area from Elwill Bay eastwards to Foreland Point. This unit contains Wringapeak, an important auk colony, which holds *c.*10% of the county's Guillemot. Nearby, Elwill Bay holds *c.*10% of Razorbill. Sixteen percent of the county's Fulmar also breed along this stretch of coast.

Discussion

Seabird 2000 has provided valuable up-to-date information regarding seabird breeding populations and distribution in Devon. Coverage in this survey was comprehensive, but this was not the case in the two earlier surveys. Comparative population estimates should therefore be treated with due caution.

A substantial decline of Fulmar is perhaps the most surprising find. It will be interesting to see if Fulmar has declined in neighbouring South West counties and indeed nationally. The key area for Fulmar is Lynmouth-Heddon's Mouth where counts have gradually increased to the mid 1990s. The large decline in *Seabird 2000* is difficult to explain and perhaps casts some doubt on the accuracy of estimates between the national surveys. There has also been a continued decline in Kittiwake numbers, but its colonies have always shown highly variable trends in terms of population and distribution. Puffin maintains only a tenuous foothold in the county and its future is far from certain.

On the plus side, Lesser Black-backed Gull, Herring Gull and Guillemot have higher estimated populations than 1985–87. Perhaps only the former has undergone a real increase, in common with the regional trend. The greater coverage of roof-nesting birds in the latest census may be the main factor in the apparent increase in Herring Gull.

The recent survey of Manx Shearwater on Lundy has provided the most accurate estimate of its population to date. Differing methodologies probably explain the large difference in recent population estimates. Bearing in mind differences in coverage, it appears that the populations of the remaining species have remained relatively stable since 1985–87.

Acknowledgements

The Devon Bird Report is grateful to Ian Mitchell and Rod Mavor (JNCC) for supplying the raw data and to JNCC for permitting its use to write this report. I am grateful to Dave Price and Pete Reay for commenting on an earlier draft, and to Helen Booker (RSPB) for supplying survey results for Lundy Island, and lending other relevant literature.

Bibliography

GRANT, K. R. (ed.) (1995). *The Devon Bird Report, 1995*. Devon Bird Watching & Preservation Society.

LLOYD, C., TASKER, M. L. & PARTRIDGE, K. (1991). *The status of seabirds in Britain and Ireland.* T & A. D. Poyser Ltd.

MANNING, C. (1991). North Devon seabirds along the Exmoor coast – Lynmouth to Heddon's Mouth. *Devon Birds* **44**(2):46-48.

MAVOR, R. A., PICKERELL, G., HUEBECK, M. & THOMPSON, K. R. (2001). *Seabird numbers and breeding success in Britain and Ireland*, 2000. Peterborough, Joint Nature Conservation Committee. (UK Nature Conservation, No.25.)

PRICE, D. & BOOKER,H. (2001). Lundy Island Manx Shearwaters: Breeding population & Distribution Survey, May 2001. Unpublished RSPB Report, RSPB, Exeter.

PRICE, D. & BOOKER, H. (2001). Survey of breeding seabirds on Lundy, June 2000. Unpublished RSPB Report, RSPB, Exeter.

SITTERS, H. P. (ed) (1988). *Tetrad Atlas of the Breeding Birds of Devon*. Devon Bird Watching & Preservation Society.

TAYLOR, A. M. (1985). Manx Shearwaters on Lundy: Ringing Studies and other observations. *Lundy Field Society Report* **36**: 23-34.

TAYLOR, A. M. (1989). Manx Shearwaters on Lundy: Further ringing studies and observations. *Lundy Field Society Report* **40**: 31-33.

WALSH, P. M., HALLEY, D. J., HARRIS, M. P., del nevo, A., Sim, I. M. W. & TASKER, M. L. (1995). *Seabird monitoring handbook for Britain and Ireland.* Peterborough, JNCC, RSPB, ITE, Seabird Group.

WOODLAND, J. (1996). The 1994 survey of gulls nesting on buildings. *Devon Birds* **49**(1): 27-28.

Shag - *Mike Langman*

PREY SELECTION BY URBAN PEREGRINE FALCONS IN 2001

Nick Dixon and Ed Drewitt

Introduction

A pair of Peregrine Falcons *Falco peregrinus* have been successfully breeding on a church tower in Exeter since 1997. A single bird was first noticed at the church in 1987 and a pair have been present since 1989. They first bred in 1997, when they took over a stick nest, newly built by Ravens. The nest was constructed on an east-facing ledge at the base of the spire, about 30m above ground level. Having taken over the nest site, they reared three young in 1997 and have successfully fledged young each year since then. The old Ravens nest has since been replaced with a purpose made tray by DBWPS, which was again used in 2001 with the pair producing three young. The adult Peregrines are present on the church all year and can be seen perching on the various pinnacles by day.

There are a wide variety of habitats within a four-mile radius of Exeter. These include four freshwater river systems and associated water meadows, tidal reaches and mudflats on the estuary, urban, suburban, industrial and mixed agricultural land as well as deciduous and coniferous forestry. The proximity of these habitat types to the nest site on the church tower provides ample opportunities for the Peregrines to prey on a wide selection of species, both resident and migrant, throughout the year.

The pair return to the church with prey all year round, and so it is an ideal site to study prey selection. Having returned with prey, the Peregrines will plume it on the pinnacles or on the flat ledge at the base of the spire before feeding. The fallen feathers are blown over a wide area dependant on the wind speed, and so many will be lost, but heavier items will either fall to the ground or onto one of the roofs.

Prey species have been recorded since 1997 and this report records the findings and changes in seasonal selection during the 12 months of 2001.

Method

Weekly collections of dropped or discarded material, including whole or part carcasses, heads/skulls, feet, humeri (wing bones), feathers and pellets are retrieved from the base of the church and other buildings in the immediate surroundings. These are bagged, dated and subsequently identified as to species, sex and age using personal knowledge, experience or a specialised identification book (Brown *et al.* 1992).

Feathers proving particularly difficult to identify to species level are checked against skins or complete specimens at The Natural History museum at Tring, Herts or Bristol University.

An annual roof, gutter and drainpipe clearance is undertaken by steeplejacks in the winter, to remove the build-up of autumn leaves, windblown rubbish and airborne debris. Amongst the considerable amount of soil cleared are many more remains of bone and feather, which often confirm species recorded from feather identification during the course of the year.

Regular weekly collections of fallen items are an ideal way of determining seasonal changes in prey brought back to the church. The Peregrines are unconcerned by human activity below so there is no disturbance to the birds, as would be the case at a more traditional site such as sea cliff or inland quarry. However, it is accepted many items could be removed or missed due to street cleaning, scavenging by foxes, cats and rats and being obscured amongst ground vegetation. It may also be possible that many more birds, particularly smaller ones, were taken but not found or consumed away from the church.

Results

Throughout 2001, the remains of no less than 145 individual birds were found and identified. These consisted of 30 different species, out of a total of 52 recorded since 1997 (Table 132). In 2001, 6% of all prey species were found to be juvenile birds. **Feral Pigeon** was by far the most important prey species comprising 40% by number and, because of its relatively large size, well over 50% by weight. Due to the large numbers of pigeons and doves taken, it is not sufficient to rely on feather alone to determine actual numbers, so humeri are removed as found and counted at the end of the year. Of the 130 humeri recovered, only one was **Woodpigeon**, 15 were **Collared Dove** and 114 were **Feral Pigeon** (59 were from the right wing and 55 from the left). Although **Woodpigeons** are very common birds locally, they are rarely taken by the Peregrines. The only one recorded in 2001 was a juvenile taken in early September.

The relative abundance of the different prey species varies throughout the year, and the main seasonal changes are described as follows.

The New Year

High water levels due to January floods, and colder weather often cause increased numbers of **Snipe** and **Teal** to move down into the SW, which make them more likely to be taken by Peregrines, and indeed, the remains of at least four **Teal** were found early in the year. This coincides with the peak Devon WeBS counts for these species. Many winter migrants were still in the area and although the greatest number of **Redwing** and **Fieldfare** were taken between October and December 2000, fresh feather remains of adults were still being found at the start of 2001. Interestingly, in the late autumn, many of the **Redwings** taken were found to be first-winter birds. Although thousands of **Black-headed Gulls** winter around Exeter, surprisingly only the remains of one was found, in January 2001. However, it happened to carry a ring and had been ringed as a nestling on 6 July 1998 at Holmsund, in Sweden (pers comm., Riksmuseum, Stockholm).

As the winter migrants disappeared, the more sedentary songbirds were busy setting up territories, singing and displaying. The Peregrines were incubating eggs from the end of March, and smaller prey such as **Chaffinch**, **Greenfinch**, **Song Thrush** and **Blackbird** were recorded, mainly caught by the tiercel to feed himself and his mate (Ratcliffe, 1995). Many of the beaks of finches, particularly **Greenfinches**, were found apart from the rest of the skull as individual remains. **Teal** continued to be taken through to the end of March, while one late **Fieldfare** was taken during the middle of April.

Spring and Hatching

As the eggs were hatching at the end of April, summer migrants were arriving into the country, establishing territories and getting ready to breed. A **Whitethroat** was an unusual find, whereas the taking of **Swifts** became regular, although the remains of five were down on last year's eleven. As the young chicks start to grow fast and require more food, the falcon increased the amount of time she spent hunting. At around the same time there was an increase in the numbers of **Feral Pigeons** being caught. Around this time, inexperienced young birds were also taken, particularly locally fledged **Starlings**. However, out of 17 taken in 2001, only 12% were juveniles, compared to 52% in 2000, while mainly adults were taken early in the season. Throughout April and May the diet also included such varied species as **Greenfinch**, **Chaffinch**, **Goldfinch**, **Bullfinch**, **Moorhen**, **Collared Dove**, **Blackbird** and **Song Thrush**.

Summer In The City

Throughout June, the focus was on common town species, particularly **Feral Pigeon**, **Starling**, **Collared Dove**, **Swift**, **Blackbird** and **Song Thrush**. This situation continued throughout July and August, together with some more unusual exotica in the form of **Cockatiel** and **Budgerigar**, perhaps easily taken due to their vunerability. The first return of migrant waders in late summer was reflected in the third record since 1997 of **Whimbrel** in the prey remains.

Departures and Arrivals

As autumn and winter drew closer and the young Peregrines dispersed, the adults were hunting less. As the summer visitors left and winter birds came in to replace them, the pair were catching more birds associated with the estuary such as **Lapwing**, **Golden Plover**, **Dunlin**, **Bar-tailed Godwit**, **Snipe** and **Teal**. A surprising find was **Little Grebe**, two of which were recorded in 2001. These possibly become vulnerable when travelling between stretches of water at night. **Woodcock**, another secretive species recorded, may also become vulnerable during nocturnal migration. Peregrines could be hunting these two species at dusk on light evenings, or may actually be utilising ground light from Exeter to hunt during the night. Although night hunting has yet to be fully proven in the UK, increasing evidence suggests that it does occur (unpublished observations, R. Tully, 1998; Dixon, in prep.). Perhaps **Moorhen** are also taken at night. The pair in Exeter took at least one in December and another back in April, and this prey species also occurred at the end of the year at two other urban Peregrine sites in the South-West (unpublished observations, E. Drewitt).

The November gutter clearance revealed many carcasses, skulls and bones. The heads and skulls usually relate to feathers found earlier in the year, and for 2001, included many **Feral Pigeon**, **Blackbird**, and **Redwing** skulls. There were also skulls of two **Moorhen**, two **Little Grebe**, **Whitethroat** and **Lapwing**, a **Whimbrel** beak, and the breastbones of **Lapwing** and **Golden Plover**.

Table 132: A summary of Peregrine prey remains found at the Church during 2001, by numbers and weight. (Average unit weight data from BWP).

Species	Total number	% Frequency	Average weight (g)	Total weight.(g)	% by weight
Feral Pigeon *Columba livia*	59	*40.69%*	300	17700	*56.90%*
Starling *Sturnus vulgaris*	17	*11.72%*	75	1275	*4.10%*
Collared Dove *Streptopelia decaocto*	10	*6.90%*	205	2050	*6.59%*
Teal *Anas crecca*	7	*4.83%*	325	2275	*7.31%*
Blackbird *Turdus merula*	7	*4.83%*	102.5	717.5	*2.31%*
Swift *Apus apus*	5	*3.45%*	43.5	217.5	*0.70%*
Lapwing *Vanellus vanellus*	4	*2.76%*	230	920	*2.96%*
Golden Plover *Pluvialis apricaria*	4	*2.76%*	220	880	*2.83%*
Snipe *Gallinago gallinago*	3	*2.07%*	110	330	*1.06%*
Fieldfare *Turdus pilaris*	3	*2.07%*	100	300	*0.96%*
Greenfinch *Carduelis chloris*	3	*2.07%*	75	225	*0.72%*
Moorhen *Gallinula chloropus*	2	*1.38%*	330	660	*2.12%*
Little Grebe *Tachybaptus ruficollis*	2	*1.38%*	190	380	*1.22%*
Song Thrush *Turdus philomelos*	2	*1.38%*	82.5	165	*0.53%*
Redwing *Turdus iliacus*	2	*1.38%*	62.5	125	*0.40%*
Carrion Crow *Corvus corone*	1	*0.69%*	510	510	*1.64%*
Woodpigeon *Columba palumbus*	1	*0.69%*	480	480	*1.54%*
Whimbrel *Numenius phaeopus*	1	*0.69%*	449	449	*1.44%*
Bar-tailed Godwit *Limosa lapponica*	1	*0.69%*	340	340	*1.09%*
Woodcock *Scolopax rusticola*	1	*0.69%*	300	300	*0.96%*
Black-headed Gull *Limosa limosa*	1	*0.69%*	300	300	*0.96%*
Jackdaw *Corvus monedula*	1	*0.69%*	220	220	*0.71%*
Cockatiel *Nymphicus hollandicus*	1	*0.69%*	90	90	*0.29%*
Dunlin *Calidris alpina*	1	*0.69%*	47.5	47.5	*0.15%*
Bullfinch *Pyrrhula pyrrhula*	1	*0.69%*	32.5	32.5	*0.10%*
Unidentified passerine	1	*0.69%*	32	32	*0.10%*
Budgerigar *Melopsittacus undulates*	1	*0.69%*	28	28	*0.09%*
Chaffinch *Fringilla coelebs*	1	*0.69%*	23.5	23.5	*0.08%*
Whitethroat *Sylvia cummunis*	1	*0.69%*	16.5	16.5	*0.05%*
Goldfinch *Carduelis carduelis*	1	*0.69%*	15.5	15.5	*0.05%*
Total	**145**	**100%**		**31104.5**	**100%**

References

BROWN, R., FERGUSON, J., LAWRENCE, M., and LEES, D. 1992. *Tracks and Signs of the Birds of Britain and Europe. An Identification Guide*. Christopher Helm, London.
RATCLIFFE, D.A. 1995. *The Peregrine Falcon*. Second Edition. Poyser, London.
SNOW, D.W. and PERRINGS, C.M. 1998. *The Birds of The Western Palearctic*. Concise Edition. OUP Oxon.

Further Information

The authors would be interested in hearing from anyone who may be able to contribute with past or current information regarding prey selection at this site. Please contact Nick Dixon by email: nickdixondevon@aol.com or by post to: N. Dixon, The Hawk and Owl Trust, c/o Zoological Society of London, Regents Park, London NW1 4RY

Nick Dixon *is a researcher for The Hawk and Owl Trust, and is currently undertaking a three year nationwide survey, investigating the increasing trend of Peregrine Falcons nesting on man-made structures and in urban environments. Prey selection at different sites is one of the factors being investigated during this study.*

Ed Drewitt *is currently the RSPB's Bristol Schools Officer, having recently completed a degree course in Zoology at Bristol University. He has been researching urban Peregrine prey selection at two other city sites in the South-West over the past three years, with funding from Bristol Zoo and the Millenium Commission.*

Peregrine - *Steve Young*

OBSERVATIONS OF VISIBLE MIGRATION ON THE SW DEVON COAST

Simon Geary

Introduction

Although visible migration is a well-known phenomenon, there is a dearth of quantifiable information on this subject for Devon. During autumn 2001, however, the opportunity arose to investigate the 'vis mig' potential of a section of the South West Devon coast in more detail.

All observation was from Staddon Point, located just outside of the Plymouth City boundary at the western extremity of the Staddon Heights plateau. At Staddon Point the land rises steeply to 100 m above sea level, and the coastline is orientated north-south forming the eastern shore of Plymouth Sound, thus making it a natural 'jumping off' area for daytime migrants 'coasting' west into Cornwall. A 360° panorama offers distant views over Plymouth Sound into Cornwall to the west, Heybrook and Wembury Point and the Western Approaches to the south, the City of Plymouth and Dartmoor to the north, and the South Hams to the east.

Data were collected over 70 hours on 43 days between 15 Sep and 3 Dec. Most recording occurred from 15 mins before local sunrise and continued for two to three hours, when passage either ceased, or diminished very considerably, for most species.

Species and numbers

Counts of the commonest and most regular species are presented in Table 1. Just over 100,000 birds of 41 species were counted flying over the site during the study period. All approached from vectors between north and southeast, most from the east, and many followed the coastline from Wembury Point. The majority of birds continued on vectors between west and northwest.

By far the most abundant species was **Woodpigeon**, with the majority passing on just seven days, including a massive 23,885 in two hours on 10 Nov. By contrast, **Meadow Pipit**, **Chaffinch** and **Linnet** had the most protracted passage, lasting the whole survey period, and with peak day counts of just 310 (late Sep), 553 (mid Oct) and 111 (mid Nov), respectively. The peak passage period for each species is highlighted by the italicised figures in the table. **Meadow Pipit** and **Chaffinch**, together with **Stock Dove**, **Woodpigeon**, **Fieldfare, Redwing**, **Jackdaw** and **Starling** comprised 95.2 % of all birds recorded. The remaining species were rather few in numbers, and some were very scarce.

TABLE 133: Numbers of the most abundant regular species passing Staddon Point in each half-monthly period, with peak passage for each species in bold.

	Half-monthly period					
Species	***15-30 Sep***	***1-16 Oct***	***17-31 Oct***	***1-15 Nov***	***16-30 Nov***	***Autumn total***
Stock Dove	-	-	85	**1270**	6	***1361***
Woodpigeon	-	-	158	**80358**	319	***80835***
Skylark	31	84	122	**388**	-	***625***
Swallow	**793**	11	-	-	-	***804***
House Martin	**375**	1	-	-	-	***376***
Tree Pipit	**26**	-	2	-	-	***28***
Meadow Pipit	**1232**	1109	246	137	40	***2764***
Grey Wagtail	**58**	-	1	-	-	***59***
Pied Wagtail	†33	**75**	22	3	-	***100***
Blackbird	-	6	**97**	34	-	***137***
Fieldfare	-	-	176	**951**	109	***1236***
Song Thrush	7	21	**101**	48	6	***183***

Redwing	-	-	733	**1744**	52	*2529*
Mistle Thrush	-	-	2	**27**	-	*29*
Jackdaw	-	-	519	**1625**	-	*2144*
Starling	-	26	1111	**1242**	176	*2555*
Chaffinch	108	1033	**1227**	1100	74	*3542*
Brambling	-	-	8	**36**	2	*46*
Greenfinch	16	74	**223**	125	13	*451*
Siskin	**350**	98	66	48	1	*563*
Linnet	119	155	191	190	**263**	*918*
Redpoll	-	13	5	**29**	1	*48*
Reed Bunting	-	2	**7**	5	-	*14*
Total birds	**3115**	**2708**	**5102**	**89360**	**1062**	**101347**
				Grand Total (all species)		**101814**

*Grand total includes species unlisted in table. †Includes two *M. a. alba*

Scarcer species

Occurrence on several dates (sf = single figures; df = double figures)

Sparrowhawk - several singles W over Plymouth Sound; **Lapwing** - sf W; **Golden Plover** - low df W; **Common Snipe** sf W; **House Sparrow** several singles and small groups NW; **Goldfinch** low df W and a 'resident' flock; **Bullfinch** low df NW; **Yellowhammer** sf NW. Although not considered as a typical migrant in the UK, the occasional **Sparrowhawk** moving west, including two on several dates, may indicate passage, or at least a limited movement.

One-day wonders (number-flight direction)

Black-throated Diver (1-NW over Plymouth), **Little Egret** (1-E), **Grey Heron** (flock 7-W), **Merlin** (1-NW), **Grey Plover** (1-W), **Dunlin** (flock 4-N), **Jack Snipe** (1-N), **Greenshank** (1-W), **Short-eared Owl** (2-W), **Collared Dove** (flock 5-NW), **Woodlark** (1-W), **Crossbill** (flock 7-NW).

Smaller numbers of migrants, comprising 19 species, were recorded on site, and were mainly chats, thrushes, warblers and crests. A thorough search of the adjacent bushes was undertaken only occasionally, and therefore this aspect of migration was under-recorded. In most cases the numbers of these species was very low and often disappointing. Noteworthy, however, were influxes of **Robin** on 25 Sep and 20 Oct with c.20 on site both dates. Only three **Black Redstart** and few **Goldcrest** (in keeping with a generally accepted poor migration this year in Devon) and the total absence of **Whinchat** and **Ring Ouzel** were surprising. The site is usually quite good for **Black Redstart** and **Ring Ouzel** (Vic Tucker, pers comm).

Changes in species assemblage over autumn

September

The peak month for hirundine, pipit and **Grey Wagtail** passage, although the majority of **Tree Pipit** would have passed before the survey began. Relatively few hirundines migrated during the morning compared to later in the day. On several dates, large easterly movements occurred during the afternoon over nearby Wembury (arriving from the direction of Staddon Heights) after the morning watches had counted fewer.

Siskin was the most abundant finch (127 on 16th and 128 on 17th), with fewer **Chaffinch**, **Goldfinch** and **Greenfinch**. The first two **Song Thrush** and a **Firecrest** were on 17th, and the only **Wheatear** and **Redstart** of the autumn were on the 23rd.

October

Unsettled weather limited observations, especially during the first week. **Pied Wagtail** passage continued from late Sep. The first **Golden Plover** passed westward on the 3rd. During mid month, the species composition changed apace, with the first **Redwing** (14th), **Jackdaw** (20th), **Reed Bunting** (18th), **Redpoll** (9th), and **Brambling**, surprisingly not until 24th but noted much earlier at the main South Devon coastal promontories (per V R Tucker). Increasing numbers of **Song Thrush**, **Chaffinch**, **Greenfinch** and **Linnet**, but decreasing **Siskin** towards the end of the month, and the last five days saw influxes of pigeons, thrushes and **Starling**.

November

Apart from a couple of days in Oct, the first 11 days of November were the most prolific and exciting period, as it was when the majority of birds moved through the area, comprising mainly **Stock Dove**, **Woodpigeon**, **Skylark**, **Fieldfare**, **Redwing** and **Jackdaw**. Finch passage continued throughout the month, tailing off slowly, but not before a surprisingly late pulse of **Linnet** during last ten days.

The numbers of **Fieldfare** and **Redwing** were disappointing in comparison to those consistently recorded over Central Plymouth between 1994 and 2000 (personal observations). Observations from Staddon Point, however, suggest that **Fieldfare** and **Redwing** flocks tend to fly W or NW further inland, effectively cutting off this corner of the coast. As usual, passage was concentrated into a few mornings. Small numbers of **Mistle Thrush** passed through between 28 Oct-14 Nov, with a maximum day count of 12 on 10 Nov.

And December?

The last count on 3 Dec recorded a few straggling **Meadow Pipit**, **Starling**, **Chaffinch**, **Linnet**, with single **Woodpigeon**, **Brambling** and **Greenfinch**. At a time when autumn migration has virtually ceased in northern Europe and north and eastern UK, the lateness of these birds demonstrates the protracted nature of autumn migration in the far South West of England.

Flight direction and behaviour

Flight direction

There were clear interspecific differences in flight direction, although in most cases this may have been a local variation on the typical westward movement. **Stock Dove**, **Woodpigeon**, **Starling**, **Jackdaw** and thrushes all flew WNW or even NW. **Jackdaw** and **Starling** made strong flights NW over Plymouth Sound toward the mouth of the River Tamar. Some may have had their sights on the southeast edge of Bodmin Moor, which is clearly visible 23 km distant.

Woodpigeon mainly approached from the east over Netton Point (Yealm Estuary), many reorienting around Wembury Point into Plymouth Sound, over toward Mt Edgcumbe and then probably west out to Whitsand Bay. Some of the larger flocks stayed out over the sea, heading toward Penlee Point and Rame Head. It was suspected that some **Woodpigeon** arrived at the South Devon coast from inland, possibly following river valleys to the coast before reorienting west along the coast.

Many **Meadow Pipits** flew in a south or southwest direction. Using optics, these could be followed out to sea until they disappeared from sight. Others flew west toward Penlee Point or continued northwest toward Plymouth.

Skylarks always flew due west toward Cornwall, many moving straight out from Heybrook and Bovisand along the route of Plymouth Breakwater toward Penlee Point and Cawsand.

A **Short-eared Owl** used Ramscliff Point, the westernmost extremity of the coast here, to circle and gain height in typical raptor fashion, before launching high over Plymouth Sound toward Penlee Point.

Virtually all finches flew west or mainly northwest. Under the influence of easterly winds, flight directions became more complex, with pipits moving south and northwest, and **Chaffinch** moving northwest and southeast.

Flight behaviour

Birds flew in patchy light drizzle and fog, maintaining the typical northwest direction. Patchy radiation fog drifting off the coast sometimes disorientated birds and caused them to temporarily alight or circle overhead before moving off northwest.

Almost daily, single or small groups of **Chaffinch** and **Woodpigeon** temporarily came off passage, to alight in nearby treetops. **Chaffinch** constantly gave excited contact calls, often attracting other individuals or small groups before eventually moving off together northwest over Plymouth Sound.

Jackdaw flocks exhibited the same flight behaviour as noted in 2000: approaching from the southeast, on reaching the cliffline they would bunch tightly, gaining altitude by circling in active flight or using the updrafts created by the cliffs. After gaining several hundred feet above the clifftop they would set out on a northwest flightpath over Plymouth Sound toward Devonport and the Tamar Estuary. The circling behaviour was probably initiated in order to re-orient on reaching a break in the predominatly east-west lie of the coastline.

On clear, calm mornings, the flight altitude of many **Chaffinch** and **Starling** was much higher by 09.00 h than during more cloudy and breezy conditions, and many birds may have past high overhead undetected.

Multispecies flocks

Several combinations occurred. Occasionally, **Starling** would flock with **Jackdaw** or **Woodpigeon** as they circled above the area; this may have benefits for both species in terms of predator detection and successful reorientation. **Stock Dove** regularly flew with **Woodpigeon** but individuals were scattered throughout the large **Woodpigeon** flocks. **Starling** and **Common Snipe** occasionally occurred amongst the larger pigeon flocks. **Starling** also flew with **Jackdaw** flocks, as did the occasional **Woodpigeon**.

Reaction to aerial predators

Peregrines were ever present in the area. Flocks of **Woodpigeon** and **Starling** avoided stooping **Peregrine** by dividing and plummeting earthwards. Out over the sea, **Woodpigeon** flocks quickly regrouped and resumed their westward flightpath. Sometimes pigeon flocks would circle and regain altitude before moving off. Close to the coast, in particular over Staddon Point, several flocks alighted in nearby trees after avoiding a stooping **Peregrine**. One such flock of several hundred scattered and descended all around me, creating a rather disconcerting rush of air as they passed close by. Later, such flocks always resumed their migration.

Discussion

Most birds were undergoing 'coasting' migration but smaller numbers arrived off the landmass and some continued south out over the sea. Some species were using geographical cues for orientation. Most used the westernmost tip of the coast to launch out over Plymouth Sound towards Cornwall. The **Woodpigeon** migration was truly amazing. This study has recorded both the largest day and autumn totals of migrating **Woodpigeon** ever in Devon. The *Devon Bird Reports* 1946-2000 contain sporadic references to migrating flocks along the whole south coast, and personal observations from the Plymouth area since 1994 have confirmed that many pigeons annually follow the South Devon coast westwards. There are also a few inland records, which together with personal observations, suggest that some flocks cross the county in a south or southwesterly direction, perhaps following river valleys before reaching the south coast. During the survey period, other observers recorded a few west, south or northward moving flocks of **Woodpigeon** and **Stock Dove** at Prawle Point and Soar area, indicating that pigeons were moving along the South Devon Coast. Although some of these records indicated immigration in off the sea, it is unknown if these birds had arrived direct from the continent, or were simply re-joining the Devon coastline from further up country.

The origin of the species

The origins, flightpath through the region, and ultimate destination of such large numbers of **Woodpigeon** warrants further investigation. The following simplistic calculation gives a crude estimate of the origins. The peak daily passage was 0800-0930h. Given that pigeons are diurnal migrants, perhaps initiating flight shortly before local sunrise, we can estimate that the birds had been travelling for 1-2 h before passing Staddon Point. Flying at an average speed of 60 kph (Mead, 2001), this would put their origin on the day in question between 60 to 120 km distant, and therefore direct immigration from the continent was unlikely (although this cannot be discounted, depending on the exact time of departure from their roost). More likely, these birds had started the day in East Devon, Dorset/Somerset. Observations from the Dorset coast during the late 1990s to present have recorded large numbers passing west over East & West Bexington and Christchurch Harbour from late Oct-early Nov (*Dorset Bird Report* 1994-2000). By contrast, Portland Bird Observatory always misses out on this movement (M Cade, pers comm).

Finches are also diurnal migrants and using the same criteria, but allowing for slightly slower flight speed, they almost certainly begin their journey on the day from South Hams, especially as finches were the first species to pass by, often before full daylight.

It is likely that the **Woodpigeon** and many other birds passing in late autumn were from continental populations, which had winged their way to and across the UK during previous days. The immediate destination of these birds was Cornwall, but their actual wintering area remains undetermined – France and Ireland are obvious possibilities.

Whilst far from being the Falsterbo of South West England, Staddon Point is never the less an interesting location to observe 'vis mig'. The late H G Hurrell must be acknowledged for this, as it was the Staddon area that he chose for his pioneering 'vis mig' watches in the 1920s (Vic Tucker, pers. comm.). Hopefully, this report will stimulate others to record or even co-ordinate their observations on this somewhat neglected subject.

Acknowledgement

Vic Tucker and Pete Reay commented on an earlier draft.

Bibliography

Devon Bird Report (1946-2000).
Dorset Bird Report (1994-2000). Dorset Ornithological Society.
MEAD, C. (2001). Scientific Discoveries in 2001. *Bird Watcher's Yearbook and Diary* 2002: 12-21.

FOOT AND MOUTH DISEASE - IMPACT ON BIRDS & BIRDING

Richard Hibbert

2001 will be remembered by Devon birdwatchers as the year in which the outbreak of Foot and Mouth disease (FMD) resulted in the effective closure of the countryside. For wildlife enthusiasts, as for a great many others, this development had a number of negative consequences. Without wishing to overlook the suffering of the farming community and the column inches already devoted to this sorry subject, it is interesting to take a look at circumstances from the perspective of birds and birders.

FMD and bird recording

Despite the access restrictions imposed during the crisis and the constriction that this placed on the activities of birdwatchers, Devon's observers managed to produce more records than usual: a total of 48,217 records compared to 31,940 in 2000 – a 51% increase. The annual crop of records (for these purposes a record is defined as an entry, or row, in the DBWPS database) has been steadily on the increase in recent years, but this is a considerable leap. Without exhaustive analysis of this and other years' data sets it is impossible to say precisely where and how such an increase occurred, but it is tempting to speculate on the reasons for this marked glut of records, and a comparison of the number of records in 2000 and 2001 for 100 annually occurring species gives some interesting results (see table 134).

As we will see below, the number and nature of those rows in the database is as much about the occurrence and behaviour of observers as it is about the birds. Changes in recording activity by groups of observers, or even individuals, can significantly affect the amount of records received. For instance, one development in 2001 was the *c.*7,500 records from Dawlish Warren collated by Ivan Lakin. Without these records the percentage increase in 2001 would be somewhat less, but it should be remembered that a considerable number of records have of course been received from Dawlish Warren in previous years.

Other factors should be taken into account, such as technical glitches and duplicate records affecting the totals. Such an extensive set of information is seldom free of such impurities, and for that reason as well as due to the rather random character of 'casual' records, analysis should be undertaken with caution. Nevertheless it is worth putting these reservations aside to examine a few of the possible reasons for the increase in records in 2001.

One is that observers concentrated on common and/or obvious species that were easy to see without leaving home or inhabited areas, and took the opportunity to fill their notebooks with records of birds that they would not normally bother with – a sort of comfort bird recording. The greatest increase in records between 2000 and 2001 was made by Pheasant, which had 266% more records this year than last. Jackdaw's 187% increase would also appear to bear this theory out, and the records of Song Thrush, Starling, Coal Tit, Pied Wagtail, Rook, Reed Bunting, Nuthatch and Meadow and Rock Pipits all increased by over 100%. Were they suddenly popular, or suddenly more worth recording by default?

It is always intriguing (if scientifically dubious) to examine the number of records that each species attracts. Generally speaking, the scarcer the bird the greater the percentage of observations that end up being sent in as records, and the number of records for each species in the database can be seen as a rough indication of popularity as well as occurrence. However in 2001 FMD introduced an entirely new set of factors to the league table of species recording. Pure popularity was outweighed by practical considerations, and factors such as the timing and duration of the crisis (spring/summer in many areas), as well as its geographical bias, affected bird recording. Thus it might be expected that a summer visitor to mainly inaccessible rural habitats (e.g. Redstart, Cuckoo) will have been less well recorded than another frequenting inhabited areas (e.g. Swallow), and that the recording of winter visitors will have been less affected than that of residents (see the big increases in Redwing and Fieldfare records – so far so good). Jokers in the pack like the two whitethroat species and Reed Warbler at one end of the scale and Magpie at the other (fairly static at only 11%) put paid to this neat reasoning, however.

Of the 100 species looked at, 21 improved on their 2000 record tally by over 100%, 33 by 50-100%, and only six (Woodlark, Nightjar, Lesser Spotted Woodpecker, Cuckoo, Little Owl and Red Grouse) had fewer records in 2001 than 2000. In a year of such widespread restriction of access to the countryside this is surely remarkable, though the latter six fit the preconception that FMD will have directly hindered the recording of certain species, and it is interesting that all of them are relatively scarce birds difficult to record in one way or another – Cuckoo and Red Grouse specifically because their Dartmoor stronghold was largely out of bounds. But how does one explain the massive increase in records for Lesser Whitethroat, for instance?

Another possible reason for the increase in records is that having been confined for so long during the crisis, the county's observers took to the field full of energy and made up for their dearth of birding earlier in the year by making a greater than usual recording effort once released. This might be thought to account for the surge in Yellowhammer records – until one consults the database and finds that survey work by R Jutsum produced 226 out of a total of 353 records for that species! If the input of one observer can swing the picture so thoroughly, what can we learn from looking at these figures? Perhaps simply what we knew before – that the annual gamut of county bird recording is highly variable and touches on as many human issues as avian. FMD just added another set of variables to what was already a jumbled scene.

The impact of FMD – some direct examples

Despite the increase in the number of casual records available for the production of DBR 01, we are lacking in certain kinds of information that would be of great interest were they available. Survey work was greatly disrupted, with the lack of CBC and WBS results (owing to the discontinuation of those surveys) being compounded by the cancellation of BBS work and the postponement of much other fieldwork. Frustratingly we will never know much about the effects that FMD had on breeding success and behaviour, but a few records give a fascinating taste of what might have happened. It is likely that the lack of disturbance suited many species, particularly ground nesters, and it is pleasing to think of scarce breeders such as Ring Ouzel, Red Grouse, Golden Plover, Dunlin, Curlew and Lapwing enjoying an undisturbed breeding season in their upland habitats. Whether a season's respite has benefited these species in terms of breeding success and range expansion will be difficult to quantify and impossible to prove.

There were, however, some direct positive results of FMD that are worth noting here. Coastal ground nesters prospered in the north of the county, with Ringed Plover breeding at Northam Burrows and Shelduck likely to have done so at Saunton Sands – both probably as a result of reduced disturbance. Likewise at Stover Country Park a pair of Sparrowhawks raised two young from a nest above a footpath closed to the public. On Dartmoor there was a rumour of Ring Ouzels nesting on a farm building by a footpath – but it remained a rumour, and as such is an apt illustration of the confinement mingled with unbearable curiosity that many birders must have felt. Otherwise the story of spring and summer 2001 is a litany of surveys not done, nestboxes unchecked, guilty roadside observations, dazzling rarities unseen in ungrazed pastures and frustrated birdwatchers with an itch to record Pheasants.

Table 134 - The record tallies of 100 species ranked by magnitude of change

Species	No. records 2000	No. records 2001	% change
Pheasant	47	172	266
Yellowhammer	118	353	199
Jackdaw	53	152	187
Lesser Whitethroat	49	139	184
Skylark	196	517	164
Whitethroat	153	404	164
Dunlin	202	524	159
Song Thrush	161	376	134

Starling	108	248	130
Coal Tit	77	175	127
Pied Wagtail	151	332	120
Kestrel	216	469	117
Reed Warbler	81	175	116
Rook	80	171	114
Redwing	113	238	111
Lapwing	225	472	110
Meadow Pipit	163	342	110
Reed Bunting	149	312	109
Fieldfare	103	212	106
Nuthatch	67	137	105
Rock Pipit	73	147	101
Grey Heron	268	524	96
Linnet	156	299	92
Goldfinch	183	349	91
Stock Dove	101	192	90
Kingfisher	213	388	82
Greenfinch	135	241	79
Bullfinch	142	253	78
Cetti's Warbler	72	127	76
Feral Pigeon	8	14	75
Tawny Owl	51	89	75
Raven	177	306	73
Chaffinch	178	302	70
Little Egret	821	1397	70
Swallow	320	542	69
Curlew	307	515	68
GS Woodpecker	167	279	67
Mistle Thrush	73	121	66
Black-headed Gull	217	358	65
Garden Warbler	98	162	65
Chiffchaff	415	679	64
Water Rail	148	243	64
Redshank	247	399	62
Swift	172	279	62
Buzzard	271	434	60
Collared Dove	85	134	58
Marsh Tit	62	98	58
Merlin	129	204	58
Sparrowhawk	220	344	56
Coot	242	375	55
Green Woodpecker	149	230	54
Woodcock	38	58	53
Great B-b Gull	128	195	52
House Sparrow	99	149	51
Canada Goose	351	528	50
Woodpigeon	117	176	50
Moorhen	191	284	49

Red-legged Partridge	31	46	48
Pied Flycatcher	71	104	47
House Martin	186	271	46
Robin	147	214	46
Stonechat	354	515	46
Blackcap	370	532	44
Cirl Bunting	120	173	44
Long-tailed Tit	150	216	44
Mallard	437	631	44
Siskin	205	290	42
Wheatear	362	511	41
Wren	118	166	41
Willow Tit	26	36	39
Oystercatcher	239	326	36
Carrion Crow	103	136	32
Jay	89	117	32
Snipe	228	301	32
Willow Warbler	213	281	32
Spotted Flycatcher	159	207	30
Wood Warbler	45	58	29
Golden Plover	198	252	27
Treecreeper	64	81	27
Great Tit	138	174	26
Peregrine	355	446	26
Blue Tit	163	204	25
Blackbird	181	220	22
Dunnock	147	179	22
Grasshopper Warbler	69	84	22
Grey Wagtail	244	296	21
Dipper	86	103	20
Tree Pipit	102	121	19
Goldcrest	239	282	18
Hobby	174	205	18
Whinchat	120	142	18
Dartford Warbler	94	106	13
Magpie	124	138	11
Redstart	117	116	0
Woodlark	29	26	-10
Nightjar	65	57	-12
LS Woodpecker	20	17	-15
Cuckoo	136	101	-26
Little Owl	54	37	-32
Red Grouse	19	13	-32

FOOT AND MOUTH AND BIRDS - Two Personal Perspectives

DARTMOOR
Richard Hibbert

Between 26 February and 4 October 2001 much of the Hexworthy area disappeared from the map. The local emergency originated at Dunnabridge, approximately 2km up the West Dart valley. All adjacent farms had to cull their stock and access restrictions persisted longer here than in surrounding areas. Whilst the area was off limits to all but dead livestock and television anchormen hawking the grimness of it all, the wildlife of the area continued with a sigh of relief. Roe Deer broke cover and roamed freely, with several sightings from high open moorland as soon as access restrictions were lifted. Despite light returning to much of the crippled region during June, the hamlet of Huccaby was still under the heel of what became in the end, for a selfish resident in need of exercise and birds, something of a ridiculous bore with a bad sense of humour. Having moved all the way to Dartmeet, we were still on the edge of the abyss of barred access and were not allowed through the garden gate. Lurid orange-red tape became an inescapable feature.

Meanwhile the eerily cattle-free field on the other side of the hedge beckoned, as did the lush oaks waving in the wood in Huccaby Cleave, temptingly visible from the window. After an excruciating summer of hearing inaccessible Pied Flycatchers singing in no man's land over the river, Yellowhammers clamoured for the seed I put down in hopes of attracting some birds, and when the surrounding fields became accessible once more, were to be found enjoying the rank and seed-rich grasses. This was the only noticeable change in bird occurrence here that I was able to attribute to FMD. Green and Great Spotted Woodpeckers bounded around over the no-fly zone and sometimes touched down over the border with us. But they seemed to have done just fine in the deserted surrounding woods – fine without any disturbance and without the unreal feeling of something having gone absurdly wrong.

Rewind to March and to the sound of gunfire all day long one Friday, to the stench of putrefying carcasses and to the interminable ballet of contractors' machines on the hillside, carving trenches and building expensive bonfires. My Huccaby recording area was reduced to 23 acres of private land and whatever territory could be viewed from there and from surrounding roads. This necessarily affected bird-recording, meaning that much of my 'patch' went unobserved for a significant part of the year. This resulted in the complete lack of coverage of one species - Whinchat. The constrained access situation actually meant that attention was focused on this core area more than it might otherwise have been, as by necessity more time was spent here than in a normal year. Few observers were fortunate enough to spend time on Dartmoor from February through to June. On that score, FMD might be said to have had some benefits – and with species such as Hen Harrier, Goshawk, Woodlark, Black Redstart, Ring Ouzel and Firecrest enlivening a grey year, who could complain?

THE GRAND WESTERN CANAL
Ray Jones

The FMD outbreak, which virtually closed the canal (with the exception of the urban areas of Tiverton and Sampford Peverell) from 28 February, originated some 40+ miles away in the Holsworthy area. However, Devon County Council deemed it desirable to close all public footpaths in Devon in the belief that the disease could be spread by walkers. It was hoped that the closure of the footpaths would only be of a few weeks' duration. In the event footpaths were closed for over two months. As time progressed towards the end of March and on into April, it was hoped that the lack of disturbance by the normal activities along the canal would, at least, allow some wildlife to take advantage of the closure.

By the end of April, FMD in Devon seemed to be under some sort of control and the areas around the canal were thankfully clear of the disease. This being the case, it was decided to view a few selected parts of the canal from bridges where there was road traffic access: i.e. Battens Bridge and Ebear Bridge.

It should be noted that not more than one bridge was visited in any one day, and that disinfectant routines were also carried out.

The one thing that was immediately noticeable was the fact that birds were making much more use of the towpath than normal due to the lack of disturbance. This particularly applied to the major species of the canal - Coot, Moorhen and Mallard, although other species such as Grey Heron and Mute Swan were also using the towpath to some extent.

On the re-opening of the canal on Friday 4th May (with the exception of an area between Ayshford and the access at Westcott) prior to the first Bank Holiday in May, I walked some length along it to try to determine some of the effects of the closure. The two facts that were immediately noticeable were the large amount of vegetation growth, and the number of birds breeding on the towpath side of the canal. This was particularly applicable to Sedge and Reed Warblers, which normally (but not always) nest on the far side of the canal; this year breeding had taken place on both sides. This also applied to Coot, Moorhen and, in one case, to a pair of Mute Swans which nested almost on the towpath near Rock. However, there did not appear to be any difference in the breeding activities of Mallard.

There were also Sedge and Reed Warblers breeding in areas of the canal outside of their normal breeding areas, i.e. Rock to Sampford Peverell. Sedge Warblers were breeding as far down the canal as Holly Dam, and the Reed Warblers as far north as Ebear. This expansion of breeding areas may be pure coincidence, but it could also be that the lack of disturbance allowed them to expand into areas that do not normally hold breeding warblers. It is to be hoped that the new territories settled in 2001 will continue to be used in the coming breeding season.

It would appear from these casual observations that the normal activities of the canal do affect its breeding birds, in that under normal circumstances they nest on the far side, with very few nests being normally found in the towpath side. Unfortunately, due to the restrictions caused by FMD, the normal annual counts of breeding birds were unable to take place, and it was not possible to determine the total numbers of pairs, which would have enabled a clearer comparison to be made with previous years. Therefore the true effects of the closure will probably never be accurately known.

It was a great shame that, due to political pressure (and not wildlife considerations), the whole of the canal was re-opened for the Bank Holiday at the beginning of May, as I have no doubt that some disturbance was thereby caused to those birds still nesting on the towpath side of the canal, together with a more limited disturbance to those on the far side. If the re-opening had been delayed until the Bank Holiday at the end of May, almost all birds would have finished breeding and any disturbance would have been of a minimal nature. However, as usual, wildlife lost out.

ROSE-BREASTED GROSBEAK ON LUNDY

Steve Cooper (Cambridgeshire)

Finally 6 October had arrived, and as the annual bout of Lundy sickness had reached fever proportions, the relief of once again walking around the wooded slopes of Millcombe Valley came as the perfect cure. Fellow Cambridgeshire birders Ade Cooper, Richard Patient and myself slowly made our way down to the walled gardens, high with the usual fantasy rarity expectations, where every call and movement was met with a rapid trio of birding reflexes. On approaching the last walled gardens near the pond, Richard and I had drifted ahead, engaged in a conversation on the possibilities for the forthcoming week when, as we moved closer to look over a wall, we were met with an explosion of head stripes and wing bars as a robust passerine shot to the back and alighted in a small elder tree. There, in all its glory, sat a Rose-breasted Grosbeak! Fantasy had turned into reality after only 80 minutes on the Island, and the adrenalin rush caused by the sight of such a familiar American vagrant put us into orbit, especially when the bird moved to the top of the tree to reveal the stunning crimson underwings of a male bird.

Such is the highly distinctive appearance of a Rose-breasted Grosbeak that a long list of written details of its plumage would not do the bird any justice, but fortunately, being a very confiding and photogenic individual, the photograph (see page 7 of the centre colour pictures) of the Lundy bird reproduced in this report portrays what a handsome species this is. The coarse, heavy streaking on the underparts, as well as the crimson colouration of the underwing and a few underlying breast feathers of male, separate this species from its more western cousin the Black-headed Grosbeak in the highly unlikely event of one ever turning up here.

Over the next four days the Grosbeak remain in the valley, increasingly to be found on the seed that we had strategically provided on the grass in front of our cottage. Here he was usually first bird down in the glow of pre-dawn and would often remain for up to two hours at a time, finding the split maize, sun flower seeds and wheat much to his liking. It is perhaps of interest to briefly describe other food items taken. On one occasion a large black beetle made an ill-fated landing adjacent to the Grosbeak and was promptly dispatched with some relish. In the walled gardens, a shallow trickle of water flows narrowly through, and is covered by a blanket of a crowfoot-type water plant - this was where the Grosbeak was often to be found, as when first discovered. The assumption that it was feeding on seeds was dismissed when, on closer inspection, many lentil-sized, shiny, bottle-green beetles were found, and assumed to be providing an interesting change of diet. Of the four previous Rose-breasted Grosbeaks that I have had the fortune to see this side of the Atlantic, all have found blackberries to their liking, and this one was no exception. Elderberries too proved popular. These were popped and the tiny seed then eaten, the cracking of the seed on one occasion being clearly audible at a range of just eight feet.

The seed that we had provided attached a gathering of up to 70 birds including a Common Rosefinch that, when spooked by various noises, Peregrines or sheep, flew into nearby bushes. This induced the Grosbeak to call on only a very few occasions, and the call could not be better described than by the words in the excellent new Sibley guide: 'like sneakers on a gym floor', a very accurate description! The presence of most species within the feeding flock caused no distraction whatsoever, but the Grosbeak showed a distinct dislike for the humble Blackbird whose presence in close proximity was met with a Rambo-like aggression, the Grosbeak always being very much the dominant bird (as well he would be with that bill!). The only reason I can suggest for this targeted lack of tolerance could be the vague resemblance of Blackbirds to the parasitic Cowbirds found back in the Grosbeak's homeland.

As with many of the other migrants present around that time, the Grosbeak departed on the starry night of 9 October, thus bringing to an end a Lundy episode that we all felt privileged to have witnessed. This was the second record for Lundy and Devon, the first being in 1985, and only the 26th record for Great Britain.

Rose-breasted Grosbeak - Lundy, 12 Oct 2001 - *Ade Cooper*

Rose-breasted Grosbeak - Lundy, Oct 2001 - *Richard Patient*

BLACK-FACED BUNTING ON LUNDY - FIRST FOR DEVON

Richard Patient (Cambridgeshire)

Steve and Ade Cooper and myself arrived on Lundy on 6 October 2001 for our fourth week-long visit to this wonderful island in successive autumns. Our 2001 trip got off to a flying start – within 90 minutes of our arrival on the island, Steve and I found a fine first-winter male Rose-breasted Grosbeak in Millcombe Valley (see separate article), but as the week progressed, the number of migrants dwindled. On a foggy 11 October, the only new species for the trip list was a Lesser Whitethroat, but as we made our way back from The Marisco Tavern that evening, we noted that the wind had turned to a southeasterly for the first time during the week, and it remained as such the next morning.

On 12 October I took my telescope with me as I wanted to have a closer look at the Lesser Whitethroat. I met Ade in Millcombe Valley and we stood at the top of the valley watching the Lesser Whitethroat until it disappeared. At 09:15h I heard the 'tik' of a rare bunting, and glancing around I noted a small bird flying towards us with Meadow Pipits. It had white outer tail feathers and luckily it landed among House Sparrows in a nearby bush. Raising my binoculars, I could clearly see a bunting, superficially rather like a Reed Bunting, and indicated it to Ade. Changing to 'scope views I quickly checked a few features – the legs were pink, as was much of the lower mandible, there was a bold dark malar, the primary projection was short with three primary tips visible beyond the tertials, the rear flanks were buff with heavy dark streaks and the belly was white in contrast to the greyish cast to the breast and much of the head. I said to Ade "I think it's a Black-faced", to which he replied "So do I!" I then added "I'm not joking" and he responded "Nor am I!" The bird sat for several minutes and then flew to a more distant tree before dropping down out of sight.

Although we were both fully confident of the identification, we wondered what a records committee might make of such an extreme rarity seen relatively briefly by just two observers, and were keen for others to see the bird, and to get a fuller description and photos. I was also concerned that the skulking nature of this species could make this difficult, although we were fortunate in that it was probably the best day of the week, with excellent visibility and no rain. A sea mist meant that we could not see the Devon coast and this was probably responsible for keeping some birds on the island; although, despite this, Chaffinches, Swallows and Siskins in particular were confidently heading out to sea in numbers during the day.

Ade left to summon other observers and fetch his 'scope. Frustratingly I had not managed to contact Steve via his mobile phone whilst the bird was in sight, but did manage to get him just as it dropped down out of sight - he soon appeared running up the valley! Andy Jewels happened to walk past within a minute of this and I told him that we had just seen a Black-faced Bunting. AMJ and SLC then headed to check the area into which it had disappeared whilst I remained where I was, so that we had an extra chance to keep track of it. Soon a party of Meadow Pipits was flushed, and as they flew overhead towards the church, I noticed that the bunting was with them - but unfortunately the others had not seen it. We headed after it but failed to find it until I located it at the top of St John's Valley, under a gorse bush. I whistled to SLC and had it ready in my 'scope for him – finally a third person had seen it! However, it was not very obliging and only gave brief glimpses. I ran off to fetch AMJ and sent him back to join SLC – he also had very brief views before it again flew off towards the Castle, at which point Ian Kendall finally appeared.

We split up but could not relocate it around the Castle – we were hoping that it had not headed out to sea. IK wished this even more fervently when he realised that the message he had received that we were looking for a Black-*headed* Bunting was incorrect on one crucial word! I headed back to the top of St John's Valley just as the bunting flew over the stone wall, again landing out of sight, so I waited here until the others appeared – luckily more or less simultaneously - a few minutes later. The bunting then hopped into view at the base of a stone wall and all five of us managed some 10 minutes constant viewing of the bird in the open as it worked its way along the wall base before flying off again. At 60x magnification the views were good, all agreed with the identification, and the news was released nationally. The bird then showed intermittently until early afternoon with a final sighting at 16:00h. Unfortunately there was no sign of it on 13 October, but a small fall of crests (including a number of Firecrests) and probably six different Yellow-browed Warblers seen then perhaps give a clue as to its origin.

The following provides a brief description of this Asian vagrant.

Behaviour. The bird was only easy to locate when in flight, in part due to its typical bunting shape and flight manner, in part due to the absence of other buntings on the island, and also aided by flight calls when given. At other times it would grub around unobtrusively under bushes and could be extremely dif-

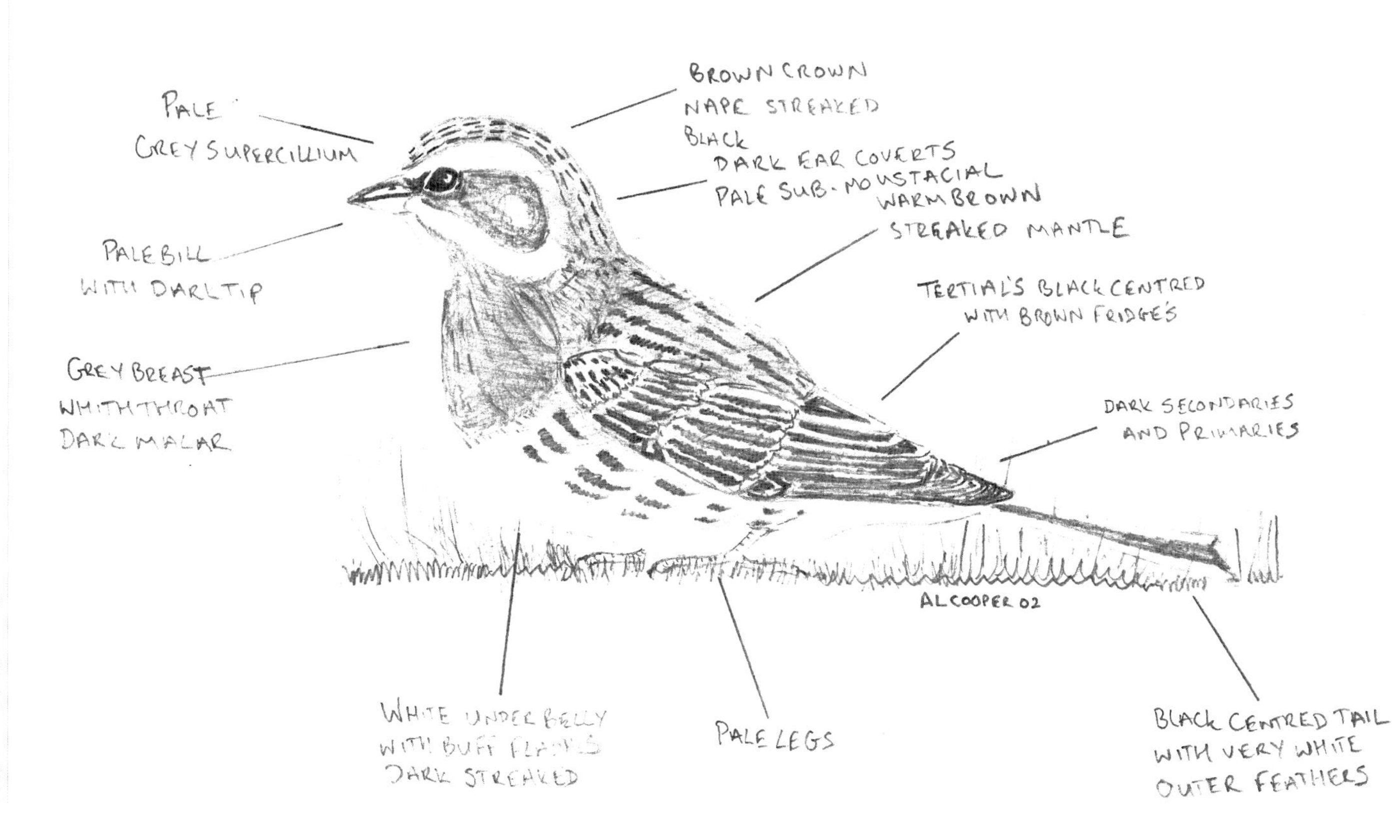
BLACK-FACED BUNTING, LUNDY, 12TH OCT 2001
PALE
GREY SUPERCILLIUM
BROWN CROWN
NAPE STREAKED
BLACK
DARK EAR COVERTS
PALE SUB-MOUSTACIAL
WARM BROWN
STREAKED MANTLE
PALE BILL
WITH DARK TIP
TERTIAL'S BLACK CENTRED
WITH BROWN FRIDGE'S
GREY BREAST
WHITH THROAT
DARK MALAR
DARK SECONDARIES
AND PRIMARIES
AL COOPER 02
WHITE UNDER BELLY
WITH BUFF FLANKS
DARK STREAKED
PALE LEGS
BLACK CENTRED TAIL
WITH VERY WHITE
OUTER FEATHERS

ficult to locate. The bird proved very flighty throughout the day and was often spooked by Meadow Pipits flying up and calling. When apparently alarmed by the presence of observers the crown feathers were raised, giving a pointed rear to the head rather reminiscent of Rustic Bunting.

Bare Parts. The bill was short and pointed, with a straight culmen. The whole of the upper mandible and the tip of the lower mandible were dark grey, contrasting noticeably with the pink base to the lower mandible. The latter was the first point I checked upon seeing the bird at rest and it was conspicuous from the outset, as were the pink legs.

Size. A small bunting, judged to be about the size of a Reed Bunting.

Plumage:

Head: contrastingly patterned. Median crown stripe greyish, contrasting with dark brown / blackish lateral crown stripes. Supercilia greyish from the bill to the (greyish) nape. Ear coverts greyish and surrounded by a dark brown border. Paler brownish spot in lower rear of ear coverts behind eye. Bold dark (blackish) malar stripe contrasting with the white submoustachials, the latter tracing the lower edge of the ear coverts and kinking upwards at the rear, ending underneath the pale spot referred to above. Throat whitish.

Upperparts: Mantle unremarkable – brownish feathers with dark brown centres. There was no sign of the paler 'tramlines' often shown by Reed Bunting. Wings surprisingly similar to those of Reed Bunting. Brownish double wing bar formed by tips to median and greater coverts. Tertials classic *Emberiza* pattern – each feather had chestnut fringes contrasting with a dark brown internal pattern reminiscent of two lobes of an oak leaf. Primary projection rather short – three primary tips counted as extending beyond the tertials. Rump unstreaked grey brown – seen clearly when looking down on the bird as it fed below us in St John's valley.

Underparts: Breast obviously greyish, with quite a lot of fine darker browner streaks within this. The streaks also continued down the flanks, increasing in size towards the rear flanks, where they were particularly bold. The grey stopped sharply on the lower breast, almost in a pectoral band. This was not a clean division however as towards the breast sides some brown streaking extended downwards from the grey. The belly was off-white and contrasted strongly with the grey breast. The flanks were noticeably washed with buff, and the vent a clean white.

Tail: A typical bunting tail – largely blackish, browner in the centre, with white outer tail feathers. There were two white outer tail feathers on each side, giving the impression of much white when the bird was in flight. The white also showed well when the bird was hopping around as it had a habit of flicking its tail open sideways.

Call. A very useful feature. The call was a 'rare bunting 'tic'', but not as hard in tone as that of other rare buntings such as Little or Rustic. The bird was heard to call several times during the day, but the calls did not vary.

This vagrant from eastern Asia has only two accepted records in Great Britain to date – in Greater Manchester March-April 1994 and in Northumberland in October 1999 (Rogers *et al*. 2000). Another was seen on Fair Isle, Shetland, just eight days after the Lundy occurrence (French 2001), while one in Humberside in May 2000 may have been an escape (*Birding World* **13**: 182-183).

There are three subspecies: *spodocephala*, *sordida* and *personata*. The last inhabits the extreme Far East and has a yellow belly at all times. The Manchester bird was attributed to the most westerly form *spodocephala* (D Parkin, in Alker 1997), to which it is likely that the Lundy record also refers. On geographical grounds this is the most likely, and additionally, *sordida* has a more yellow belly and greenish head (Bradshaw 1992) not shown by the Lundy individual.

Ageing and sexing is complex. The very noticeable greyish cast to the head and breast of the Lundy bird, shown well in the photographs, suggests a male, and as adult winter males recall those in breeding plumage (Bradshaw 1992), it seems reasonable that the Lundy bird was a first winter male *Emberiza spodocephala spodocephala*.

References

ALKER, P J (1997). Black-faced Bunting: new to Britain and Ireland. *British Birds* **90**: 549-561.
BRADSHAW, C (1992). Field identification of Black-faced Bunting. *British Birds* **85**: 653-665.
FRENCH, P (2001). The Black-faced Bunting on Fair Isle. *Birding World* **14**: 437-438.
ROGERS, M J *et al*. (2000). Report on rare birds in Great Britain in 1999. *British Birds* **93**: 512-567.

SIBERIAN STONECHAT AT START POINT - FIRST FOR DEVON
(Saxicola torquata maura/stejnegeri)
Mike Langman

On 26 September 2001, Mike Langman (ML) and Bill Macdonald (BMc) had gone to Start Point after seeing the previous night's forecast – overnight rain with light - moderate SE winds, which all looked very promising for migrants. Arriving at dawn, the first stop was the lighthouse and its tamarisk bushes, but these were surprisingly birdless, and, as even the seawatching was not up to much, ML and BMc decided to head back up to the denser vegetation of Start Farm.

At about 08:45h, and halfway up the road from the lighthouse, both observers noticed some bird movement on the lower slope of bracken involving several Stonechats and a Whitethroat, and then ML picked out a small pale buffy bird about 150 feet away partially hidden by the bracken. He could see a whitish throat, buffy orange upper chest and a fairly bland face with a slight dark mask. Initial thoughts were of a small Black-eared Wheatear, but the bird's size and habitat were both wrong. ML quickly directed BMc onto the bird and both observers set up their telescopes. The bird was fairly inactive and sat for two or three minutes facing up the slope. ML suggested to BMc that the bird was probably a Siberian Stonechat, but much better views, and all plumage features, particularly the rump, needed to be seen. During the next hour and 15 mins that ML and BMc watched the bird, it was observed down to perhaps less than 40 feet for lengthy periods through both telescopes and binoculars.

Description

Eventually the bird started to move around. As it turned, the most striking feature was the paleness of the bird when compared with the surrounding Stonechats; it was much more the colour of a Whinchat, though without the broad supercilium. The bird also had whitish edges to tertials, secondaries, primary tips, median and lesser coverts, alula and primary coverts, all of which were black-centred. The greater coverts were also black-centred with pale whitish tips and pale buffy edges, though the innermost ones appeared completely white, as on most Stonechats. The mantle and scapulars were a buffy greyish colour, with brownish streaks created by dark centres to the feathers.

The head was rather plain, and although showing a stronger buffy supercilium than the nearby Stonechats, this was nowhere near as striking as a Whinchat's. The supercilium was most prominent in front of the eye and extended up over the top of the bill. More obvious was a short blackish mask, mostly behind the eye, but extending underneath it to a small area in front. The eye had a whitish eye-ring. The cheeks were buffy-grey with darker streaks. The crown had a buffy-grey ground colour with darker speckles. The nape was a buffy-grey with only very faint darker streaks. The throat was pure white, with no dark smudges or centres to any of the feathers.

A couple of times the bird flew up to catch flies and we could then see an extensive pale buffy rump which also had a hint of orange, but it was impossible to establish the extent of this area. The tail of the bird was entirely black with a thin whitish outer edge to the entire length of the outer tail feathers on both sides of the tail. More extensive were the pale whitish/buff tips to all the tail feathers. The tail did not have pale bases to the feathers like Whinchat and the *variegata* race of Stonechat. The primaries extended to cover the upper tail coverts so, again, the observers could not see the rump.

Eventually, however, excellent views were obtained of the rump when the bird hovered long enough for us to 'scope' it, but even better when the bird sat with its back to us, drooped its wings and preened. The rump and upper tail coverts were indeed completely pale buffy-orange in colour, with perhaps a little more orange on the uppertail coverts. There were no darker marks or centres to any of these feathers.

The underparts were mostly whitish with a richer orange-buff upper chest that was strongest coloured on the sides of the breast where it met the mantle and scapulars. The flanks were also pale orange-buff where they disappeared up under the overlying primaries. The belly was whitish, as were the under tail coverts and vent area. The bill, legs and feet were completely jet-black.

The shape, size and character of the bird appeared to match that of the surrounding Stonechats, and if it were not for its paleness it could easily have been overlooked. News of the bird was released at about 08:50hrs but unfortunately it disappeared at about 10:00h and could not be relocated.

Both ML and BMc are 100% confident that this bird was a Siberian Stonechat. This is Devon's first accepted record of this race of Stonechat, and it is a race that is likely to gain full species status in the near future (Wink *et al*. 2002).

Note:

Another record of Siberian Stonechat, at Lannacombe on 13 Oct 2001 (ARHS, DS), has now been accepted by BBRC.

References

WINK, M, SAUR-GURTH, H, & GWINNER, E. (2002). Evolutionary relationships of stonechats and related species inferred from mitochondrial-DNA sequences and genetic fingerprinting. *British Birds* **95**: 349-355.

Siberian Stonechat - *Mike Langman*

REDPOLL SHOWING CHARACTERISTICS OF MEALY REDPOLL

Matthew Knott

In early February 2001 I found a redpoll on Topsham Playing Fields. It appeared greyer and more coldly toned than accompanying Lesser Redpolls, but I only had brief views of the bird front-on. The features I noted were enough for me to consider the possibility of Mealy Redpoll, but I was unable to find the bird on subsequent visits over the next two weeks. Redpoll taxonomy was very much on my mind because the British Ornithologist's Union had just, on 1 January 2001, split Lesser Redpoll *Carduelis flammea cabaret* from the Common Redpoll group (Mealy *C. f flammea*, Greenland *C. f rostrata* and Icelandic *C. f islandica*).

On the afternoon of 16 February, I again found the bird, feeding on what I assume were buds in a bare tree. This time I was able to note striking white tramlines, decent white wing bars, a broad white rump streaked black/grey and the overall pale colouration lacking strong ochre/buff tones. The undertail coverts at times appeared wholly white, though there did prove to be a narrow grey streak on the longest undertail covert and limited streaking on other undertail covert feathers. In beautiful late afternoon sunshine the bird appeared very pale and not unlike photographs I've seen of Arctic Redpolls! It was associating with Goldfinches and no other redpolls were around for comparison.

According to all the literature I have on redpolls, this bird was showing all the characteristics of a Mealy Redpoll. I made a few phone calls that evening in order to get the opinions of other birders. One of the problems of living down in Devon is that we are not even used to seeing Lesser Redpolls, let alone Mealies! On the occasions when we do see redpolls they are invariably flying high overhead. Most observers who have seen Mealies saw them when twitching Arctics and then they were probably scrutinising the largest, palest Mealies.

Over the weekend the bird was seen by a number of observers, in gardens bordering the playing fields and coming to seed along a stretch of footpath, always associating with Goldfinches. The bird was however frequently disturbed by dog-walkers. Most observers agreed that the features pointed towards Mealy Redpoll, but so far we hadn't had a side-by-side size comparison with Lesser Redpoll but after much deliberation the news was released that it was indeed a Mealy Redpoll.

On the morning of Monday 19 Feb, I again went to see the bird. This time it was feeding on seed with Lesser Redpolls and I had an uncomfortable sinking feeling as I realised it was apparently the same size as the Lessers and, in actual fact, didn't appear markedly different in overall colouration. It was slightly paler than the accompanying Lessers but I was expecting the latter to be much smaller, browner and dingier. Additionally, in duller light the bird looked duller and not 'frostier' as I had anticipated. Any similarity to Arctic Redpoll had vanished, though it always looked more striking perched up in a tree.

Over the next two weeks I saw the bird, by itself, several times. Interestingly it remained in company with Goldfinches and was only sporadically and briefly joined by Lesser Redpolls. Many observers were still favouring it being a Mealy and Peter and Carole Leigh's photographs seemed to support this.

On the morning of 4 March, after a light covering of snow, John Gale and I were able to watch the bird coming to seed with a small number of Lesser Redpolls. I again had the sinking feeling as it 'blended in' with the others. There were two or three 'classic' buff and brown Lessers but there were at least a couple of others that approached our bird in ground colour. The small gathering of Lessers between them showed grey-white tramlines, pale rumps (exact colours not discerned), pure white wing bars and the 'trousers' noted on our bird. What's more, John felt that one bird was larger and greyer than our bird! It had to be said, however, that no other individual showed all these features together.

In short, all the features that made our bird a Mealy are exhibited by Lesser Redpolls to some degree. Neither John nor myself (and subsequently other observers) feel that our bird can confidently be identified as a Mealy. There is the highly unlikely possibility that there are even more poorly marked Mealies

Various Lesser Redpolls - (Topsham Playing Fields - Feb 11, 2001) - *Matthew Knott*

present, but, given the rarity of the species in the South-West, and the fact that 2001 was not an 'invasion year', this seems improbable. None of the texts we've looked at hint at the variability of Lesser Redpoll, but skins, kindly photographed at the Natural History Museum in Tring by John Gale, show them to be anything but straightforward.

After the bird's departure, some experienced, high-profile birders were sent copies of the photographs. Each, *without exception*, has identified the bird as a Mealy Redpoll. Most recently, Lars Svensson has commented:

> *"In my experience, February and early March is too early for a Lesser Redpoll to get its plumage bleached and abraded enough to look so Mealy-like greyish and white. It should have had more of the tawny or ochrous tinge associated with cabaret left at this date. I think it is less reliable to base an ID on size and proportions than on plumage colours and if it is not an undisputed Mealy female, it should at least be labelled 'unidentified intermediate Mealy/Lesser.'"*

In December 2001 my brother and I travelled to Titchwell in north Norfolk to look at a mixed flock of Arctic and Mealy Redpolls. I was instantly struck by their stockiness and overall pale colouration that simply didn't tally with my impression of the Topsham bird. I spent the 2001/2002 winter period tracking down small parties of Lesser Redpolls that all looked typically ochre-toned. In short, I came to the conclusion that I could, personally, reach no conclusion! In my records the bird remains a 'Redpoll sp.'. The whole episode sparked a lot of useful controversy and served to highlight some of the difficulties with redpoll taxonomy, but hopefully it won't be long before an *undisputed* Mealy Redpoll is discovered in Devon.

Redpoll species - (Topsham Playing Fields - Feb 16, 2001) - *Matthew Knott*

SWIFTS IN A PLYMOUTH TOWER

Leonard Hurrell

When the outside of our house was re-rendered in 1961, nest boxes suitable for Swifts were installed inside the tower, with holes and tunnels made to allow access through the thick stone walls. However, it was 38 years before Swifts made use of these purpose-built facilities! When a pair eventually became established, our next move was to install an infra-red camera above the nest box to facilitate detailed observations in subsequent years. Although the equipment transmits a monochrome image, the Swift's behaviour during both day and night is not affected, and is thus entirely natural. All activity can be watched in one of our living rooms, and video recordings made. Some of the highlights of the observations in 2001 are described below.

The first excited party of Swifts, screaming as they passed the site, was noted on 12 May. This was also the day when both adults first entered the nest box, one at 15.00h and the other at 18.00h. Two days later, nest-building activity started, being observed at 05.15h. During the next few days they were not in the site much during the day, but usually came in for the night between 20.00 and 20.45h. Sometimes there was a period of quite frenetic activity with nest building, preening, allopreening and occasional intervals of dozing. During the night there were longer periods of sleep, but even then, there were occasional episodes of vigorous preening. Swifts do not seem to sleep soundly for long.

The first egg was laid on 22 May and on 23rd a pair was display-flighting past the tower with fast, shallow wing beats. This could well have been the resident pair. A second egg was laid on 24th and a third on 26th. On this day the incubating bird received a small item of nesting material brought in by the other adult and added it to the shallow nest cup with the rapid up and down head movements associated with salivary secretion. While engaged in this activity, the incoming adult shuffled beneath the first one and took over incubation. Every fragment of grass and all feathers used in the construction of the nest are collected on the wing, and the addition of such material goes on through the nesting period.

Occasional screaming parties continued to race past the tower during the incubation period. On 12 June, a party of three appeared about to enter one of the unoccupied sites, flying up close to the entrance hole but then veering away again. On 13 June, during a change-over of adults, some broken eggshell and the first youngster to have hatched could be seen in the nest cup. By 16 June, all three had hatched and a feed was observed.

During the next three days it gradually became evident that while two youngsters were thriving, it was doubtful if the third would survive. It was much smaller than its siblings and its food requests were becoming weaker. It was simply not competing effectively. Each food bolus delivered by the parents was fed to the nearest and most vociferous open mouth, and the smallest had expired by 21 June. The sky had been overcast, with a cool wind blowing and periods of rain during the days since the hatch. An unusual observation at this time involved one of the adults coming in at roosting time with a fairly large winged insect, probably a cranefly, which it then appeared to swallow itself rather than feeding to the young.

On 9 July, the weather was windy and cool with occasional drizzle, but did not seem to present particularly unfavourable feeding conditions for insectivorous birds. However, both adults remained in the nest until 13.30h, before both went out for a period of hunting.

On 24 July at 21.00h, a flight of 19 Swifts was high over Peverell, and since one of our juveniles had departed that afternoon, we wondered if this was a party of non-breeders and juveniles on the move. Early on 26 July we watched outside the tower since no Swifts were visible through the camera, and it seemed that the second juvenile was in the twenty-four inch tunnel that leads from the nest box through the stone wall to the outside world. There was a considerable amount of calling from the tower between 07.40 and 08.20h, before the pale throat and forehead of the juvenile appeared at the exit hole. After a little hesitation, it launched forward in a dive, passing between the high evergreen trees and clearing the dense shrubbery. The parent followed soon afterwards, accelerating more confidently down the driveway close to the ground, the usual route taken by the adults. This was a day when there was a massive hatch of flying ants, which must have provided a generous food supply for young Swifts on their incredibly steep learning curve. Both adults returned to the nest to roost that night, but not the juveniles; it is well known that unlike Swallows for example, juvenile Swifts do not return to the nest site once they have left it.

Thereafter, only one adult returned to roost on both 27 and 28 July, before it also set off for that long trek to southern Africa. The difference in the adults' departure dates makes it seem unlikely that they accompany each other on the migration. It is, therefore, all the more remarkable that they had managed to arrive at their nest site at the beginning of the nesting season within three hours of each other.

Swifts - *Len Hurrell*

LOOKING BACK to the 1951 and 1976 Devon Bird Reports

Peter Goodfellow

It is vital that the Society's records are well-maintained year by year. Unless they are, we shall never know for sure what changes in status or behaviour have occurred, so our records and annual reports are essential resources for both historical and conservation reasons.

Devon Bird Report 1951

The *Devon Bird Report* for 1951 was the Society's 24th. Compared to the current report its 72 pages seem very slim, but there are nevertheless many fascinating records, and these are arranged by geographical area rather than the chronological sequence typical of recent reports. I suppose it is because I was an English teacher that I so enjoyed re-reading the President's Forward. S D Gibbard's turn of phrase is virtually nonexistent in present day ornithology:

> *"Here's riches! Garnered from part of the great number of records sent in by our members, part of the harvest of yet another year of patient observation and pleasant toil…Treasures too are here, including many Buntings, Black-headed, Red-headed, Lapland and Ortolan, Icterine and Melodious Warblers, Red-breasted Flycatcher…Of Spoonbills, a sight of eleven together [on the Teign estuary] is richness indeed, but even that is exceeded by a view of fourteen of these birds [on the Oar Stone]."*

Plenty there to worry our Rarities Committee! Among other rarities were the seven Avocets on the Tamar, and just one Little Egret ("so pleasingly described") seen on the Axe from June to August – but the Collared Dove had not even arrived yet. How things have changed!

Besides the Systematic List, there were reports on Swift migration and surveys on Black Redstart and Spotted Flycatcher. The Secretary's Report mentions the improved Wild Birds Protection Order for Devon on which DBWPS had been consulted throughout, and the great shock when the Order came before Devon County Council for final approval – Buzzard and Raven had been removed from the fully protected list and Heron was not protected at all! The DBWPS and RSPB protested strongly, and the Council rescinded their decision at a later meeting. We are *still* battling with other conservation problems.

It is striking how many behavioural records were published in 1951 compared with more recent *DBR*s: for example, the Willow Warblers that fed their chicks 29 times in an hour; the Song Thrush that deliberately soaked paper and lined its nest with the resulting *papier maché*; the analysis of the 161 feathers in a Wren's nest; the Buzzard that called 112 times in eight minutes from one perch; and the underwater swimming technique of a Moorhen. Are we too busy today, 'atlasing,' 'twitching' or doing counts for surveys, to notice such 'treasures'?

Devon Bird Report 1976

Twenty-five years ago, the 1976 *DBR* contained only 64 pages, and although the Systematic List ran to barely 40 pages, there were the now familiar reports on ringing, wildfowl counts (not yet enlarged to include waders and gulls), heronries and review of the year.

The Report starts spectacularly with a black and white photograph of a Quail on its nest – two pairs bred successfully that year, as did a pair each of Shoveler, Montagu's Harrier and maybe Wryneck. On the other hand, all 12 pairs of Dartford Warblers were confined to the East Devon Commons, and only one pair of Cetti's Warbler was proved to breed, at Slapton. The status of Cirl Bunting was maintained in the E, SE and S divisions, but there was a marked lack of records from the north.

To many modern birdwatchers, perhaps the most startling difference between the 1976 and 2001 Reports, is that then there were only *three* pages devoted to skuas, gulls and terns. Since then, our interest in, and understanding of, seabird numbers and movements has increased enormously.

It was interesting to read that Canada Goose numbers were "remarkably static." I expect many people wish that were still true. On the other hand, Avocets were now wintering on the Exe *and* the Tamar, the latter still winning the numbers game with a maximum of 70 against the Exe's 35. Whitethroats had "almost returned to pre-1969 status" after the dreadful population crash resulting from the drought in the African Sahel. That species' status still applies today, but how I wish we could now see the huge flocks of Linnets recorded in the south in the autumn of 1976 (300, 400, 500, 600, and biggest of all, *c.*2,000 flying south from Prawle Point on 24 October) and enjoy the Tree Sparrows seen in all areas of the county in the second winter-period, including a flock of *c.*100 at Colyton on Boxing Day. What a Christmas present that would be today!

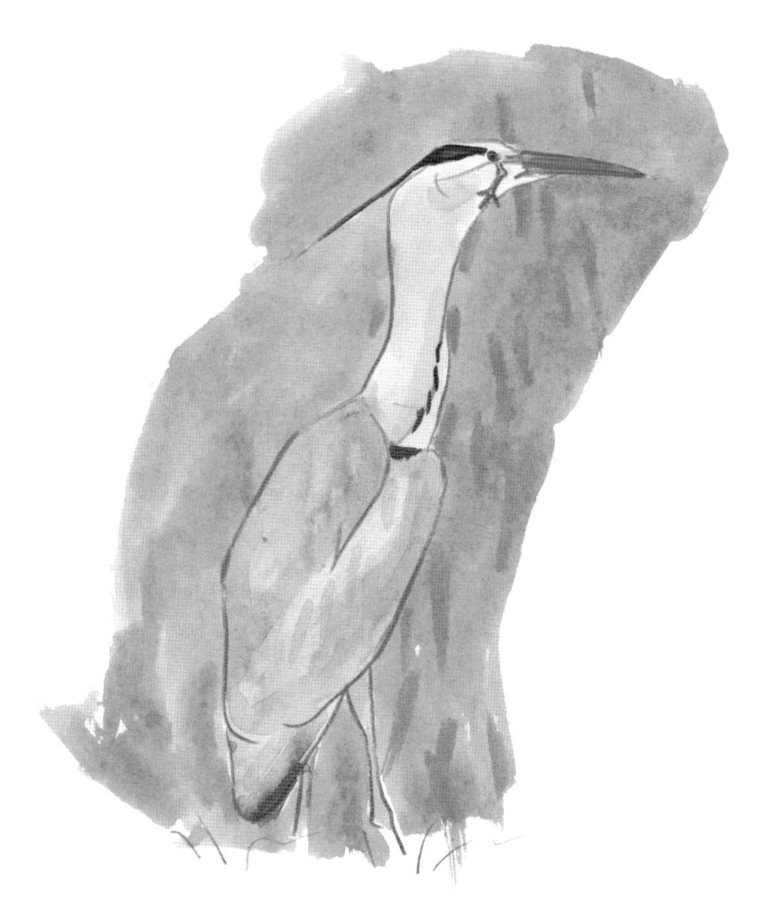

Grey Heron - *John Walters*

Devon Ringing Report for 2001

Compiled by Roger Swinfen

My thanks are once again extended to all those ringers who have sent me details of birds they have ringed in Devon during the year. Like any fieldwork, the necessary statutory and voluntary restrictions of movement and access imposed during the outbreak of Foot and Mouth disease affected ringing. The reduced number of nestlings ringed illustrates this most clearly. The number of species ringed is also low but Tufted Duck was added to our list. Hopefully, next year's report will show a renewed effort. Ringing is an extremely valuable tool for the study of birds. Not many animal species are so easily and effectively marked. I believe that it is at its best when the data gained is used in conjunction with other work, such as counts or detailed observations, to strengthen the case for conservation.

I wish to highlight one control that does illustrate a number of interesting points. Jon Avon, as Warden of the Stover Country Park, has closely monitored the breeding Nightjars on his patch. The birds had been found to be taking advantage of the ground clearance beneath the high voltage pylons and overhead cables to breed. Most members of our society know that when he was working at Haldon, Robin Khan kept a close watch on Nightjars in that area. Observations at other forestry sites on Dartmoor by Roger Smaldon and the Dartmoor Study Group, show how these birds are able to use other cleared areas at the right stages of regeneration. In this climate of increased interest in the species, Nik Ward decided to try and catch some adult birds in the dusk at Haldon. It was a surprise when one of the first birds he caught was found to be already ringed, and an even greater surprise when he found that it had been ringed as a chick at Stover, just a few kilometres along the A38, some 6 years earlier by Jon Avon.

The submission of ringing data to the BTO by computers "on line" has almost completely superseded paper schedules. No applications for an upgrade to an A Permit will be accepted from a ringer who does not have access to a computer. Any existing ringer without a computer is encouraged to seek another who does have a computer with the relevant software and is willing to enter both sets of data. As if this were not enough, the preferred programme for data submission has changed from the earlier B-ring to the new Integrated Population Monitoring Reporter or IPMR. For those of us who may be less fluent in computer operation, there has been a steep learning curve. The belief is that once data is held electronically, it will be much more readily available to be used by BTO scientists. The quality of this data is obviously of paramount importance, so there is a close monitoring of the standard of ringing and the training of ringers in the latest techniques. The British ringing scheme is considered a leader in the field and the competence of British ringers amongst the highest. Not bad for a group of volunteers.

Good ringing in 2002

Table 135; Contributors to the 2001 Report.

Ringer	Adults	Pulli	Total
Jon Avon. JDA	173	61	234
Greg Conway. GJC	57	0	57
Peter Ellicott. PWE	352	0	352
Dennis Elphick. DE	125	0	125
Peter Goodfellow. PFG	36	0	36
Keith Grant. KRG	511	275	786
John High. JAH	281	126	407
Dave Jenks. DGJ	260	0	260
Tony John. AWGJ	267	78	345
Harvey Kendall. FHCK	22	216	238
David Morgan. DHWM	85	131	216
Patrick Moore. PCHM	1	0	1
David Price. DJP	715	206	921
Alan Searle. AS	241	0	241
Paul Stubbs. PLS	216	0	216

Ringer	Adults	Pulli	Total
Roger Swinfen. RCS	198	32	230
Jerry Tallowin. JRBT	98	20	118
John Turner. JNVT	3	59	62
Jon Turner. JT	741	0	741
Michael Tyler. MWT	109	111	220
Nik Ward. NCW	279	2	281
Richard Whiteside. RJW	54	17	71
Heather Woodland. HAW	77	57	134
Slapton Bird Observatory. SBO	538	4	542
Totals	**5439**	**1395**	**6834**
Other Contributors			
South Devon Seabird Trust. SDST			
Barn Owl Trust. BOT			
Gordon Vaughan. GAV			
Barrie Whitehall. BW			

Table 136; Devon annual ringing totals

Year	Adult	Pulli	Total	No. of Species
1990	10208	3579	13787	109
1991	8374	3989	12363	88
1992	10317	4171	14488	92
1993	8319	4158	12477	93
1994	8479	3161	11640	88
1995	9191	2882	12073	89
1996	10685	2380	13065	93

Year	Adult	Pulli	Total	No. of Species
1997	8293	2481	10774	86
1998	4960	2138	7098	73
1999	5080	2288	7368	71
2000	6417	2250	8667	75
2001	5439	1395	6834	71

Table 137; Birds ringed in Devon during 2001.

Species	Adult	Pulli	Year Total	71 to 2001 G.Tot
Storm Petrel	6	0	6	189
Cormorant	0	30	30	1251
Mute Swan	13	0	13	601
Tufted Duck	1	0	1	1
Sparrowhawk	5	0	5	271
Buzzard	1	0	1	60
Water Rail	1	0	1	35
Moorhen	4	2	6	47
Dunlin	4	0	4	3499
Wood Pigeon	9	2	11	165
Collared Dove	4	0	4	190
Barn Owl	8	87	95	2279
Little Owl	2	3	5	129
Nightjar	2	6	8	73
Swift	0	2	2	227
Green Woodpecker	1	0	1	55
G.S. Woodpecker	19	0	19	306
Swallow	23	97	120	14684
House Martin	22	0	22	1610
Grey Wagtail	2	0	2	636
Pied Wagtail	9	7	16	3298
Wren	86	0	86	6142
Dunnock	152	0	152	6603
Robin	199	27	226	9754
Black Redstart	2	0	2	19
Redstart	0	12	12	765
Stonechat	2	0	2	549
Wheatear	1	0	1	335
Blackbird	278	14	292	9402
Fieldfare	4	0	4	310
Song Thrush	40	4	44	2388
Mistle Thrush	2	0	2	215
Redwing	5	0	5	1824
Cetti's Warbler	15	0	15	825
Grhopper Warbler	1	0	1	221
Sedge Warbler	66	0	66	13316
Reed Warbler	105	0	105	14124
Lesser Whitethroat	1	0	1	315
Whitethroat	22	0	22	2191
Garden Warbler	8	0	8	1131
Blackcap	174	0	174	7909
Chiffchaff	187	10	197	10559
Willow Warbler	87	0	87	11904
Goldcrest	106	0	106	4998
Firecrest	17	0	17	232
Spotted Flycatcher	11	9	20	1286

Species	Adult	Pulli	Year Total	71 to 2001 G.Tot
Pied Flycatcher	39	548	587	22065
Long-tailed Tit	129	0	129	4351
Marsh Tit	11	8	19	965
Coal Tit	177	0	177	3363
Blue Tit	920	255	1175	45861
Great Tit	332	152	484	17883
Nuthatch	30	92	122	3196
Treecreeper	17	0	17	802
Jay	2	0	2	218
Magpie	3	0	3	180
Jackdaw	0	23	23	453
Rook	2	0	2	50
Crow	1	0	1	121
Starling	92	0	92	4971
House Sparrow	260	3	263	1888
Chaffinch	228	0	228	12053
Brambling	2	0	2	803
Greenfinch	1014	2	1016	25646
Goldfinch	125	0	125	2550
Siskin	232	0	232	5654
Linnet	5	0	5	872
Redpoll	2	0	2	603
Bullfinch	88	0	88	3725
Yellowhammer	13	0	13	1639
Cirl Bunting	2	0	2	370
Reed Bunting	6	0	6	4506
Totals	5439	1395	6834	

Table 138; Birds ringed in Devon during 1977/2000 but not in 2001.

SPECIES	Grand Total
Little Grebe	1
Great Crested Grebe	1
Shag	689
Grey Heron	6
Canada Goose	422
Brent Goose	70
Shelduck	2
Teal	1
Mallard	86
Eider	1
Kestrel	249
Merlin	1
Hobby	23
Grey Partridge	4
Pheasant	9
Spotted Crake	1
Corn Crake	1
Coot	6
Oystercatcher	3093
Ringed Plover	49
Grey Plover	10
Lapwing	37
Knot	73
Sanderling	7
Little Stint	2
Curlew Sandpiper	6
Purple Sandpiper	22
Ruff	2
Jack Snipe	24
Snipe	96
Woodcock	26
Black-tailed Godwit	10
Bar-tailed Godwit	82
Whimbrel	54
Curlew	223
Redshank	271
Greenshank	7
Green Sandpiper	2
Wood Sandpiper	1
Common Sandpiper	27
Turnstone	418
Black-headed Gull	278
Common Gull	14
Herring Gull	455
Great Black-backed Gull	148
Kittiwake	2
Sandwich Tern	34
Common Tern	8
Little Tern	5
Guillemot	115
Razorbill	6
Stock Dove	130
Turtle Dove	4
Cuckoo	16
Tawny Owl	229
Long-eared Owl	1
Kingfisher	338
Wryneck	3
Lesser Spotted Woodpecker	10
Woodlark	4
Skylark	180
Sand Martin	2129
Richard's Pipit	1
Tree Pipit	93
Meadow Pipit	651
Rock Pipit	123
Water Pipit	2
Yellow Wagtail	586
Dipper	1349
Nightingale	30
Bluethroat	9
Redstart	753
Whinchat	183
Ring Ouzel	51
Savi's Warbler	2
Aquatic Warbler	43
Great Reed Warbler	1
Icterine Warbler	4
Melodious Warbler	2
Subalpine Warbler	2
Barred Warbler	1
Pallas's Warbler	5
Yellow-browed Warbler	6
Radde's Warbler	1
Dusky Warbler	1
Wood Warbler	180
Red-breasted Flycatcher	1
Bearded Tit	81
Willow Tit	215
Penduline Tit	1
Red-backed Shrike	1
Great Grey Shrike	1
Woodchat Shrike	1
Raven	15
Tree Sparrow	27
Hawfinch	7
Blackpoll Warbler	1
Little Bunting	3

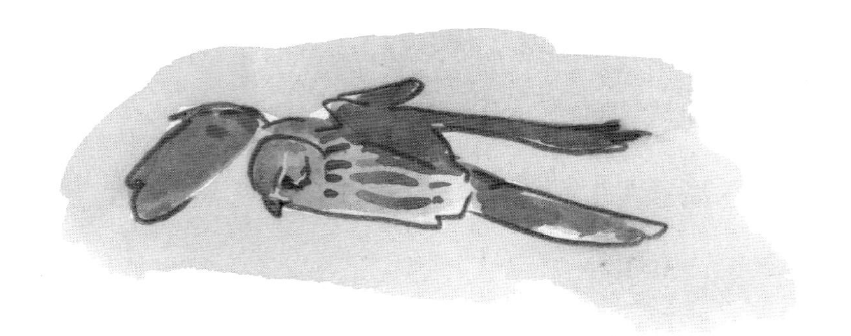

Kestrel - *John Walters*

Recoveries and Controls - Codes.

This is a selected list of birds controlled (i.e. ringed birds caught and released again by a ringer) or recovered (i.e. found dead) in Devon during 2000. Some records, especially of birds found after only a very short interval or close to the place of ringing, have been omitted in the interests of space.

Age Codes.

A code to express the age of a bird is adopted by the various European ringing schemes and is summarised below.

1	pullus (nestling or chick unable to fly)
2	full grown bird, precise year of hatching unknown.
3	Definitely hatched during the same year as ringing.
4	Hatched before the year of ringing, but precise year not known.
5	Definitely hatched during the year before the year of ringing.
6	Hatched two or more years prior to ringing, but precise year not known.
7	Definitely hatched two years before ringing.
8	Hatched 3 or more years prior to ringing.

Where it is known, the sex of the bird is given by M = male or F = female.

Codes for manner of recovery.

v	a ringed bird caught and released by a ringer.
+	shot or killed otherwise by man.
x	found dead or dying.
FR	ring number read in the field, e.g. by telescope.
cr	a bird carrying a plastic colour ring.

Table 139 - Recoveries and Controls

Species Ref. No.	Age Recovery Code	Ringing Date - - Date -	and Site and Site	Days Difference	Distance Travelled	Bearing Travelled
Cormorant	*Species: Many young birds do move across the Channel and return in their second or third year to breed.*					
Cormorant						
5176990	1	28/04/96	Great Mewstone, Wembury (RCS)			
	x	27/04/01	Newton Ferrers, River Yealm	1825 days	4km	62°
Cormorant	-					
5189452	1	06/05/96	Great Mewstone, Wembury (RCS)			
	x	30/12/99	Bouin, Vendee, France 46.58N:01.59W	1333 days	402km	157°
Cormorant	-					
5120623	1	03/05/98	Great Mewstone, Wembury (RCS)			
	x	24/02/01	Noss Mayo, nr.Plymouth	1028 days	4km	90°
Cormorant	-					
5120660	1	13/06/99	Great Mewstone, Wembury (RCS)			
	x	01/05/00	St.Andre des Eaux, Loire Atlantique, France 47.18N: 02.18W	323 days	358km	158°
Shag	-					
1342290	1	30/05/99	Great Mewstone, Wembury (RCS)			
	x	05/02/00	Petit Port, Guernsey, Channel Islands	251 days	149km	131°
Mute Swan	*Species: Due to the lack of submerged vegetation in Slapton Ley birds move to Shaldon to join the wintering flock there.*					
Mute Swan						
Z73926	5	02/01/94	River Yealm, Newton Ferrers (RCS)			
	C	06/02/01	Over Stratton, South Petherton, Somerset (RSPCA)	2592 days	111km	52°

Species Ref. No.	Age Recovery	Ringing Date - Code - Date -	and Site and Site	Days Difference	Distance Travelled	Bearing Travelled
Mute Swan	-					
Z73931	6	24/03/94	Sutton Harbour, Plymouth (RCS)			
	x	14/08/01	Saltash, Cornwall	2700 days	8km	298°
Mute Swan	05/10/01 released Chard Reservoir, fighting					
Z75449	6	10/08/95	Slapton Ley (RCS)			
	x	01/09/01	Kingsbridge Estuary (RSPCA)	2214 days	9km	246°
Mute Swan	-					
Z75463	5M	28/01/96	Kingsbridge (RCS)			
	x	09/01/01	Salcombe	1808 days	4km	198°
Mute Swan	-					
Z75486	6M	19/08/96	Slapton Ley (RCS)			
	C	21/02/01	Shaldon, nr.Teignmouth (RSPCA)	1647 days	29km	19°
Mute Swan	-					
Z75486	6M	19/08/96	Slapton Ley (RCS)			
	x water tank	17/04/01	Radstock, nr.Bath	1702 days	139km	37°
Mute Swan	-					
Z87784	3	23/01/97	Creech St.Michael, Somerset			
	v	20/01/01	Teignmouth (RCS)	1458 days	64km	212°
Mute Swan	-					
Z75470	3M	10/08/97	Slapton Ley (RCS)			
	x	29/01/01	Shaldon, nr.Teignmouth (RSPCA)	1268 days	29km	19°
Mute Swan	-					
Z87200	6F	12/02/00	Stover Lake, Newton Abbot (RCS)			
	x	02/06/01	Stover Country Park, nr.Newton Abbot (JDA)	476 days	0km	360°
Mute Swan	16/11/01 released ST8054					
Z87179	5	01/010/97	Sutton Harbour, Plymouth (RCS)			
	C	26/07/01	Sutton Harbour, Plymouth (RSPCA)	1394 days	2km	180°
Mute Swan	-					
Z87200	6F	12/02/00	Stover Lake, Newton Abbot (RCS)			
	x	02/06/01	Stover Country Park, nr.Newton Abbot (JDA)	476 days	0km	360°
Mute Swan	-					
Z94208	4	12/08/00	Slapton Ley (RCS)			
	x	09/01/01	Shaldon, nr.Teignmouth (SDST)	150 days	29km	19°
Mute Swan	-					
Z94209	4	16/09/00	Slapton Ley (RCS)			
	x road	13/12/01	Shaldon, nr.Teignmouth (SDST)	453 days	29km	19°
Mute Swan	-					
Z94220	6M	30/01/01	Kingsbridge (RCS)			
	x RSPCA	01/09/01	Kingsbridge Estuary (RSPCA)	214 days	1km	90°
Buzzard	The rescue did give it another 10 years.					
GH36192	8	22/06/90	plucked from sea off Dartmouth			
	released	23/06/90	Dartmouth (KRG)	-	-	-
Buzzard	-					
-	-	22/06/90	-			
	x storms	28/01/01	Weake Hill, Dartmouth	3872 days	2km	147°

Species Ref. No.	Age Recovery	Ringing Date - Code - Date -	and Site and Site	Days Difference	Distance Travelled	Bearing Travelled
Oystercatcher	*Species: Some more reports from this long running project indicating where our winter birds are breeding.*					
Oystercatcher						
FV63727	3	29/08/83	Dawlish Warren (DCWRG)			
	FR	20/07/01	Groote Keeten, Noord Holland, Netherlands 52.52N:04.42E	6535 days	613km	66°
Oystercatcher	-					
FV63989	8	29/09/89	Dawlish Warren (DCWRG)			
	x road	28/03/01	Brathens, Banchory, Grampian Region	4198 days	719km	5°
Oystercatcher	-					
FR92424	8	09/010/90	Dawlish Warren (DCWRG)			
	v	18/07/01	Fuglafjordur, Eysturoy, Faeroes 62.15N:06.48W	3935 days	1310km	351°
Oystercatcher	-					
FR41323	3M	05/12/96	Cockwood Harbour, River Exe (DCWRG)			
	x	02/05/01	nr.Rosemarkie, Fortrose, Highland Region	1609 days	779km	357°
Oystercatcher	-					
FC51402	8	17/11/00	Dawlish Warren (DCWRG)			
	x pred-ator	18/04/01	Vinkel, Noord-Brabant, Netherlands 51.43N:05.27E	152 days	631km	79°
BH Gull	-					
EB89348	3	30/12/96	Derriford, Plymouth (RCS)			
	FR	21/03/01	Utterslev, Copenhagen, Denmark 55.43N:12.31E	1542 days	1255km	62°
Herring Gull	*Species: The Heathfield birds were ringed as part of a DEFRA project studying gulls using landfill sites.*					
Herring Gull						
GF00445	1	04/06/92	Drakes Island, Plymouth (RCS)			
	x	30/06/01	Saltram House, River Plym, Plymouth	3313 days	5km	69°
Herring Gull	-					
GP73761	8	31/12/96	Derriford, Plymouth (RCS)			
	v	15/08/01	Raleigh Camp, Torpoint, Cornwall	1688 days	9km	232°
Herring Gull	-					
GP65158	5	20/04/00	Portland Bill, Dorset			
	v	21/05/00	Dowlands Cliffs (RJW)	31 days	45km	297°
Herring Gull	-					
GH98800	8	14/11/00	Heathfield Landfill, Newton Abbot (MAFF)			
	x garden netting	19/07/01	Teignmouth (SDST)	247 days	8km	103°
Herring Gull	-					
GH98697	8	15/11/00	Heathfield Landfill, Newton Abbot (MAFF)			
	x	26/05/01	Brixham (SDST)	192 days	22km	192°
Herring Gull	-					
GH98989	8	15/11/00	Heathfield Landfill, Newton Abbot (MAFF)			
	x	12/07/01	Shaldon (SDST)	239 days	8km	118°
Guillemot	-					
X60485	5	23/03/97	released Hope's Nose, Torquay (SDST)			
	x on beach	25/07/01	Cullen, Grampian Region	1585 days	805km	3°
Guillemot	-					
X59860	5	05/11/98	released Hope's Nose, Torquay (SDST)			
	x Erika oil spill	02/01/00	Sauzon, Morbihan, France 47.22N:03.12W	423 days	343km	176°

Species Ref. No.	Age Recovery	Ringing Date - Code - Date -	and Site and Site	Days Difference	Distance Travelled	Bearing Travelled
Guillemot	-					
X59837	4	29/04/99	released Hope's Nose, Torquay (SDST)			
	oiled	06/12/00	Thurlestone (SDST)	-	-	-
Guillemot	-					
-	-	29/04/99	-			
	released	11/04/01	Hope's Nose, Torquay (SDST)	587 days	33km	232°
Guillemot	-					
R05449	4	30/08/00	released Hope's Nose, Torquay (SDST)			
	x in care, oiled	31/01/01	Ile d'Ouessant, Finistere, France 48.27N:05.04W	150 days	249km	207°
Guillemot	-					
R05460	4	13/02/01	released Hope's Nose, Torquay (SDST)			
	x on beach	29/05/01	Trearddur Bay, Anglesey	105 days	322km	346°
Barn Owl	*Species: All the Barn Owls submitted have been from the two major projects in Devon. No reports of "wild" owls were received*					
Barn Owl						
FC12711	4F	02/07/95	nr.Yealmpton (BOT)			
	x	28/07/01	nr.Holbeton	2218 days	3km	128°
Barn Owl	-					
FC12756	1F	13/07/95	Uppacott, nr.Tawstock (BOT)			
	x	29/07/01	Umberleigh	2208 days	6km	128°
Barn Owl	-					
FC98158	1	20/06/98	Waffapool, East Putford (BOT)			
	x in barn	04/04/01	Tetcott, nr.Holsworthy	1019 days	21km	196°
Barn Owl	-					
FB00782	1F	04/06/99	nr.Awliscombe (BOT)			
	x road	17/02/01	nr.Killerton (BOT)	624 days	15km	263°
Barn Owl	-					
FC79872	1	25/06/99	Site confidential, nr.Newton Abbot (HAW)			
	x road	01/03/01	nr.Churston Ferrers	615 days	17km	131°
Barn Owl	-					
FB00852	1	04/07/99	Hutcherleigh, Halwell (BOT)			
	v	31/07/01	nr.Chillington (BOT)	758 days	8km	171°
Barn Owl	-					
FB00928	1F	14/06/00	nr.Chillington (BOT)			
	x power lines	27/02/01	West Alvington, Kingsbridge	258 days	7km	270°
Barn Owl	-					
FC83328	1F	21/06/00	nr.Lympstone (BOT)			
	? Road, released	10/01/01	Exeter area	199 days	8km	335°
Barn Owl	-					
FC97727	1	04/07/00	nr.Modbury (HAW)			
	v	22/06/01	nr.Ermington (BOT)	353 days	7km	285°
Barn Owl	-					
FC79884	1	08/07/00	nr.Meeth (HAW)			
	x	21/09/00	Okehampton (BOT)	75 days	8km	155°

Species Ref. No.	Age Recovery	Ringing Date - Code - Date -	and Site and Site	Days Difference	Distance Travelled	Bearing Travelled
Barn Owl	-					
FB13029	1	19/07/00	nr.Lewdon, Witheridge (BOT)			
	x road	28/02/01	nr.Clyst Honiton	224 days	25km	127º
Barn Owl	-					
FB13035	1F	22/07/00	nr.Modbury (BOT)			
	x road	23/07/01	A379 nr.Alphington (KRG)	366 days	45km	34º
Barn Owl	-					
FB13046	1F	26/07/00	Huddisford, Wofworthy (BOT)			
	x	14/01/01	nr.Lynton	172 days	48km	52º
Barn Owl	-					
FB13054	1F	27/09/00	Dartington (BOT)			
	x	10/01/01	Dartington (BOT)	105 days	0km	360º
Barn Owl	-					
FB13059	1F	12/06/01	Clyst Valley, Topsham (BOT)			
	x	19/12/01	Killerton	190 days	11km	360º
Nightjar	A very interesting control of this fascinating species					
RK61508	1	03/07/95	Stover Lake, Newton Abbot (JDA)			
	v	25/06/01	Great Haldon Forest (NCW)	2184 days	10km	56º
Grey Wagtail	-					
P049843	1	08/05/00	Ash Mill, South Molton (JAH)			
	x cat	03/05/01	Holcombe Regis, Wellington	360 days	26km	102º
Robin	-					
J633849	3J	17/08/00	Derriford, Plymouth (RCS)			
	x cat	29/06/01	Derriford, Plymouth	316 days	0km	360º
Blackbird	-					
RA48986	3M	03/07/94	Derriford, Plymouth (RCS)			
	v	29/01/01	Roborough, Plymouth	2402 days	2km	360º
Blackbird	-					
RK86268	6M	30/01/95	Ashridge, Sandford, Crediton (JAH)			
	x glass	29/11/00	Ashridge, Sandford, Crediton	2130 days	0km	360º
Sedge Warbler	-					
N350365	3J	15/07/00	Slapton Ley (SBO)			
	v	22/08/00	Floirac, Charente Maritime, France 45.28N:00.44W	38 days	578km	158º
Sedge Warbler	-					
N826518	3J	05/08/00	Slapton Ley (SBO)			
	v	12/08/00	Le Massereau, Loire-Atlantique, France 47.14N:01.55W	7 days	362km	159º
Reed Warbler	-					
N160197	3J	23/07/98	Greenstraight, Hallsands (AKS)			
	v	23/08/98	St-Seurin-D'Uzet, Charente Maritime, France 45.30N: 05.50W	31 days	567km	158º
Reed Warbler	-					
P260628	1	15/06/01	Chew Valley Lake, Avon			
	v	04/08/01	Slapton Ley (SBO)	50 days	135km	212º
Blackcap	-					
P110818	3J	15/07/00	Whitcombe, nr.Kenn, Exeter (KRG)			
	x	15/03/01	Beht, Khemisset, Morocco 34.25N:06.26W	243 days	1819km	187º

Species Ref. No.	Age Recovery	Ringing Date - Code - Date -	and Site and Site	Days Difference	Distance Travelled	Bearing Travelled
Blackcap	-					
P360662	4M	16/01/01	Alphington, Exeter (PWE)			
	x road	01/03/01	Alphington, Exeter (PWE)	44 days	0km	350°
Chiffchaff	-					
6M9125	3J	17/07/00	Slapton Ley (SBO)			
	v	24/03/01	Logrosan, Canceres, Spain 39.20N:05.29W	252 days	1225km	187°
Chiffchaff	-					
5Z5711	3M	01/010/00	Prawle Point (AKS)			
	v	05/01/01	Mar de Caes, Ribatejo, Portugal 38.57N:08.56W	96 days	1316km	198°
Chiffchaff	-					
6M1934	3J	17/08/01	Glencaple, Dumfries and Galloway			
	v	23/09/01	Slapton Ley (SBO)	37 days	522km	181°
Goldcrest	-					
6M9410	3M	07/010/00	Slapton Ley (SBO)			
	x	14/06/01	Manaton, Newton Abbot	250 days	39km	342°
Pied Flycatcher	*Species: A heavily edited selection from the many reports received from the vatious nest box projects in the county.*					
Pied Flycatcher						
K290042	1	02/06/96	Chittlehamholt (DHWM)			
	v	10/06/01	Castle Woods, Okehampton (FHCK)	1834 days	0km	360°
Pied Flycatcher	-					
J368822	1	12/06/96	Meldon Reservoir (FHCK)			
	v	13/06/01	Meldon Reservoir (FHCK)	1827 days	0km	360°
Pied Flycatcher	-					
K185226	1	05/06/97	Neadon Cleave (KRG)			
	v (4F)	06/06/01	Steps Bridge, Dunsford (DJP)	1462 days	9km	32°
Pied Flycatcher	-					
K906958	1	06/06/97	Dunsford Woods (DJP)			
	v	13/06/01	Meldon Reservoir (FHCK)	1468 days	0km	360°
Pied Flycatcher	-					
K979283	1	07/06/97	nr.Rifton, Stoodleigh (JAH)			
	v	24/05/00	Deerpark Wood, Arlington (PCHM)	1082 days	36km	305°
Pied Flycatcher	-					
P110006	1	03/06/99	Bovey Valley Woods (KRG)			
	v (4M)	07/06/01	Steps Bridge, Dunsford (DJP)	435 days	10km	14°
Pied Flycatcher	-					
P110006	1	03/06/99	Bovey Valley Woods (KRG)			
	v at nest box	07/06/01	Steps Bridge, Dunsford (DJP)	735 days	10km	14°
Pied Flycatcher	-					
N908190	1	08/06/99	Steps Bridge, nr.Dunsford (DJP)			
	v	19/05/00	nr.Rifton, Stoodleigh (JAH)	-	-	-
Pied Flycatcher	-					
-	-	08/06/99	-			
	v at nest box	24/05/01	nr.Rifton, Stoodleigh (JAH)	716 days	31km	18°

Species Ref. No.	Age Recovery	Ringing Date - Code - Date -	and Site and Site	Days Difference	Distance Travelled	Bearing Travelled
Pied Flycatcher	-					
K574193	1	10/06/99	Cwm-Brith-Bank, Llandrindod Wells, Powys			
	v	09/06/00	Ash Mill, South Molton (JAH)	365 days	139km	190°
Pied Flycatcher	-					
P049884	4F	19/05/00	nr.Rifton, Stoodleigh (JAH)			
	x pred-ator	21/08/01	Cwm Lasgarn, Cwmavon, Gwent	459 days	95km	23°
Pied Flycatcher	-					
N254871	1	05/06/00	Hittisleigh (HAW)			
	v	26/06/01	Coombeshead Farm, nr.Culm Davy (RJW)	386 days	42km	61°
Pied Flycatcher	-					
P110562	1	05/06/00	Yarner Woods (KRG)			
	v (4F)	03/06/01	Steps Bridge, Dunsford (DJP)	363 days	11km	6°
Pied Flycatcher	-					
P110509	1	05/06/00	Yarner Woods (KRG)			
	v at nest box	11/06/01	River Walkham Weir (AWGJ)	371 days	30km	256°
Pied Flycatcher	-					
P110562	1	05/06/00	Yarner Woods (KRG)			
	v at nest box	03/06/01	Steps Bridge, Dunsford (DJP)	363 days	11km	6°
Pied Flycatcher	-					
P110664	1	07/06/00	Neadon Cleave (KRG)			
	x	17/06/01	Belstone	375 days	17km	311°
Blue Tit	*Species: Included to demonstrate the longevity of some individuals*					
Blue Tit						
F400894	3	31/08/95	Bridford (DJP)			
	v (6M)	14/02/01	Bridford (DJP)	1994 days	0km	360°
Blue Tit	-					
K184970	3JM	03/09/96	Whitcombe, nr.Kenn, Exeter (KRG)			
	x	03/05/01	Kenn, nr.Exeter	1703 days	1km	270°
Blue Tit	-					
K644708	1	23/05/97	Sidborough Nature Reserve (MWT)			
	v	11/04/01	Sidborough Nature Reserve (MWT)	1419 days	0km	360°
Great Tit	Species: Two contrasting reports to illustrate how long it may have lived but for the cat.					
J684810	4F	09/12/95	Sidborough Nature Reserve (MWT)			
	v	09/01/01	Sidborough Nature Reserve (MWT)	1858 days	0km	360°
Great Tit	-					
J633827	3J	07/08/00	Derriford, Plymouth (RCS)			
	x cat	06/05/01	Derriford, Plymouth	272 days	0km	360°
Starling	*A species that could always be relied upon to provide interesting movements, now missing from my garden.*					
Starling						
RJ06420	5F	15/05/95	Cruwys Morchard, nr.Tiverton (JAH)			
	x pred-ator	02/04/01	Ostrovno, Vitebsk, USSR 55.07N:29.52E	2149 days	2283km	78°
Starling	-					
RP53827	3F	14/07/98	Othery, Somerset			
	x shot	10/04/01	Hemyock, nr.Wellington	1001 days	30km	236°

Species Ref. No.	Age Recovery	Ringing Date - Code - Date -	and Site and Site	Days Difference	Distance Travelled	Bearing Travelled
Chaffinch	-					
J684755	4M	05/010/95	Sidborough Nature Reserve (MWT)			
	v	24/02/01	Sidborough Nature Reserve (MWT)	1969 days	0km	360°
Greenfinch	-					
VN70179	3JF	11/06/96	Landguard Point, Suffolk			
	v	15/03/97	Sandford, Crediton (JAH)	277 days	368km	250°
Greenfinch	-					
VX72365	5M	13/02/00	Bridford (DJP)			
	x	04/12/01	Hennock	660 days	8km	162°
Siskin	*Species: More reports of wintering birds in Devon. We need to know more about local breeding birds.*					
Siskin						
P049475	3F	26/12/99	Sandford, Crediton (JAH)			
	x	01/05/01	Ballyvoy, Ballycastle, Antrim, Northern Ireland	492 days	517km	341°
Siskin	-					
E615400	4M	23/03/00	Alpington, Exeter (PWE)			
	v	14/03/01	Alpington, Exeter (PWE)	356 days	0km	360°
Siskin	-					
P483358	5M	18/02/01	Sandford, Crediton (JAH)			
	x glass	10/03/01	Uplyme, Lyme Regis, Dorset	20 days	50km	99°

Little Gull - *Ren Hathway*

DEVON COUNTY LIST

This list, of all species and recognised sub-species recorded in Devon, has been derived from Ashleigh Rosier's (1995) *A Checklist of the Birds of Devon*. This list has now been revised and updated by Mike Langman.

Taxa occurring in the current year (2001) are shown in **bold**. For all others, the year when last recorded is given in parentheses. An asterisk (*) indicates a BBRC rarity, and (A) and (B) indicate taxa on the Devon A & B lists (for further details, refer to the Guidance on Record Submissions in this report). (E) indicates an escapee during the current year. Sub-species are indented and italicised and appear under the parent or most familiar taxon and are referred to by their most familiar name.

Red-throated Diver
Black-throated Diver (B)
Great Northern Diver
White-billed Diver (1993) *
Pied-billed Grebe *
Little Grebe
Great Crested Grebe
Red-necked Grebe (B)
Slavonian Grebe
Black-necked Grebe (B)
Black-browed Albatross (1965) *
Fulmar
Cory's Shearwater (A)
Great Shearwater (A)
Sooty Shearwater (B)
Manx Shearwater
Balearic Shearwater (B)
Fea's/Zino's Petrel *
Wilson's Storm-petrel (1887) *
Storm Petrel
Leach's Storm-petrel (A)
Gannet
Cormorant
    'Continental' Cormorant (A)
Shag
Bittern (B)
American Bittern (1875) *
Little Bittern (1999) *
Night Heron (A)
Squacco Heron (1997) *
Cattle Egret * (E)
Little Egret
Great White Egret (2000)*
Grey Heron
Purple Heron (A)
Black Stork *
White Stork (A)
Glossy Ibis (1987) *
Spoonbill

Mute Swan
Bewick's Swan (B)
Whooper Swan (B)
Taiga' Bean Goose (1997) (A)
    'Tundra' Bean Goose (1998) (A)
Pink-footed Goose (A) (E)
European White-fronted Goose
    'Greenland' White-fronted Goose (A)
Greylag Goose
Snow Goose (E)
Canada Goose
Barnacle Goose (E)
Dark-bellied Brent Goose
    Pale-bellied Brent Goose
    Black Brant (1988) *
Red-breasted Goose *(E)
Egyptian Goose
Ruddy Shelduck (E)
Shelduck
Mandarin
Wigeon
American Wigeon (2000) (A)
Gadwall
Eurasian Teal
Green-winged Teal (A)
Mallard
American Black Duck *
Pintail
Garganey(B - ♀ & eclipse)
Blue-winged Teal (1997) *
Shoveler
Red-crested Pochard
Pochard
Ring-necked Duck (A)
Ferruginous Duck* (E)
Tufted Duck
Scaup (B - not ad ♂)
Eider
Long-tailed Duck

Common Scoter
Surf Scoter (A)
Velvet Scoter (B)
Bufflehead (1999) *
Goldeneye
Smew (B)
Red-breasted Merganser
Goosander
Ruddy Duck (B)
Honey Buzzard (A)
Black Kite (1999) *
Red Kite (B)
White-tailed Eagle (1946) *
Marsh Harrier (B)
Hen Harrier
Montagu's Harrier (A)
Goshawk (A)
Sparrowhawk
Buzzard
Rough-legged Buzzard (1997) (A)
Osprey (B)
Kestrel
Red-footed Falcon (1999) *
Merlin
Hobby
Gyr Falcon (1998) *
Peregrine
Red Grouse
Black Grouse (1997) (A)
Red-legged Partridge
Grey Partridge
Quail (B)
Pheasant
Golden Pheasant (E)
Water Rail
Spotted Crake (A)
Sora (2000) *
Little Crake (1983) *
Baillon's Crake (1995) *
Corncrake (2000)(A)
Moorhen
Coot
Crane (A)
Little Bustard (1912) *
Great Bustard (1871) *
Oystercatcher
Black-winged Stilt (1995) *
Avocet
Stone Curlew (A)
Cream-coloured Courser (1959) *
Collared Pratincole (1956) *
Little Ringed Plover (B)
Ringed Plover
Killdeer (1999) *
Semipalmated Plover (1998) *
Kentish Plover (A)
Greater Sand Plover (1988) *
Dotterel (B)
American Golden Plover (1995) *
Pacific Golden Plover (1992) *
Golden Plover
Grey Plover
Sociable Plover (1963) *
Lapwing
Knot
Sanderling
Semipalmated Sandpiper (1994) *
Western Sandpiper (1973) *
Little Stint
Temminck's Stint (A)
Least Sandpiper (1966) *
White-rumped Sandpiper (1999) *
Baird's Sandpiper *
Pectoral Sandpiper (A)
Curlew Sandpiper
Purple Sandpiper
Dunlin
Broad-billed Sandpiper (2000) *
Buff-breasted Sandpiper (2000)(A)
Ruff
Jack Snipe
Snipe
Great Snipe (1956) *
Long-billed Dowitcher (1990) *
Woodcock
Black-tailed Godwit
Hudsonian Godwit(1982) *
Bar-tailed Godwit
Whimbrel
Curlew
Upland Sandpiper (1986) *
Spotted Redshank
Redshank
Marsh Sandpiper (1990) *
Greenshank
Greater Yellowlegs (1996) *
Lesser Yellowlegs *
Green Sandpiper
Wood Sandpiper (B)
Terek Sandpiper (1996) *

Common Sandpiper
Spotted Sandpiper (1999) *
Turnstone
Wilson's Phalarope (1994) *
Red-necked Phalarope (1995) (A)
Grey Phalarope (B)
Pomarine Skua (B)
Arctic Skua
Long-tailed Skua (A)
Great Skua
Great Black-headed Gull (1859) *
Mediterranean Gull (B)
Laughing Gull (1996) *
Franklin's Gull *
Little Gull
Sabine's Gull (B)
Bonaparte's Gull *
Black-headed Gull
Ring-billed Gull (A)
Common Gull
Lesser Black-backed Gull
    fuscus *LBB Gull* *
    intermedius*LBB Gull* (B)
Herring Gull
    argentatus *H Gull* (2000) (B)
    Yellow-legged Gull (B)
    'American' Herring Gull (1998) *
Iceland Gull (B)
    Kumlien's Gull (1995) *
Glaucous Gull (B)
Great Black-backed Gull
Ross's Gull (1996) *
Kittiwake
Ivory Gull (1853) *
Gull-billed Tern (1988) *
Caspian Tern (1966) *
Lesser Crested Tern (1985) *
Sandwich Tern
Roseate Tern (A - except ad Apr-Jul then B)
Common Tern
Arctic Tern (B)
Bridled Tern (1977) *
Sooty Tern (1979) *
Little Tern
Whiskered Tern (1988) *
BlackTern
White-winged Black Tern (1994) *
Guillemot
Razorbill
Black Guillemot (B)
Ancient Murrelet (1992) *
Great Auk (1829)
Little Auk (B)
Puffin
Pallas's Sandgrouse (1888) *
Rock Dove
Stock Dove
Woodpigeon
Collared Dove
Turtle Dove
Rose-ringed Parakeet (E)
Great Spotted Cuckoo (1998) *
Cuckoo
Black-billed Cuckoo (1982) *
Yellow-billed Cuckoo (1989) *
Barn Owl
Scops Owl (1985) *
Snowy Owl (1972) *
Little Owl
Tawny Owl
Long-eared Owl (2000) (B)
Short-eared Owl
Nightjar
Swift
Alpine Swift *
Little Swift (1985) *
Chimney Swift (1999) *
Kingfisher
Blue-cheeked Bee-eater (1987) *
Bee-eater (2000)(A)
Roller (1989) *
Hoopoe (B)
Wryneck (B)
Green Woodpecker
Great Spotted Woodpecker
Lesser Spotted Woodpecker
Eastern Phoebe (1987) *
Bimaculated Lark (1962) *
Short-toed Lark (1999) (A)
Crested Lark (1959) *
Woodlark (B)
Skylark
Shore Lark (1998) (A)
Sand Martin
Swallow
Red-rumped Swallow *
House Martin
Richard's Pipit (A)
Tawny Pipit (1998) (A)
Olive-backed Pipit(1997) *

Tree Pipit
Meadow Pipit
Red-throated Pipit (1997) *
Rock Pipit
  Scand. Rock Pipit (2000)(A)
Water Pipit(B)
Yellow Wagtail
  Blue-headed Wagtail (B)
  Ashy-headed Wagtail (A)
Citrine Wagtail (1998) *
Grey Wagtail
Pied Wagtail
  White Wagtail (B)
Waxwing (1999) (B)
Dipper
  'Black-bellied' Dipper (A)
Wren
Dunnock
Alpine Accentor (1993) *
Rufous Bush Chat (1980) *
Robin
Thrush Nightingale (1981) *
Nightingale (B - non-breeding)
Red-spotted Bluethroat (1995) (A)
  White-spotted Bluethroat (1995) (A)
Black Redstart
Redstart
Whinchat
Stonechat
  Siberian Stonechat *
Wheatear
  Greenland Wheatear (B)
Pied Wheatear (1983) *
Black-eared Wheatear (1984) *
Desert Wheatear (1997) *
Rock Thrush (1985) *
White's Thrush (1984) *
Swainson's Thrush (1995) *
Gray-cheeked Thrush (1986) *
Veery (1997) *
Ring Ouzel
Blackbird
Fieldfare
Song Thrush
Redwing
Mistle Thrush
American Robin (1982) *
Cetti's Warbler
Grasshopper Warbler
Savi's Warbler (1998) *
Aquatic Warbler (1999) (A)
Sedge Warbler
Marsh Warbler (1997) (A)
Reed Warbler
Great Reed Warbler (1992) *
Booted Warbler (1994) *
Icterine Warbler (1999) (A)
Melodious Warbler (A)
Dartford Warbler
Spectacled Warbler (1999) *
Subalpine Warbler *
Sardinian Warbler (1992) *
Ruppell's Warbler (1979) *
Desert Warbler (1992) *
Barred Warbler (A)
Lesser Whitethroat
  Sib. Lesser Whitethroat (1994) (A)
Whitethroat
Garden Warbler
Blackcap
Greenish Warbler (1992) *
Arctic Warbler (1993) *
Pallas's Warbler (2000) (A)
Yellow-browed Warbler (A)
Hume's Warbler (1992) *
Radde's Warbler (2000) *
Dusky Warbler (1999) *
Western Bonelli's Warbler (1986) *
Wood Warbler
Common Chiffchaff
  abietinus *Chiffchaff* (1999) (A)
  tristis *Chiffchaff* (1999) (A)
Iberian Chiffchaff (1999) *
Willow Warbler
Goldcrest
Firecrest
Spotted Flycatcher
Red-breasted Flycatcher (A)
Collared Flycatcher (1948) *
Pied Flycatcher
Bearded Tit (1998) (B)
Long-tailed Tit
Marsh Tit
Willow Tit (B)
Crested Tit (1947) (A)
Coal Tit
  Cont.Coal Tit (1998) (A)
Blue Tit
Great Tit
Nuthatch

Treecreeper
Penduline Tit (1996) *
Golden Oriole (B)
Isabelline Shrike *
Red-backed Shrike (B - ♀/juv)
Lesser Grey Shrike (1992) *
Great Grey Shrike (B)
Woodchat Shrike (2000)(A)
Jay
Magpie
Nutcracker (1969) *
Chough (A)
Jackdaw
Rook
Carrion Crow
Hooded Crow (A)
Raven
Starling
Rose-Coloured Starling (A)
House Sparrow
Spanish Sparrow (1966) *
Tree Sparrow(B)
Red-eyed Vireo (2000) *
Chaffinch
Brambling
Serin (A)
Greenfinch
Goldfinch
Siskin
Linnet
Twite (A)
Lesser Redpoll
Mealy Redpoll (1997) (A)
Two-barred Crossbill (1990) *
Crossbill
Parrot Crossbill (1892) *
Scarlet Rosefinch (A)
Bullfinch
Hawfinch (B)
Black-and-White Warbler (1987) *
Chestnut-sided Warbler (1995) *
Yellow-rumped Warbler (1960) *
Blackpoll Warbler (1984) *
Ovenbird (1985) *
Common Yellowthroat (1954) *
Rufous-sided Towhee (1966) *
Lapland Bunting (A)
Snow Bunting (B)
Black-faced Bunting *
Yellowhammer
Cirl Bunting
Ortolan Bunting (A)
Rustic Bunting (1997) *
Little Bunting (A)
Yellow -breasted Bunting (1993) *
Reed Bunting
Black-headed Bunting *
Corn Bunting (A)
Rose-breasted Grosbeak *
Bobolink *
Northern Oriole (1967) *

MAP OF DEVON

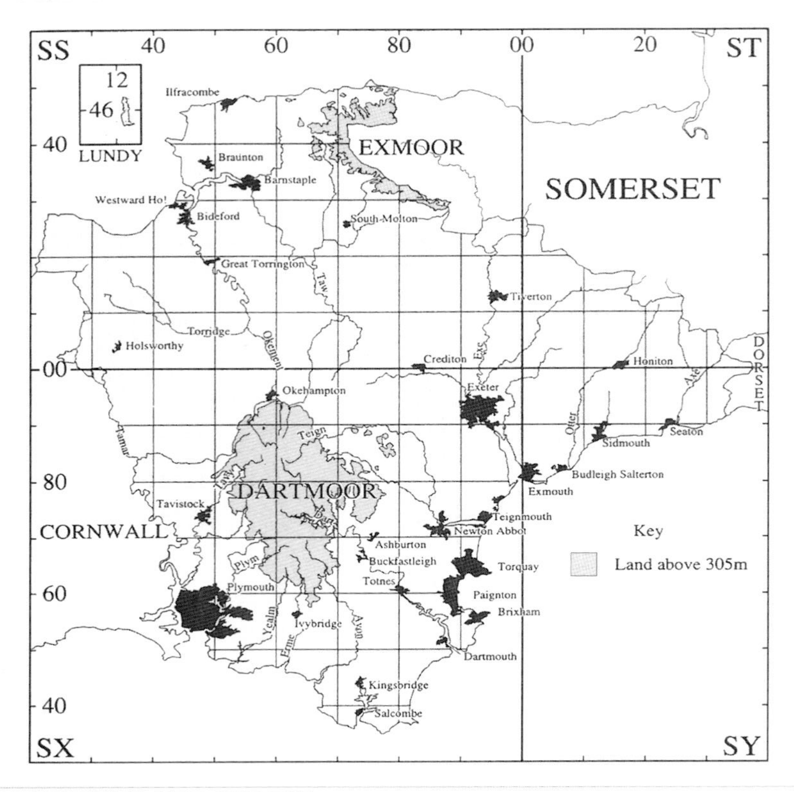

ROSEATE TERNS AT DAWLISH

One of the impressive features of 2001 was the number of Roseate Terns recorded at Dawlish Warren. Careful observations of plumage and rings by JEF has enabled him to gain insight into the number of individuals involved and their duration of stay, and his study provides a valuable lesson in the merits of making detailed descriptions of individual birds. His drawings appear opposite, and the following is based on his account of events.

Spring records from DWNNR Logbook. All in May, with singles on 1, 2, 4, 8 & 16 May, three on 15th, two on 17th, three on 18th and four on 22nd. Without descriptions, it is impossible to know the numbers involved, but it must have been at least four, and could have been up to 17.

Autumn. One or two seen most days, 4 Jul – 12 Aug. Five or six were present on 10 Jul, and, although distant, appeared to be summer adults. Usually only one present until 22 Jul, but there were three (all adults) on 14th. Initially roosting by the railway, and too far away to observe closely, bird A began to come close to the hide from 19th. On 21st, birds B and C joined A, and then five on 22nd included these three plus two new birds, D and E. On 27 Jul, four birds included A again, but also three new individuals, birds F, G and H. Three on 30th included birds A and B, but also another new bird, I, bringing the total to nine individuals present in nine days. A tenth individual, bird K, joined A and B on 8 Aug, and just A on 12th.

Subsequently, two adults seen by LC on 15 Aug probably included at least one new adult, and four recorded by GV on 25th were probably new because of the long gap, but impossible to be sure without descriptions.

The short stay of most birds suggest that more pass through than the records suggest, and for example, it is likely that all birds on 10 Jul were different to those subsequently described. On the other hand, it is interesting that one bird (A) remained in the area for at least 25 days.

According to KRy, Roseate Terns on the Farne Islands are ringed on both legs with BTO rings. Observations at Dawlish would be much easier if the birds were colour-ringed as well.